Functional Equations in Applied Sciences

This is volume 199 in
MATHEMATICS IN SCIENCE AND ENGINEERING
Edited by C.K. Chui, *Stanford University*

A list of recent titles in this series appears at the end of this volume.

Functional Equations in Applied Sciences

Enrique Castillo
UNIVERSIDAD DE CANTABRIA
SANTANDER, SPAIN

Andrés Iglesias
UNIVERSIDAD DE CANTABRIA
SANTANDER, SPAIN

Reyes Ruíz-Cobo
UNIVERSIDAD DE CANTABRIA
SANTANDER, SPAIN

2005
ELSEVIER
Amsterdam – Boston – Heidelberg – London – New York – Oxford
Paris – San Diego – San Francisco – Singapore – Sydney – Tokyo

ELSEVIER B.V.
Sara Burgerhartstraat 25
P.O. Box 211, 1000 AE Amsterdam
The Netherlands

ELSEVIER Inc.
525 B Street, Suite 1900
San Diego, CA 92101-4495
USA

ELSEVIER Ltd
The Boulevard, Langford Lane
Kidlington, Oxford OX5 1GB
UK

ELSEVIER Ltd
84 Theobalds Road
London WC1X 8RR
UK

First edition 2005

Library of Congress Cataloging in Publication Data
A catalog record is available from the Library of Congress.

British Library Cataloguing in Publication Data
A catalogue record is available from the British Library.

ISBN: 0-444-51788-x
ISSN (Series): 0076-5392

∞ The paper used in this publication meets the requirements of ANSI/NISO Z39.48-1992 (Permanence of Paper).

Transferred to Digital Printing 2008

DEDICATION

To Janos Aczél with admiration.

Note from the Editor

Founded about four decades ago by the visionary distinguished mathematician, the late Richard Bellman, this "red series" Mathematics in Science and Engineering (MISE) has the entrepreneurial tradition of being one of the very first to publish interesting monographs of mathematical topics that may have the potential to make a significant impact to the advancement of sciences and engineering. Since the unfortunate early departure of Professor Bellman, a lot has been happening in mathematics that is beyond the fascination of solutions of "Big Problems" and settlement of well-known conjectures. Various exciting research areas and directions that have more direct applications in the scientific and engineering fields have been introduced. Even the tradition of pursuing individual mathematical research has gradually been adapting to the more common way of carrying out collaborative work in other scientific disciplines. Indeed, it has been a very interesting period of changes, but these changes are both natural and necessary.

The invention of semiconductors, integrated circuits (IC), as well as the exponential rate of technological advancement in IC functionalities and "chip" size, called Moore's law, has been the core and source of this "industrial revolution." Most notably, the unbelievable escalation in computing power, along with the most significant miniaturization of computing devices, not only leads to rapid advancement of all areas in sciences and engineering, but also becomes the source of creating new fields and research directions. At the same time, the tremendous IC capabilities have also significantly advanced such technologies as intelligent sensors, which, in turn, rely on computing. The enormous computing power has also played the key role in advancement and creation of new fields and new directions in other subject areas beyond the traditional science and engineering subjects, including economics, commerce, etc. Hence, the scope of science and engineering applications for MISE, as originally envisioned by Professor Bellman, is to be broadened accordingly as well.

In taking over the editorship of this book series of MISE from Professor William Ames, I am committed to preserving the entrepreneurial spirit of the founder, Professor Bellman, by encouraging publication of mathematics monographs that have the potential to make an impact on the advancement of sciences and engineering technologies, now under a broader scope. It is still the "red series", only with a new artistic design to reflect the closer and more direct relationship between the advancement of mathematics and that of other scientific and engineering fields to be interpreted in broader horizons. We welcome submission of book proposals and manuscripts from all fellow mathematical scientists who share a similar vision of MISE. We need our readers' support to make a lasting impact on the advancement of sciences and engineering technologies.

Charles Chui
Editor-in-Chief
Stanford, California
July, 2004

Note from the Publisher

Having been established in the 1960's by Academic Press, the *Mathematics in Science and Engineering*, "red series", became well-known under the leadership of the founding editor Richard Bellman, many pioneering works being produced.

Almost two hundred volumes were published, up to volume 198 by Igor Podlubny in November 1998, the Editor-in-Chief at this stage being Professor William Ames at Georgia Tech.

After the acquisition of Harcourt General by Elsevier in 2001, which included Academic Press, responsibility for publication of the "red series" passed from the San Diego office of the former AP to Elsevier in Amsterdam, as part of the merging of the two individual publishing programs.

Following the completion of the detailed merger, we are very happy to announce the continuation of the MISE "red series" under the editorship of Prof. Charles Chui, Stanford, USA.

It is our intention to publish around 3 volumes per year, of the highest level of mathematical sciences scholarship, starting with the present vol. 199 by Castillo, Iglesias & Ruíz-Cobo

Keith Jones
Publisher
PMCA-- Physics, Mathematics, Computer Science & Astronomy
Elsevier

Contents

Preface

Functional equations is one of the most powerful and beautiful fields of Mathematics we have encountered in our professional life. It was during the summer of 1983, on the occasion of a stay at the ETH (Zürich), when E. Castillo together with A. Fernández-Canteli discovered for the first time the real importance of functional equations. We were trying to model the influence of length and stress range on the fatigue life of longitudinal elements and, when analyzing the inconsistencies of some tentative models, we found a compatibility equation written in terms of a functional equation. Immediately, the 1966 Aczél book on functional equations came to our minds (two or three years earlier, somebody in our library had ordered the book and so it was only by chance that we had the opportunity of taking a look at it without realizing, at first glance, its real importance, yet noting that some powerful methods were behind it). Since then, we have completely changed our minds and incorporated the functional equations' philosophy and techniques to our daily procedures. Even though many years were required to find our first functional equation, many others have appeared since then in our work, and, in fact, today we cannot think of building models or stating problems without using functional equations.

Our experience is that model building in science and engineering is frequently performed based on selecting simple and easily tractable equations that seem to reproduce reality to a given quality level. However, on many occasions these models exhibit technical failures or inconsistencies, such as those we discovered in our fatigue models when we obtained the compatibility equation, and which make them unacceptable. Functional equations is one of the main tools that prevent arbitrariness and allow a rigorous and consistent selection and design of models. In fact, conditions required by many models to be adequate replicas of reality can be written as functional equations.

Functional equations arise in many fields of Applied Science, such as Mechanics, Geometry, Statistics, Hydraulics, Economics, Engineering, etc. However, though the theory of functional equations is very old, not only technicians but many mathematicians are still unaware of the power of this important field of Mathematics. As J. Aczél and J. Dhombres indicate in the preface of their book: *"from their very beginnings, functional equations arose from applications, were developed mostly for the sake of applications and, indeed, were applied quite intensively as soon as they were developed"*. However, most of the recent advances in the theory of functional equations have been published in mathematical journals which are not written in a language that many engineers and scientists can easily understand. This fact, which is common to many other areas of Mathematics, has been the reason why many engineers and applied scientists are still unaware of a long list of these advances and, consequently, they have not incorporated common functional equation techniques into their daily procedures.

Our experience with functional equations was so positive and relevant to applications that we became engrossed in this still relatively unknown field of Mathematics. Impressed by its importance and wishing to share with others this discovery, we decided to write the present book.

One of the aims of this book is to provide engineers and applied scientists with some selected results of functional equations which can be useful in applications. We are aware that this is not an easy task, and that any effort to bring together mathematicians and engineers, as experience shows, has many associated difficulties. We have, intentionally, omitted or simplified many proofs and details of theorems in order to make the text more readable to engineers. However, we wish to go even further, trying to offer the readers a different point of view and offer them a new way of thinking in mathematical modelling. Traditionally, engineers and scientists state practical problems in terms of derivatives or integrals, which lead to differential or integral equations, respectively. With this book we want to offer them the possibility of using functional equations too, as one more alternative, which is at least as powerful as either of the other two.

This book, which is based on lectures delivered by the authors at the University of Cantabria and in the book "Functional Equations in Science and Engineering", published by Marcel Dekker in 1992, focuses primarily upon applications and includes many examples of applications aiming to illustrate how functional equations are the ideal tool to design mathematical models. Thus, special attention is given to the analysis and discussion of the functional equations, in the light of their physical meaning, and to practical examples.

The book is organized in two parts. The first part is devoted to functional equations in general. Chapter 1 is an introduction to functional equations. In it, we use several simple problems to motivate functional equations. The beauty of functional equations becomes apparent when some formulas, such as the area of a rectangle or a trapezoid, or the interest formulas, arise as the only expressions that satisfy some natural conditions. Furthermore, we discover generalized formulas showing that the standard formulas are not sufficient to deal with all practical cases. In Chapter 2 an important effort has been made to identify a list of methods to solve functional equations, and give some illustrative examples to facilitate its understanding. We know of no other book giving this general methodology to solve functional equations. In Chapters 3 to 6 several functional equations in one or several functions in one or several variables are discussed, and several examples of applications are given. In Chapter 7 we discuss the problem of equivalence of functional, difference, and differential equations and use this equivalence to solve functional equations. The possibility of stating problems as functional equations, as an alternative to the usual statement of problems, based on differential or difference equations, is a new and powerful alternative that deserves special attention. To end this part, Chapter 8 deals with vector and matrix equations.

In the second part we apply functional equations to solve a wide range of practical problems. In Chapter 9 we introduce functional networks, a powerful generalization of neural networks. It is shown how every functional equation or system of functional equations leads to a functional network, and how it can be exploited to solve functional equations numerically. Functional networks have proven to be a powerful technique that allows simple and very efficient networks to be built. In Chapter 10 we deal with some applications to engineering, including the laws of Science, models for fatigue life, and beam equations. In Chapter 11 some applications to Geometry and to computer aided design are presented. Chapter 12 is devoted to applications in the Economic field: taxation functions, price indices, interest formulas, and many other material, including monopoly and duopoly models, are analyzed. Finally, in Chapter 13 some applications to Probability and Statistics are presented. In particular, several families of distributions are characterized.

We would like to thank A. Fernández-Canteli, J. Galambos, Barry C. Arnold, and J.M. Sarabia, with whom we have done joint work related to functional equations, for their invaluable stimulus and encouragement.

We also thank José Antonio Garrido and Iberdrola for partial support of this book.

Special recognition must be given to Janos Aczél. As mentioned before, his 1966 book drew the attention of the authors to the field of functional equations and made possible all their work in this interesting area of Mathematics. Professor Aczél has marked the path to follow for all those who love functional equations. We must also mention the remarkable book of Eichhorn (1978), where extremely interesting applications to Economics were presented.

Finally, we wish to mention the scientific community, mainly those included in the bibliography and those who were, surely but unintentionally, omitted. They, through their life's work, have made this book possible. To all of them, our most sincere thanks.

Enrique Castillo
Andrés Iglesias
Reyes Ruíz–Cobo

Santander, June 10, 2004.

Part I

Functional Equations

CHAPTER 1

Introduction and motivation

1.1 Introduction

Mathematical modelling is one of the basic techniques for solving problems and analyzing reality in Physics and Engineering. Experienced engineers and scientists know how a successful analysis or design depends on an adequate selection of the model and method of analysis. The modelling or idealization of the problem under consideration (structure, road, harbor, water supply system, etc.) should be sufficiently simple, logically irrefutable, admitting a mathematical solution, and, at the same time, represent sufficiently well the actual problem. The selection of the idealized model should be achieved by detecting and representing the essential first-order factors, and discarding or neglecting the inessential second-order factors.

Model building is based on an adequate selection of simple equations that seem to represent reality to a given quality level. However, on many occasions these models exhibit technical failures or inconsistencies which make them inadmissible. Functional equations are a tool that prevents arbitrariness and allows model selection to be based on adequate constraints.

Though the theory of functional equations is very old (some examples of functional equations appear in Oresme (1347, 1352), Napier (1614, 1617, 1620), Kepler (1624), Galileo (1638), Abel (1823, 1826b,a), etc...), it is not only technicians but many mathematicians too, who are still unaware of the power of this important field of Mathematics.

Functional equations arise in many fields of Applied Science (see Aczél (1984)), such as *Mechanics*: D'Alembert (1747, 1750, 1769), Lagrange (1788, 1799), *Geometry*: Aczél (1966), Rassias (1994), *Geometric Design*: Castillo and Iglesias (1995, 1997), Monreal and Santos (1998), *Statistics*: Alsina and Bonnet (1979), Alsina (1981a,b), Arnold et al. (1992, 1993), Castillo and Galambos (1987a), Castillo et al. (1987, 1990b), *Economics*: Aczél (1966, 1975, 1987b,

1988), Aczél and Eichhorn (1974), Eichhorn (1978a,b,c), Eichhorn and Kolm (1974), Eichhorn and Gehrig (1982), Young (1987), *Artificial Intelligence*: Castillo et al. (1990c, 1999b), *Engineering*: Aczél (1966, 1987b), Castillo and Ruiz-Cobo (1992), Castillo and Galambos (1987b), Kahlig (1990), etc.

One of the most appealing characteristics of functional equations is their capacity for model design. In fact, those conditions required by many models in order to be adequate replicas of reality can be written as functional equations. Thus, the engineer finds there an appropriate tool for his design purposes. In this manner, functions are not arbitrarily chosen; on the contrary, they appear as the only solutions to the adequate set of requirements.

In Section 1.2 we introduce some simple motivating examples of functional equations, such as the formula for the area of a rectangle, the simple interest, the sum of the internal angles of a polygon, or the associativity equation. In Section 1.3 we introduce some definitions and basic concepts that are needed in order to understand the rest of the book.

1.2 Some examples of functional equations

This section introduces some illustrative examples of how functional equations can be applied to solve some interesting problems related to many different fields.

1.2.1 First example: Area of a rectangle (Legendre (1791))

Assume that the formula of the area of a rectangle is unknown but given by $f(a, b)$, where f is an unknown function, b is its basis and a is its height. Consider Figure 1.1(left) in which the rectangle of basis b and height a has been horizontally divided in two different sub-rectangles with the same basis b and heights a_1 and a_2, respectively. According to our assumptions, the areas of the sub-rectangles and the initial rectangle cannot be calculated, but they can be expressed in terms of our unknown f function as $f(a_1, b)$, $f(a_2, b)$, and $f(a_1+a_2, b)$, respectively. Similarly, we can perform the division vertically, as shown in the right rectangle of the same figure, and write the areas of the resulting rectangles as $f(a, b_1)$, $f(a, b_2)$, and $f(a, b_1 + b_2)$, respectively.

Stating that the areas of the initial rectangles must be equal to the sum of the areas of the sub-rectangles, we get the functional equations

$$\begin{aligned} f(a_1 + a_2, b) &= f(a_1, b) + f(a_2, b) \\ f(a, b_1 + b_2) &= f(a, b_1) + f(a, b_2). \end{aligned} \tag{1.1}$$

Because b is constant in the first equation and a is constant in the second, both equations become Cauchy's Equation (3.7) (to be discussed in Section 3.4) for non-negative f and then, because of Theorem 3.3(b), we have

$$f(a, b) = c_1(b)a = c_2(a)b,$$

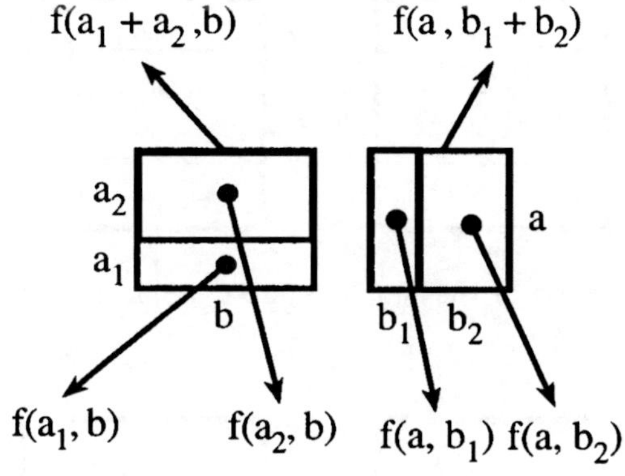

Figure 1.1: Basic rectangles.

where $c_1(b)$ and $c_2(a)$ are initially arbitrary functions, but due to the second identity, they must satisfy the condition

$$\frac{c_1(b)}{b} = \frac{c_2(a)}{a} = c,$$

which implies

$$f(a,b) = cab, \tag{1.2}$$

where c is an arbitrary positive constant.

As a consequence, the area of a rectangle is the product of its basis a, its height b and a constant c.

This proves that the area of a rectangle is not the well known "basis × height", but "a constant × basis × height". The constant takes care of the units we use for the basis, the height and the area. This means that if b is measured in inches, h in feet, and we want f in square miles, the constant must be different from the constant required for the case of b measured in meters, h in kilometers, and f in square meters.

The interesting result is that functional equations discover the need to consider the units of measure.

1.2.2 Second example: Simple interest

Let $f(x,t)$ be the interest we receive from the bank after a deposit of an amount x during a period of duration t. Then, if the assumptions of simple interest hold, the function $f(x,t)$ must satisfy the following conditions:

1 At the end of the time period t, we receive the same interest in the following two cases (see Figure 1.2):

  (a) We deposit the amount $x + y$ in one account.

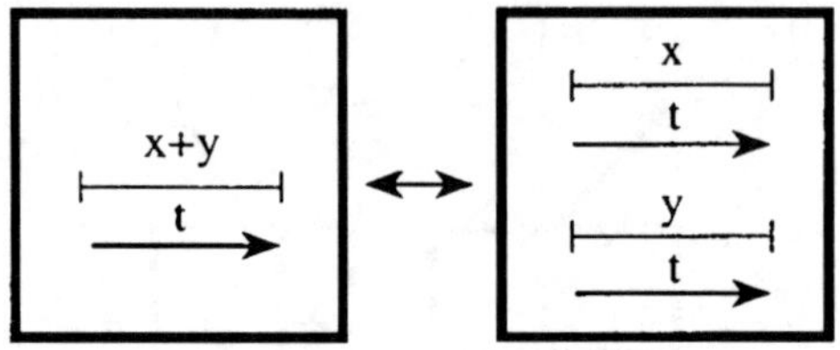

Figure 1.2: Illustration of the simple interest problem. Splitting the amount into two parts.

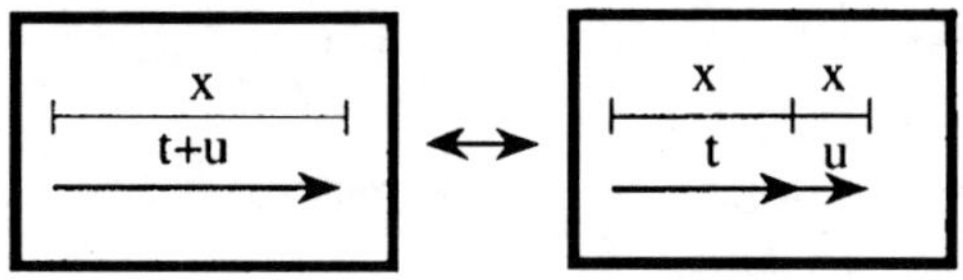

Figure 1.3: Illustration of the simple interest problem. Splitting the total deposit duration into two parts.

(b) We deposit the amount x in one account, and the amount y in another account.

Thus, we have

$$f(x+y,t) = f(x,t) + f(y,t).$$

2 At the end of the time period $t+u$, we receive the same interest in the following two cases (see Figure 1.3):

(a) We deposit the amount x during a period of duration $t+u$.

(b) We deposit the amount x first during a period of duration t and later for a period of duration u.

Thus, we have

$$f(x,t+u) = f(x,t) + f(x,u).$$

That is, the following equations hold

$$\left.\begin{array}{l} f(x+y,t) = f(x,t) + f(y,t) \\ f(x,t+u) = f(x,t) + f(x,u) \end{array}\right\}; \quad x,y,t,u \in \mathbb{R}_+ \tag{1.3}$$

According to Theorem 3.3 and Example 2.14, the solution of the first equation is given by

$$f(x,t) = c(t)x,$$

and back-substitution into the second leads to

$$c(t+u)x = c(t)x + c(u)x \quad \Rightarrow \quad c(t+u) = c(t) + c(u) \quad \Rightarrow \quad c(t) = Kt,$$

and then, we finally obtain

$$f(x,t) = kxt, \tag{1.4}$$

where the constant k is the interest rate. Expression (1.4) is the well known formula of the simple interest.

It is important to note here that the above assumptions do not hold in reality, but they are the simple interest assumptions. It can be seen from the bank office that if we deposit a larger amount or we do so for a longer period the interest rate increases.

We note that the bank policy has to be such that:

$$f(x+y,t) \geq f(x,t) + f(y,t).$$

Otherwise, the bank is inviting its clients to deposit their money in many accounts (a low amount in each account).

In addition, we must have:

$$f(x,t+u) \geq f(x,t) + f(x,u).$$

Otherwise, the bank is inviting its clients to withdraw the money every day and deposit it again in a new account.

Consequently, *simple interest* is the optimal way of keeping account stability by giving the least possible interest.

Fortunately, the actual bank policy does not follow the simple interest rule.

A comparison of the system of equations of the rectangle area and of the simple interest examples shows that, apart from notation, the two systems of functional equations (1.1) and (1.3) are identical. This means that we have two physical problems: one geometric and one economic, leading to exactly the same mathematical model.

It is interesting to point out that the commutativity of a and b in (1.2) results as a consequence of the assumptions in (1.1), but it is not axiomatic. This implies that a rectangle can be rotated through an angle of $\pi/2$ radians (exchange basis and height) without changing its area.

The commutativity also holds for the simple interest case; that is, we get the same interest if we deposit \$1 for 2 years as if we deposit \$2 for 1 year.

1.2.3 Third example: Sum of the internal angles of a polygon

Let $f(n)$ be the function giving the sum of the internal angles of a polygon with n sides. In order to obtain this function, we consider the following fact: if we split one of the sides, of a given polygon with n sides, into two sides, the sum of the internal angles of the new polygon corresponds to the sum of the internal angles of the initial polygon plus the sum of the angles of a triangle (see Figure 1.4).

Therefore, we have

$$f(n+1) = f(n) + \alpha + \beta + \gamma = f(n) + f(3), \tag{1.5}$$

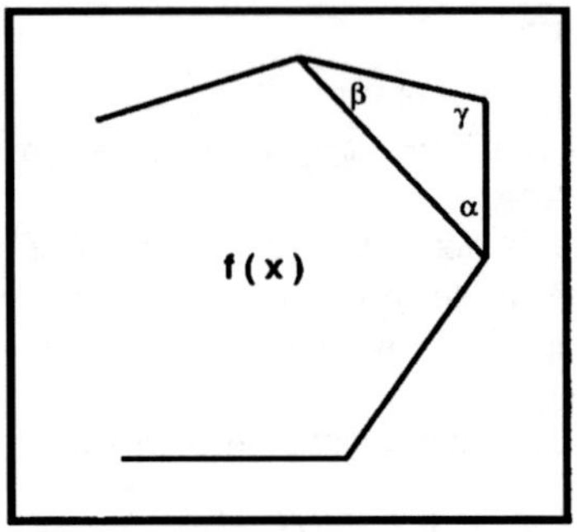

Figure 1.4: Elemental perturbation of a polygon.

which is a difference equation. As we shall show in Example 3.13, its general solution is:

$$f(n) = f(3)(n-2). \tag{1.6}$$

Thus we have obtained the expression for f as a function of the number of sides, n, and an initial value corresponding to the sum of the internal angles of a triangle. If this value is assumed to be known, say π, Expression (1.6) leads to the well known formula:

$$f(x) = \pi(x-2).$$

1.2.4 Fourth example: The associativity equation

Let us consider three real numbers x, y, and z, and an operation $\oplus$. The associative property establishes that

$$(x \oplus y) \oplus z = x \oplus (y \oplus z), \tag{1.7}$$

that is, the same result is obtained operating x with y and then with z, as operating x with the result obtained by operating y with z.

If we define the function $F(x, y) = x \oplus y$, Equation (1.7) becomes

$$F(F(x,y),z) = F(x, F(y,z)), \tag{1.8}$$

which is the well known *associativity functional equation* (see Section 6.5).

Under certain assumptions (see Theorem 6.6), its more general continuous and invertible solution is given by

$$F(x,y) = f^{-1}[f(x) + f(y)], \tag{1.9}$$

where f is an arbitrary continuous and strictly monotonic function.

It is interesting to remark that this solution characterizes every associative operation under the previous conditions. In fact, it is sufficient to choose an arbitrary continuous and strictly monotonic function in order to obtain an associative operation.

Example 1.1 (Two associativity operations). Taking $f(x) = \log x$ Equation (1.9) becomes

$$F(x, y) = f^{-1}[f(x) + f(y)] = \exp[\log x + \log y] = \exp[\log(xy)] = xy, \quad (1.10)$$

which shows that the product is an associative operation.

Alternatively, we can take $f(x) = x^n$, and then Equation (1.9) becomes

$$F(x, y) = f^{-1}[f(x) + f(y)] = \sqrt[n]{x^n + y^n}, \quad (1.11)$$

which is another associative operation, for any value of n. ■

1.3 Basic concepts and definitions

It is not easy to give a precise definition of functional equations. However, before starting the discussion of functional equations and systems, some definitions are required.

Definition 1.1 (Functional equation). *In a broader sense, a functional equation can be considered as an equation which involves independent variables, known functions, unknown functions and constants; but we exclude differential equations, integral equations and other kinds of equations containing infinitesimal operations. In our equations, the main operation is the substitution of known or unknown functions into known or unknown functions.* ■

Example 1.2 (Functional equations). The following equations are typical examples of functional equations:

1. Cauchy's Equation (see Section 3.4):

$$f(x + y) = f(x) + f(y); \quad x, y \in \mathbb{R} \quad (1.12)$$

2. Pexider's Equation (see Section 4.2):

$$f(x + y) = g(x) + h(y); \quad x, y \in \mathbb{R} \quad (1.13)$$

Note that this equation involves three unknown functions. As we shall see, this is a characteristic of functional equations: a simple functional equation can determine several unknown functions.

3. Homogeneous Equation (see Section 5.3):

$$f(zx, zy) = z^n f(x, y); \quad x, y \in \mathbb{R}_+, \quad z \in \mathbb{R}_{++} \quad (1.14)$$

Note that the unknown function f depends on several variables.

4. Transformation Equation (see Section 6.8):

$$f(f(x,y),z) = f(x,g(y,z)); \quad x,y,z \in \mathbb{R} \tag{1.15}$$

   This is an example of an equation with several functions of several variables (see Chapter 6).

In the equations above, $\mathbb{R}, \mathbb{R}_+$ and $\mathbb{R}_{++}$ are the sets of real numbers, the set of non-negative real numbers and the set of positive real numbers, respectively. ■

Definition 1.2 (System of functional equations). *A system of functional equations is a set of $n \geq 2$ functional equations.* ■

Example 1.3 (Systems of functional equations). Examples of systems of functional equations are

$$\left.\begin{aligned} f(x+y-z,p+q-z) &= f(x,p)+f(y,q)-2 \\ f(x+z,p+1) &= f(x,p)+z-1 \end{aligned}\right\}; \quad x,y,z,p,q \in \mathbb{R} \tag{1.16}$$

$$\left.\begin{aligned} f(x_1+x_2,y_1+y_2,z) &= f(x_1,y_1,z)+f(x_2,y_2,z) \\ f(z,z,x_1+x_2) &= f(z,z,x_1)+f(z,z,x_2) \end{aligned}\right\}; \quad x_1,x_2,y_1,y_2,z \in \mathbb{R} \tag{1.17}$$

Other systems of functional equations have appeared in the previous Examples 1.2.1 and 1.2.2. ■

Difference equations (as Equation (1.5)), which appear in many practical cases, will be considered here as particular cases of functional equations, even though only very simple cases will be studied.

Definition 1.3 (Domain of a functional equation). *Given a functional equation, the set of all values of the variables, on which it is supposed to hold, is called its domain (not to be confused with the domain of definition of each known or unknown function appearing in it).* ■

Example 1.4 (Domain of a functional equation). The functional Equation (1.12), which includes two variables, x and y, can be considered to be valid on many domains such as, for example $x,y \in \mathbb{R}$, $x,y \in \mathbb{R}_+$ or $(x,y) \in T$, where $T = \{(x,y) \in \mathbb{R}^2 | x \geq 0, y \geq 0, x+y \leq 1\}$. The domains of the functional equation in these three cases are $\mathbb{R}^2$, $\mathbb{R}^2_+$ and T, respectively. ■

If the functional equation comes from a physical problem we can talk about its *natural domain*, as the set of values of the variables with a physical sense. For instance, since function f of Example 1.2.1 represents an area, it has the set of non-negative real numbers, $\mathbb{R}_+$, as a natural domain. In other cases the so called natural domain could be considered somewhat artificial. In this book, unless stated otherwise, we assume that the functional equations are defined by their natural domains.

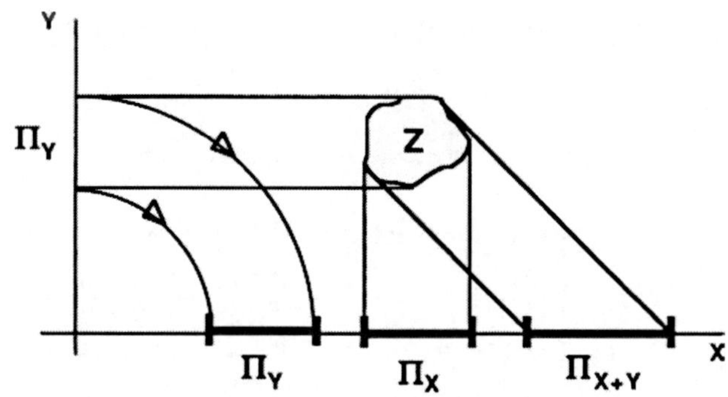

Figure 1.5: Graphical illustration of the sets $\Pi_X(Z)$, $\Pi_Y(Z)$ and $\Pi_{X+Y}(Z)$ in Example 1.6.

Sometimes we find functional equations which are stated on a *restricted domain*, that is, restricted when compared with their natural or initial domain. In this case two different names have been proposed: *"functional equations on restricted domains"*, Kuczma (1978), and *"conditional functional equations"*, Dhombres and Ger (1975).

It is interesting to point out that the domain of the functional equation can be independent of the unknown functions or dependent on them. The domains in Example 1.2 are independent of the unknown functions. One case of dependence is given in the following example.

Example 1.5 (Domain dependent on the unknown function). The domain of the functional equation

$$f(x+y) = f(x) + f(y), \quad f(x+y) \neq 0 \tag{1.18}$$

is clearly dependent on the unknown function. ∎

It is also important to note that the domains of the known and unknown functions are dependent on the domain of the functional equation. We illustrate this fact by the following example.

Example 1.6 (Domain of the unknown function). Let us assume that the domain of the functional Equation (1.12) is the set Z in Figure 1.5. Then the domain of the function $f(x)$ must be, at least, (see Figure 1.5) the set $\Pi_X(Z) \cup \Pi_Y(Z) \cup \Pi_{X+Y}(Z)$, where $\Pi_X(Z)$, $\Pi_Y(Z)$ and $\Pi_{X+Y}(Z)$ are the projections of Z onto the X axis, the Y axis, and the X-axis parallel to the line $x + y = 0$, respectively. ∎

Definition 1.4 (Particular solution). *We say that a function or a set of functions is a particular solution of a functional equation or system if, and only if, it satisfies the functional equation or system in its domain of definition.* ∎

Example 1.7 (Particular solution). The following functions or sets of functions

$$f(x) = 3x, \tag{1.19}$$

$$f(x) = 2x + 3, \;\; g(x) = 2x + 1, \;\; h(x) = 2x + 2, \tag{1.20}$$

$$f(x, y) = \begin{cases} \left(\sqrt{xy}\right)^n \exp\left(\dfrac{x}{y}\right) & \text{if} \quad xy \neq 0, \\ 2x^n & \text{if} \quad y = 0, \; x \neq 0, \\ 3y^n & \text{if} \quad x = 0, \; y \neq 0, \\ k \; (kn = 0) & \text{if} \quad x = y = 0, \end{cases} \tag{1.21}$$

$$f(x, y) = g(x, y) = x + y, \tag{1.22}$$

$$f(x, y) = x - y + 2, \tag{1.23}$$

$$f(x, y, z) = zx, \tag{1.24}$$

are particular solutions of equations (1.12) to (1.17), respectively. Note that the corresponding equations are identically satisfied by these functions. ∎

Definition 1.5 (General solution). *Given a class of functions $\mathcal{F}$, the general solution of a functional equation or system is the totality of particular solutions in that class.* ∎

Example 1.8 (General solution). The general solution of the system (1.16) in the class of continuous functions is

$$f(x, y) = x - y + 2. \tag{1.25}$$

∎

To obtain the general solution of a functional Equation (or system), the following considerations must be taken into account:

1. The general solution of the functional equations can depend on one or more arbitrary constants.

2. In addition to arbitrary constants, arbitrary functions can appear in the general solution. Thus, an infinite number of point conditions could be necessary to get a unique solution.

3. Unlike any other kind of equations, a single equation can determine several unknown functions.

4. To have a well defined equation, its domain of definition (integer, real, complex, etc.) and the domains and ranges of the functions appearing in the functional equation or system should be clearly established. It is important to mention that the general solution of a given functional equation is strongly dependent on its domain of definition.

5. To have a well defined equation, the class (continuity, measurability, differentiability, integrability, etc.) of admissible functions should be given.

The following examples illustrate these cases.

Example 1.9 (General solution with arbitrary constants). The general solution of the functional Equation (1.12) in the class of continuous functions is

$$f(x) = cx, \tag{1.26}$$

where c is an arbitrary constant (see Theorem 3.3). ∎

Example 1.10 (General solution with arbitrary functions). The general solution of Equation (1.14) in the class of all real functions of real variables is

$$f(x,y) = \begin{cases} \left(\sqrt{xy}\right)^n \phi\left(\dfrac{x}{y}\right) & \text{if } \ xy \neq 0, \\ ax^n & \text{if } \ y = 0, x \neq 0, \\ by^n & \text{if } \ x = 0, y \neq 0, \\ c\ (cn = 0) & \text{if } \ x = y = 0, \end{cases} \tag{1.27}$$

where ϕ is an arbitrary function and a, b and c are arbitrary constants.

The general solution of the system (1.17) in the class of continuous functions is

$$f(x,y,z) = \varphi(z)(x-y) + azy, \tag{1.28}$$

where φ is an arbitrary continuous function and a is an arbitrary constant. ∎

Example 1.11 (Single equation determining several unknown functions). The general solution of the functional Equation (1.13) in the class of continuous functions is

$$f(x) = cx + a + b, \quad g(x) = cx + a, \quad h(x) = cx + b, \tag{1.29}$$

where a and b are arbitrary constants (see Theorem 4.1).

The general solution of (1.15) in a certain class of functions (see Theorem 6.9) is

$$f(x,y) = \alpha[\alpha^{-1}(x) + \beta(y)], \quad g(x,y) = \beta^{-1}[\beta(x) + \beta(y)], \tag{1.30}$$

where α and β are arbitrary continuous and strictly monotonic functions.

∎

Example 1.12 (Cauchy's Equation III). The general solution of the functional equation

$$f(xy) = f(x) + f(y),$$

in the class of functions continuous-at-a-point is $f(x) = c\log|x|$ (c arbitrary real constant) if $f(x)$ is defined only for non-null real numbers, but is $f(x) = 0$ if it is defined for all real numbers. ∎

Hence, a functional equation is completely defined as soon as the following three units of information are given:

1. The equation $E[\mathbf{f}, \mathbf{x}] = 0$.

2. Its domain D.

3. The class $\mathcal{F}$ of admissible functions (including domains and ranges).

Thus, a functional equation can be considered as a triplet $(E, D, \mathcal{F})$. This means that we are interested in all functions $f \in \mathcal{F}$ such that

$$E[\mathbf{f}, \mathbf{x}] = 0, \quad \forall \mathbf{x} \in D,$$

where $\mathbf{f}$ and $\mathbf{x}$ are the vectors of unknown functions and variables, respectively.

Sometimes it is interesting to compare the general solutions of the same functional equation but with different domains and/or classes of admissible functions; that is, the solution of the equation with restricted or enlarged domains or classes of admissible functions. In other words, we want to compare the general solutions of

$$E\left[\mathbf{f}, \mathbf{x}\right] = 0, \quad \forall \mathbf{x} \in D_1, \quad \mathbf{f} \in \mathcal{F}_1 \tag{1.31}$$

and

$$E\left[\mathbf{f}, \mathbf{x}\right] = 0, \quad \forall \mathbf{x} \in D_2, \quad \mathbf{f} \in \mathcal{F}_2. \tag{1.32}$$

If the sets of all solutions of (1.31) and (1.32) are denoted by $\mathcal{S}_1$ and $\mathcal{S}_2$, respectively, we have

$$\mathcal{F}_2 \subseteq \mathcal{F}_1, \quad \text{and} \quad D_1 \subseteq D_2, \quad \Rightarrow \quad \mathcal{S}_2 \subseteq \mathcal{S}_1. \tag{1.33}$$

These implications are obvious because, on the one hand, an enlargement of the admissible functions allows an enlargement of the set of all solutions but never a reduction. On the other hand, an enlargement of the equation domain implies a more restrictive equation because this implies more restrictions on the unknown functions (conditions at more points); hence, the set of all solutions may be reduced.

This property suggests the following four methods for solving functional equations, which will be used frequently throughout the book.

Method A includes the following steps:

1. Enlarge the domain of the functional equation.

2. Find its general solution on the enlarged domain.

3. Use the obtained solutions as particular solutions of the equation on the initial domain.

Method B includes the following steps:

1. Restrict the class of admissible functions.

2. Find its general solution in the restricted class.

3. Use the obtained solutions as particular solutions of the equation in the unrestricted class.

Method C includes the following steps:

1. Restrict the domain of the functional equation.

2. Find its general solution on the restricted domain.

3. Find which of the solutions of (1.31) are true solutions of the equation with the initial or unrestricted domain.

Method D includes the following steps:

1. Enlarge the class of admissible functions.

2. Find its general solution in the enlarged class.

3. Determine which of the solutions of (1.31) belong to the initial class.

Of course, many other methods can be used, too. A description of the main methods for solving functional equations will be given in Chapter 2.

Note that methods A and B lead to *particular solutions* of the given functional equation for the initial domain and class. On the contrary, methods C and D allow the general solution to be obtained in a more restrictive situation. So, the obtained solution may not actually be a solution for the specified domain and class and an additional test is required. In such a case, we shall refer to them as *candidate solutions.* These concepts of general, particular and candidate solutions can be used to obtain computer solutions, even under changes in the domain and/or class of the equation.

Given a functional Equation $(E, D, \mathcal{F})$, it is interesting to find triplets $(E, D_1, \mathcal{F}_1)$ and $(E, D_2, \mathcal{F}_2)$ with $D_1 \subset D \subset D_2$ and $\mathcal{F}_2 \subset \mathcal{F} \subset \mathcal{F}_1$, such that they have the same general solution as the initial equation. In particular, one of the most interesting results is obtained when the sets $D_1, D_2, \mathcal{F}_1$ and $\mathcal{F}_2$ cannot be improved; that is, when they lead to an optimum (a minimum of restrictions) characterization of the general solution set.

Definition 1.6 (Equivalent functional equations). *Let*

$$E_1 [\mathbf{f}, \mathbf{x}] = 0, \quad \forall\, \mathbf{x} \in D, \quad \mathbf{f} \in \mathcal{F}_1 \tag{1.34}$$

$$E_2 [\mathbf{f}, \mathbf{x}] = 0, \quad \forall\, \mathbf{x} \in D, \quad \mathbf{f} \in \mathcal{F}_2 \tag{1.35}$$

be two functional equations. We say that equations (1.34) and (1.35) are equivalent if and only if their general solutions coincide. ■

For a detailed treatment of restricted domains see Kuczma (1978), Dhombres and Ger (1975) and Aczél and Dhombres (1989), Chapters 6 and 7.

Even though functional equations lead to a very long list of mathematical problems, as has been shown above, in this book, mainly addressed to applied scientists and engineers, we are more concerned about finding solutions to physical and engineering problems. We shall, therefore, intentionally omit discussions of many mathematical problems and details.

Exercises

1.1 State the system of functional equations that lead to the formula of the area of a trapezoid. How many arbitrary constants are expected to appear in its general solution?.

1.2 The area of a trapezoid is given by $f(b_1, b_2, a)$, where b_1 and b_2 are the bases, a is the height and f is a non-negative function to be determined. Fill the right hand sides of the incomplete system of functional equations

$$f(b_1, b_2, ka) = \dots$$

$$f(kb_1, kb_2, a) = \dots$$

based on some geometric properties of areas and the trapezoid.

1.3 Modify the interest functional equations in (1.3) to represent a more realistic behavior of the actual bank policy than the simple interest assumptions.

1.4 Find a system of functional equations to characterize the sum of complex numbers.

1.5 Write the functional equation associated with the distributivity property.

1.6 Show that the general solution

$$F(x, y) = f^{-1}[f(x) + f(y)],$$

of the associativity functional equation

$$F(F(x, y), z) = F(x, F(y, z)),$$

can be replaced by

$$F(x, y) = f^{-1}[f(x) \oplus f(y)],$$

where $\oplus$ is any associative operation.

1.7 Knowing that the general continuous-at-a-point solution of the functional equation

$$f(xy) = g(x) + h(y); \ \ Mx, y \in \mathbb{R}_{++}$$

is

$$\begin{array}{lll} f(x) = A\log(BCx), & g(x) = A\log(Bx), & h(x) = A\log(Cx), \\ f(x) = A + B, & g(x) = A, & h(x) = B, \end{array}$$

find, using method C, the general continuous-at-a-point solutions of

$$f(xy) = g(x) + h(y); \ \ x, y \in \mathbb{R} - \{0\},$$

$$f(xy) = g(x) + h(y); \ \ x, y \in \mathbb{R}.$$

1.8 The general continuous solution of the functional equation

$$f(x, yz) = f(f(x, y), z); \ \ x, y \in \mathbb{R}_{++}$$

is

$$f(x, y) = \phi\left(y\phi^{-1}(x)\right),$$

where ϕ is an arbitrary function. Determine, using the methods A, B, C and D, the kind (general, particular or candidate) of solution for this equation for the following domains and classes:

(a) Domain: $\mathbb{R}_{++}$. Class of functions: Arbitrary.
(b) Domain: $\mathbb{R}$. Class of functions: Continuous.
(c) Domain: $\mathbb{Z}_{++} = \{x \in \mathbb{Z}/x > 0\}$. Class of functions: Continuous.
(d) Domain: $\mathbb{R}_{++}$. Class of functions: Differentiable.
(e) Domain: $\mathbb{R}$. Class of functions: Differentiable.
(f) Domain: $\mathbb{Z}_{++}$. Class of functions: Arbitrary.

1.9 Under some regularity conditions, and in a certain class of functions, the general solutions of the following two functional equations

$$S(x + y, z) = H[S(x, z), S(y, z)] \tag{1.36}$$

and

$$S(x, z) = S(y, z)^{N(y,x)} \tag{1.37}$$

are

$$S(x, z) = w[f(z)x], \ \ H(x, y) = w[w^{-1}(x) + w^{-1}(y)]$$

and

$$S(x, z) = p(z)^{q(x)}, \ \ N(y, x) = \frac{q(x)}{q(y)},$$

respectively, where w is an invertible arbitrary function, f is a continuously differentiable arbitrary function, and p and q are arbitrary positive functions. Prove that the general solution of the system (1.36)–(1.37) is

$$S(x, z) = \beta(z)^{x^C},$$

where $\beta(z)$ is an arbitrary positive function and C is an arbitrary constant.

1.10 If $S(x, z)$ is the survivor function of a piece of length x, give the natural domain of the functional equations (1.36) and (1.37) and physical interpretations for them.

1.11 Give one example of two equivalent functional equations.

CHAPTER 2

Some methods for solving functional equations

2.1 Introduction

In Section 1.3 we showed some basic examples of functional equations and their solutions. However, it was not explained how these solutions were obtained. Unlike the field of differential equations, where a clear methodology to solve them exists, in functional equations such a methodology does not exist. In fact, in many cases "ad hoc" methods are required. This represents a great shortcoming and perhaps one of the reasons explaining why engineers and applied scientists have not incorporated functional equations into their daily work. To facilitate the use of functional equations we can:

- Elaborate a list including the main functional equations and their corresponding solutions.
- Identify the sets of equations which can be solved using the same methods.

This chapter is devoted to a complete description of some methods for solving functional equations.

The main methods for solving functional equations to be analyzed are:

1. Replacement of variables by given values
2. Transforming one or several variables
3. Transforming one or several functions
4. Using a more general equation
5. Treating some variables as constants
6. Inductive methods

7. Iterative methods

8. Separation of variables

9. Reduction by means of analytical techniques (differentiation, integration, etc.)

10. Mixed methods

In the following sections we give a full description of these methods and include some representative examples.

2.2 Replacement of variables by given values

If we replace one or several variables appearing in the functional equation by carefully selected values, some mathematical relations that give some of the unknown functions or simpler functional equations can be obtained. This method requires a final check of the resulting solutions because the previous replacement leads to equations associated with a set of necessary, but not sufficient, conditions for the functions to be solutions of the initial equation.

Theorem 2.1 (Homogeneous equations). *The most general solution of the equation*

$$f(yx) = y^k f(x); \quad x, y \in \mathbf{R}_+, \tag{2.1}$$

where f is a real function of a real variable and k is a given constant, is

$$f(x) = cx^k, \tag{2.2}$$

where c is an arbitrary constant. ■

Proof: With $x = 1$ in (2.1) we get $f(y) = cy^k$, where $c = f(1)$. This solution, taking into account the commutativity of the product of real numbers, satisfies (2.1) and then (2.2) is proved. Note that this proof only requires the existence of a unit element and the commutativity of the product. Thus, the same solution is valid for many other domains and classes of functions. ■

Example 2.1 (Replacing variables by constant values). To solve the functional equation

$$f(x + y) = f(y) + x; \quad x, y \in \mathbb{R}, \tag{2.3}$$

we make $y = 0$, to obtain

$$f(x) = x + k, \quad k = f(0), \tag{2.4}$$

where k is an arbitrary constant. Next we check that (2.4) satisfies (2.3). ■

Example 2.2 (Cosine function). The cosine function is the only function that satisfies the functional equation

$$f(x+y)+f(x-y)=2\cos(x)\cos(y);\quad x,y\in\mathbb{R}.$$

In fact, by setting $y=0$ it reduces to

$$f(x)+f(x)=2\cos(x)\cos(0)=2\cos(x)\quad\Rightarrow\quad f(x)=\cos(x).$$

■

Example 2.3 (Replacing variables by constant values). The functional equation

$$f(x+y)+f(x-y)=2f(x)\cos(y) \tag{2.5}$$

can be solved by making the following substitutions

$$\left.\begin{matrix} x=0 \\ y=t \end{matrix}\right\},\quad \left.\begin{matrix} x=\dfrac{\pi}{2}+t \\ y=\dfrac{\pi}{2} \end{matrix}\right\},\quad \left.\begin{matrix} x=\dfrac{\pi}{2} \\ y=\dfrac{\pi}{2}+t \end{matrix}\right\}.$$

Then, Equation (2.5) simplifies to

$$\begin{aligned} f(t)+f(-t) &= 2a\cos(t), \\ f(t+\pi)+f(t) &= 0, \\ f(t+\pi)+f(-t) &= 2f\left(\frac{\pi}{2}\right)\cos\left(\frac{\pi}{2}+t\right)=-2b\sin(t), \end{aligned}$$

with $a=f(0)$ and $b=f\left(\dfrac{\pi}{2}\right)$.

Subtracting the third equation from the sum of the first two and dividing by 2, we conclude that

$$f(t)=a\cos(t)+b\sin(t),$$

where a and b are arbitrary constants. Since it satisfies (2.5), it is the general solution of the initial equation. ■

Example 2.4 (Using Cauchy's equation I). Every solution of the equation

$$f(zx+y)=zf(x)+f(y);\quad x,y,z\in\mathbb{R} \tag{2.6}$$

is a solution of the Cauchy Equation (see Section 3.4)

$$f(x+y)=f(x)+f(y).$$

Proof: With $z=1$ in (2.6) we get the result. ■

2.3 Transforming one or several variables

By transforming one or several of the variables appearing in the functional equation we can transform the given equations in others, the solutions of which are known.

Example 2.5 (Transforming one or several variables). The general solution of equation

$$G(x+z, y+z) = G(x,y) + z \tag{2.7}$$

is

$$G(x,y) = x + g(y-x), \tag{2.8}$$

where g is an arbitrary function.

Proof: Replacing $z = -x$ in (2.7) and calling $g(x) = G(0,x)$, we get (2.8), which satisfies (2.7). ∎

Theorem 2.2 (Cauchy's equation II). *The most general continuous-at-a-point solutions of the functional equation*

$$f(x+y) = f(x)f(y); \quad x, y \in \mathbb{R} \text{ or } x,\ y \in \mathbb{R}_{++} \tag{2.9}$$

are

$$f(x) = \exp(cx) \text{ and } f(x) = 0, \tag{2.10}$$

where c is an arbitrary constant. ∎

Proof: Replacing x and y by $t/2$ in (2.9) we obtain

$$f(t) = f\left(\frac{t}{2}\right)^2 \quad \Rightarrow \quad f(t) \geq 0, \ \forall t.$$

Now, if $f(t_0) = 0$ for some t_0 then $f(t) = f(t - t_0 + t_0) = f(t - t_0)f(t_0) = 0$ for all t. Thus, either $f(t) > 0$ for all t, or $f(t) \equiv 0$.
If $f(t) > 0$ for all t, then we can take logarithms on both sides in (2.9) and get

$$\log f(x+y) = \log\left[f(x)f(y)\right] = \log f(x) + \log f(y),$$

which, using the notation $g(x) = \log f(x)$, leads to

$$g(x+y) = g(x) + g(y),$$

which is Cauchy's equation with solution (see Theorem 3.3)

$$g(x) = cx.$$

Thus, we finally get

$$f(x) = \exp(cx).$$

∎

2.4 Transforming one or several functions

Similarly, we can transform one or several functions and get some equations with known solutions.

Example 2.6 (Transforming one function). If the following equation

$$f(x+y) = f(x) + f(y) + K,$$

where K is a real constant, is satisfied for all real x and y and if the function $f(x)$ is (a) continuous-at-a-point, or (b) not smaller than K for small x, or (c) bounded in an interval, then

$$f(x) = cx - K,$$

where c is an arbitrary constant.

Proof: With $g(x) = f(x) + K$, the functional equation becomes

$$g(x+y) = g(x) + g(y),$$

which is Cauchy's Equation (see Theorem 3.3). Then, the result holds. ∎

Example 2.7 (Translation equation). The general continuous solution of the translation equation

$$F[F(x,u),v] = F(x,u+v); \quad x, F(x,u) \in (a,b), \quad u,v \in (-\infty,\infty), \tag{2.11}$$

if, for given $x = x_0$, $F(x_0,t)$ is continuous and strictly monotonous, is

$$F(x,y) = f[f^{-1}(x) + y], \tag{2.12}$$

with f an arbitrary and strictly monotonous in $\mathbb{R}$ function.
Proof: Replacing $x = x_0$ in Equation (2.11) and calling

$$f(u) = F(x_0,u),$$

we get

$$F[f(u),v] = f(u+v).$$

Due to the fact that the function f is invertible, we can make $f(u) = y$, that is, $u = f^{-1}(y)$, to get

$$F(y,v) = f[f^{-1}(y) + v].$$

∎

2.5 Using a more general equation

Assume that we know the general solution of a functional equation with say n unknown functions. Assume also that we are asked about the general solution of an equation which is a particular case of the initial equation where some of the n functions are known. The general solution of this new equation can be easily obtained by forcing the known functions to fit into the general format solution of the starting equation. Some useful equations to be used in this group of methods are

$$\sum_{k=1}^{n} f_k(x)g_k(y) = 0, \tag{2.13}$$

$$F[G(x,y), H(u,v)] = K[M(x,u), N(y,v)], \tag{2.14}$$

$$G(x,y) = H[M(x,z), N(y,z)]. \tag{2.15}$$

The first equation will be described in Section 4.4, whereas the last two can be found in Section 6.4.

Example 2.8 (A particular case of the generalized bisymmetry equation). In this example we obtain the general solution of the functional equation

$$F[G(x,y), G(u,v)] = K[x+u, y+v], \tag{2.16}$$

by noting that this equation is a particular case of the generalized bisymmetry Equation (2.14), whose general solution is given by Theorem 6.4 (see Chapter 6), with

$$H(x,y) = G(x,y), \quad M(x,u) = x+u, \quad N(y,v) = y+v. \tag{2.17}$$

Introducing these relations into the general solution of (2.14) leads to new functional equations and extra relationships. If the new equations can be solved, then the solution of the initial problem can be found. ∎

Example 2.9 (The associativity equation). The associativity equation

$$F[F(x,y), u] = F[x, F(y,u)] \tag{2.18}$$

can be solved by its reference to Equation (2.14), that is, taking into account that

$$G(x,y) = F(x,y) = K(x,y) = N(x,y), \quad H(u,v) = u, \quad M(x,u) = x.$$

∎

Example 2.10 (Another example). The following two equations

$$F(x+y,u+v) = K[M(x,u),N(y,v)]$$

and

$$F(F(x,y),z) = F(x,F(y,z)),$$

can be solved as particular cases of the generalized bisymmetry equation

$$F[G(x,y),H(u,v)] = K[M(x,u),N(y,v)].$$

■

Example 2.11 (Using a Jensen's equation). The general continuous solution of equation

$$f\left(\frac{x+y}{2}\right) = \frac{g(x)+g(y)}{2} \tag{2.19}$$

is

$$f(x) = g(x) = cx + a,$$

where c and a are arbitrary constants.

Proof: Equation (2.19) is a particular case of the generalized Jensen Equation (described in Section 4.3 and Example 2.20):

$$f\left(\frac{x+y}{2}\right) = \frac{g(x)+h(y)}{2},$$

whose general solution is

$$f(x) = cx + \frac{a+b}{2}, \quad g(x) = cx + a, \quad h(x) = cx + b.$$

Finally, we merely need to identify g and h to get the solution. ■

2.6 Treating some variables as constants

If, after considering as constants some of the variables appearing in a functional equation, we are able to solve the resulting functional equation, then, by making the arbitrary constants and/or the functions in the resulting general solution depend on those variables, we obtain the general solution of the initial problem.

Example 2.12 (Using Cauchy's equation II). The general continuous solution with respect to the first variable of the functional equation

$$f(x+y,z) = f(x,z)f(y,z); \quad x,y,z \in \mathbb{R} \tag{2.20}$$

is

$$f(x,z) = \exp[c(z)x], \tag{2.21}$$

where c is an arbitrary function.

Proof: For each value of z, Equation (2.20) is a Cauchy equation, whose solution (given by Theorem 2.2) gives (2.21). ■

Example 2.13 (A homogeneous equation). The general solution of equation

$$f(\lambda x, y, z) = \lambda^k f(x, y, z) \tag{2.22}$$

is

$$f(x, y, z) = x^k c(y, z), \tag{2.23}$$

where c is an arbitrary function.
Proof: For each value of y and z, Equation (2.22) is a homogeneous equation. Its solution (given by Theorem 2.1) gives (2.23). ■

Example 2.14 (Using Cauchy's equation I). The general continuous solution with respect to the first variable of the functional equation

$$f(x + y, z) = f(x, y) + f(y, z); \quad x, y, z \in \mathbb{R} \tag{2.24}$$

is

$$f(x, z) = c(z)x, \tag{2.25}$$

where c is an arbitrary function. ■

2.7 Inductive methods

The induction method allows us to solve some functional equations.

Example 2.15 (Using the induction principle). The general solution of the functional equation

$$f(x + y) = f(x) + f(y) - f(x)f(y) \tag{2.26}$$

is

$$f(x) = 1 - a^x.$$

Proof: Substituting successively $y = x, y = 2x, \ldots, y = (n-1)x$ we get

$$\begin{array}{l} f(2x) = 2f(x) - f(x)^2 = 1 - [1 - f(x)]^2, \\ f(3x) = 3f(x) - 3f(x)^2 + f(x)^3 = 1 - [1 - f(x)]^3, \\ \cdots \ \cdots \ \cdots \ \cdots \ \cdots \ \cdots \ \cdots \ \cdots \ \cdots \ \cdots \ \cdots \ \cdots \\ f(nx) = \sum\limits_{k=1}^{n} (-1)^{k+1} \binom{n}{k} f(x)^k = 1 - [1 - f(x)]^n. \end{array} \tag{2.27}$$

The last expression above is proved by induction: of course, it is true for $n = 2$, and assuming that it is true for n, we finally obtain

$$\begin{array}{rcl} f[(n+1)x] & = & f(x + nx) \\ & = & f(x) + f(nx) - f(x)f(nx) \\ & = & f(x) + 1 - [1 - f(x)]^n - f(x)\{1 - [1 - f(x)]^n\} \\ & = & 1 - [1 - f(x)]^{n+1}. \end{array}$$

Then, substituting $x = 0, x = 1$ and $x = -1$ in the last expression of (2.27) we have

$$\begin{array}{ll} f(0) = 1 - [1 - f(0)]^n, & f(0) = 0, \ \ or \ \ f(0) = 1, \\ f(n) = 1 - [1 - f(1)]^n = 1 - a^n, & a = 1 - f(1), \\ f(-n) = 1 - [1 - f(-1)]^n = 1 - b^n, & b = 1 - f(-1). \end{array}$$

But $f(0) = 1$ implies $f(x) = 1$ (simply make $y = 0$ in (2.26)). If $f(0) = 0$, substitution of $x = 1$ and $y = -1$ into (2.26) leads to $ab = 1$, and then, $f(x) = 1 - a^x$ for any integer x. Thus, (2.27) shows that

$$f\left(n\frac{m}{n}\right) = 1 - \left[1 - f\left(\frac{m}{n}\right)\right]^n = f(m) = 1 - a^m,$$

and then

$$f\left(\frac{m}{n}\right) = 1 - a^{\frac{m}{n}}.$$

Thus, $f(x) = 1 - a^x$ for any rational.

Consequently, the general continuous solution of the functional Equation (2.26) is $f(x) = 1 - a^x$.

Note that Equation (2.26) can be easily reduced to Cauchy's equation by replacing $g(x) = 1 - f(x)$. ■

Example 2.16 (Using the induction principle). Equation

$$f(x_1 + y_1, \ldots, x_n + y_n) = f(x_1, \ldots, x_n) + f(y_1, \ldots, y_n); \quad x_i, y_i \in \mathbb{R} \ or \ \mathbb{R}_+,$$

can also be solved by induction over n (see Theorem 5.1). ■

2.8 Iterative methods

Some techniques related to iterative methods are also useful to solve some functional equations.

Example 2.17 (Iterative method). The Abel equation

$$f(g(t)) = f(t) + 1$$

can be transformed into the following equation

$$f(g^n(t)) = f(t) + n.$$

■

Other examples of iterative methods will be described in Section 5.4.

2.9 Separation of variables

If we can force some variables to appear on the right hand side of the equation and some others on its left hand side, then neither side must depend on the non common variables. This leads to new and normally simpler functional equations.

Example 2.18 (Separate variables). The general solution of the equation

$$f^{-1}(g(x)+h(y)) = \exp x; \quad x, y \in \mathbb{R} \tag{2.28}$$

is

$$g(x) = f(\exp x) - c, \quad h(x) = c, \tag{2.29}$$

with f an invertible arbitrary function and c an arbitrary constant.

Proof: Since f is invertible, Equation (2.28) can be written as

$$g(x) + h(y) = f(\exp x); \quad x, y \in \mathbb{R}$$

and then

$$-g(x) + f(\exp x) = h(y) = c,$$

which leads to Equation (2.29). ∎

2.10 Reduction by means of analytical techniques

Some other useful techniques are:

- Transformation of a functional equation into a differential equation
- Transformation of a functional equation into an integral equation
- Finding the solution over dense sets and extrapolating solutions by continuity.

  For example, Theorem 3.3 establishes the general solution of Cauchy's equation
  $$f(x+y) = f(x) + f(y); \quad x, y \in \mathbb{R}.$$
  This theorem is proved by stating it for rational numbers (a dense subset of $\mathbb{R}$) and then extending it to $\mathbb{R}$ by continuity (see proof of Theorem 3.3).
- Use of characteristic mappings and invariants.

One theorem and some illustrative examples are given below.

Theorem 2.3 (D'Alembert's functional equation). *The functional equation*

$$f(x+y) + f(x-y) = 2f(x)f(y); \quad x, y \in \mathbb{R} \tag{2.30}$$

has as general solutions the following continuous functions

$$f(x) = 1, \quad f(x) = 0, \quad f(x) = \cosh(Bx), \quad f(x) = \cos(Bx), \tag{2.31}$$

where B is an arbitrary constant. ∎

Proof: To solve D'Alembert's equation we initially set $y = 0$ and then $x = 0$ to obtain

$$f(0) = 1 \text{ or } f(x) = 0 \text{ and } f(y) = f(-y). \tag{2.32}$$

Then we differentiate twice with respect to y and set $y = 0$ and we get

$$f''(x) = kf(x) \Rightarrow f(x) = \begin{cases} a\cosh(\sqrt{k}x) + b\sinh(\sqrt{k}x) & \text{if } k > 0, \\ a + bx & \text{if } k = 0, \\ a\cos(\sqrt{-k}x) + b\sin(\sqrt{-k}x) & \text{if } k < 0, \end{cases} \tag{2.33}$$

where we have made $k = f''(0)$. Using now (2.32) we get $a = 1$ and $b = 0$. Thus, the general differentiable solution of (2.33) becomes

$$f(x) = \begin{cases} \cosh(\sqrt{k}x) & \text{if } k > 0, \\ 1 & \text{if } k = 0, \\ 0 & \\ \cos(\sqrt{-k}x) & \text{if } k < 0. \end{cases} \tag{2.34}$$

This proves the equivalence of (2.30) and (2.33). ■

Example 2.19 (Transforming to an integral equation). Cauchy's equation

$$f(x + y) = f(x) + f(y)$$

is equivalent to the following Volterra's integral equation:

$$\int_0^t (2t - 3u)f(u)du = 0.$$

■

2.11 Mixed methods

By mixed methods we understand a combination of the previous methods, as for example:

1. Multiple replacements.
2. Transforming variables and functions.
3. Replacements and changes of variables.
4. Replacements and changes of functions.

Example 2.3 shows a typical case in which multiple replacements allow the general solution of the given equation to be obtained. Other mixed methods are applied in the theorem and examples below.

Theorem 2.4 (Cauchy's equation III). *The most general solutions, which are continuous-at-a-point, of the functional equation*

$$f(xy) = f(x) + f(y); \quad x, y \in \mathbf{T} \tag{2.35}$$

are

$$f(x) = \begin{cases} c\log(x) & \text{if} \quad \mathbf{T} = \mathbb{R}_{++}, \\ c\log(|x|) & \text{if} \quad \mathbf{T} = \mathbb{R} - \{0\}, \\ 0 & \text{if} \quad \mathbf{T} = \mathbb{R}. \end{cases} \tag{2.36}$$

■

Proof: For positive x and y in Equation (3.9), we can make the following change of variables

$$\begin{cases} u = \log(x) & \Leftrightarrow \quad x = \exp(u), \\ v = \log(y) & \Leftrightarrow \quad y = \exp(v), \end{cases} \tag{2.37}$$

and get

$$f(e^u e^v) = f(e^{u+v}) = f(e^u) + f(e^v), \tag{2.38}$$

which is equivalent to

$$g(u+v) = g(u) + g(v), \tag{2.39}$$

where $g(x) = f[\exp(x)]$.

Thus, we obtain again a Cauchy Equation (see Theorem 3.3). So, under some mild regularity conditions, we can write

$$g(x) = cx \Rightarrow f(x) = c\log(x); \quad x \in \mathbb{R}_{++}. \tag{2.40}$$

If Equation (2.35) is satisfied for $y = 0$, then $f(0) = f(x) + f(0)$, which implies $f(x) = 0$ for all x.

Finally, if Equation (2.35) is satisfied for all $x \neq 0$ and $y \neq 0$, then we have $2f(t) = f(t^2) = 2f(-t)$ and then $f(x) = f(-x) = c\log(|x|)$. ■

Example 2.20 (Changing variable). The equation

$$f\left(\frac{x+y}{2}\right) = \frac{g(x) + h(y)}{2}; \quad x, y \in \mathbb{R} \tag{2.41}$$

is known as the *generalized Jensen equation.* Its general continuous solution is:

$$f(x) = ax + \frac{b+c}{2}, \quad g(x) = ax + b, \quad h(x) = ax + c, \tag{2.42}$$

where a, b and c are arbitrary functions.

Proof With $x = 0, y = s + t$ and considering (2.41) again, we get

$$f\left(\frac{s+t}{2}\right) = \frac{g(0) + h(s+t)}{2} = \frac{g(x) + h(y)}{2} \tag{2.43}$$

which is equivalent to

$$h(s+t) = g(x) + h(y) - g(0), \tag{2.44}$$

and setting

$$u(x) = g(x) - g(0), \tag{2.45}$$

gives the expression

$$h(s+t) = u(x) + h(y), \tag{2.46}$$

which is a Pexider Equation (see Section 4.2). ∎

Example 2.21 (Changing a function). The following equation:

$$k(x+y) = g(x)l(y) + h(y)$$

by making $y = 0$ and

$$\phi(y) = \frac{l(y)}{l(0)}, \quad \psi(y) = h(y) - h(0)\frac{l(y)}{l(0)},$$

becomes

$$k(x+y) = k(x)\phi(y) + \psi(y),$$

which will be solved in Section 4.3 for the $\mathbb{R}$ domain (see Theorem 4.8). ∎

Other methods, already mentioned, are based on the extension or restriction of domains and/or classes of functions. The fact that the general solution of a functional equation is strongly dependent on the domain or class in which it is stated is clearly shown in the following example.

Example 2.22 (Restricting the class of functions). The functions

$$f(x) \equiv 0, \tag{2.47}$$

$$f(x) = cx + 1, \tag{2.48}$$

$$f(x) = \begin{cases} 1 - \dfrac{x}{x_1} & \text{for } x \le x_1, \\ 0 & \text{for } x \ge x_1 > 0, \end{cases} \tag{2.49}$$

$$f(x) = \begin{cases} 0 & \text{for } x \le x_2 < 0, \\ 1 - \dfrac{x}{x_2} & \text{for } x \ge x_2, \end{cases} \tag{2.50}$$

are the continuous solutions of the functional equation

$$f(x + yf(x)) = f(x)f(y).$$

On the other hand, only the functions (2.48) and (2.49) are the differentiable solutions of the same equation. ∎

Exercises

2.1 Use the replacement of variables by given values to solve the following functional equations:

(a) $f(xy) = f(x)^y; \;\; x, y \in \mathbb{R}_+,$

(b) $f(xy) = f(x)^{y^k}; \;\; x, y \in \mathbb{R}_+,$

where k is a constant, and $\mathbb{R}_+$ denotes the set of the non-negative real numbers.

2.2 Reduce the functional equation

$$f(x - zy) = f(x) - zf(y); \quad x, y, z \in \mathbb{R}$$

to Cauchy's equation.

2.3 Given the functional equation

$$k(xy) = g(x)\; l(y) + h(y),$$

prove that the changes

$$\phi(y) = \frac{l(y)}{l(1)}, \quad \psi(y) = h(y) - \frac{h(1)}{l(1)}\; l(y),$$

transform this equation into

$$k(xy) = k(x)\phi(y) + \psi(y),$$

to be solved in Theorem 4.9.

2.4 The following two equations arise in the study of fatigue problems:

$$G(x, ks) = G(x, s)^k$$

and

$$F(x, z) = F(y, z)^{N(x,y)}.$$

Solve these two equations using the techniques described in this chapter.

2.5 The general solution of the Pexider equation

$$f(x + y) = g(x) + h(y)$$

is

$$f(x) = Ax + B + C; \quad g(x) = Ax + B; \quad h(x) = Ax + C.$$

Use this result to solve the functional equation

$$f_1[g_1(x) + h_1(y)] = f_2[g_2(x) + h_2(y)].$$

2.6 Solve the functional equation

$$f(xy) = f(x)f(y).$$

2.7 Solve the functional equation

$$2f(\frac{x+y}{2}) = f(x) + f(y).$$

2.8 The general solution of the auto-distributivity functional equation

$$F[G(x,y),z] = H[M(x,z),N(y,z)],$$

is

$$\begin{array}{rcl} F(x,y) & = & l[f(y)g^{-1}(x) + \alpha(y) + \beta(y)], \\ G(x,y) & = & g[h(x) + k(y)], \\ H(x,y) & = & l[m(x) + n(y)], \\ M(x,y) & = & m^{-1}[f(y)h(x) + \alpha(y)], \\ N(x,y) & = & n^{-1}[f(y)k(x) + \beta(y)], \end{array}$$

where g, h, k, l, m and n are arbitrary strictly monotonic and continuously differentiable functions, $f(a) = 0$ and f, α and β are arbitrary continuously differentiable functions.

Based on the previous known result, solve the following functional equation

$$F[G(x,y),z] = H[G(x,z),F(y,z)].$$

2.9 Solve the following functional equation

$$f(x+y) + f(x-y) = 2f(x)f(y); \quad x,y \in \mathbb{R}$$

by reducing it to a differential equation. Find its natural domain of definition for the solution to be valid.

2.10 Making the adequate changes of variables and functions, solve the following functional equation

$$f(x+y) = f(x)g(y) + h(y); \quad x,y \in \mathbb{R}.$$

CHAPTER 3

Equations for one function of one variable

3.1 Introduction

In this chapter we deal with functional equations with one unknown function of one variable. One application of this kind of equation, the well known formula for the sum of the internal angles of a polygon, has already been introduced in the first chapter (see Section 1.2). This chapter starts with the functional equation for homogeneous functions and then moves on to the famous equations of Cauchy, Jensen, D'Alembert and other such important equations. To illustrate the usefulness of these equations several examples of applications from the mathematical, physical and engineering sciences are included. Among them, the characterizations of the normal, the exponential and the composed Poisson distributions or a three-parametric family of distributions for approximating extremes are given. In the final part of this chapter, the case of linear difference equations is briefly analyzed.

3.2 Homogeneous functions

Homogeneous functions play an important role in Physics and Engineering and arise very frequently in applications. In this chapter we analyze the simplest case, which will be generalized in Chapter 5, Theorem 5.9.

The general solution of the homogeneous equation

$$f(yx) = y^k f(x) \quad ; \quad x, y \in \mathbb{R}_+ \tag{3.1}$$

is given by (see Theorem 2.1):

$$f(x) = cx^k, \tag{3.2}$$

where c is an arbitrary constant.

Table 3.1: Areas of regular polygons as a function of the side length

Number of sides	Area
3	$0.433\ x^2$
4	$1.000\ x^2$
5	$1.721\ x^2$
6	$2.598\ x^2$
7	$3.634\ x^2$
8	$4.828\ x^2$
9	$6.182\ x^2$
10	$7.694\ x^2$
11	$9.366\ x^2$
12	$11.19\ x^2$

Example 3.1 (Areas and Volumes of geometric figures). Given a geometric figure such that its area depends only on a single parameter, such as a circle, a regular polygon, a regular polyhedron, etc., we call $f(x)$ the function giving its area, where x is the associated parameter (radius, side length, etc.). By dimensional analysis, it is easy to observe that these formulas must satisfy the functional equation $f(yx) = y^2 f(x)$, i.e., the area of a figure with parameter x times y is y^2 times the area of the figure with parameter x. Thus, because of Theorem 2.1, $f(x)$ must be of the form $f(x) = Cx^2$. Note that $C = \pi$ for the circle if x is the radius, $C = 1$ for the square if x is the side length and $C = 6$ for the lateral area of a cube if x is the edge length.

Table 3.1 gives the values of the constant C associated with regular polygons as a function of the number of sides.

Similarly, the volume of a family of figures depending on a single parameter (sphere, regular polyhedron, etc.) is such that $f(yx) = y^3 f(x)$. Thus, it can be written as $f(x) = Cx^3$, where C is a constant which depends on the family being considered. As an example, $C = 4\pi/3$ for the sphere of radius x.

Table 3.2 gives the lateral areas and volumes of some regular polyhedra.

∎

Example 3.2 (A general turbulent evaporation formula). (Kahlig (1990)) In the problem of turbulent evaporation from a water surface in the presence of wind, the following parameters are considered to be important (dimensions are given in brackets and expressed by length L and time T):

- V_T evaporation under turbulent conditions $[LT^{-1}]$

Table 3.2: Lateral areas and volumes of different geometric elements as a function of the edge lengths

Element	Lateral area	Volume
Tetrahedron	$1.7321a^2$	$0.11785a^3$
Octahedron	$3.4641a^2$	$0.471404a^3$
Dodecahedron	$20.6458a^2$	$7.663119a^3$
Icosahedron	$8.6603a^2$	$2.181695a^3$

- $q_s - q$ humidity difference (saturation deficit) where q is the specific humidity of the air and q_s is the specific humidity of saturation at the temperature of the water surface (both are dimensionless)
- K mean turbulent diffusion coefficient of water vapor in the air $[L^2T^{-1}]$
- S water surface area $[L^2]$
- U wind speed (mean value) $[LT^{-1}]$

From these parameters, exactly three independent dimensionless quantities can be formed, e.g.

- $\Pi_1 = V_T S^{1/2}/K$ evaporative Reynolds number
- $\Pi_2 = US^{1/2}/K$ advective Reynolds number
- $\Pi_3 = q_s - q$ saturation deficit

According to the Pi Theorem, a functional relation exists and is equivalent to

$$\Phi(\Pi_1, \Pi_2, \Pi_3) = 0 \quad \text{or explicitly} \quad \Pi_1 = f(\Pi_2, \Pi_3)$$

where Φ and f are unknown functions.
Additional physical information leads to:

- In a saturated environment ($q = q_s$) there is no evaporation, i.e., $V_T = 0$ if $\Pi_3 = 0$, therefore $f(\Pi_2, 0) = 0$.
- In general, the saturation deficit ($q = q_s$) is small, therefore a Taylor series expansion of f (with respect to Π_3) appears as feasible

$$f(\Pi_2, \Pi_3) = f(\Pi_2, 0) + \left.\frac{\partial f}{\partial \Pi_3}\right|_{\Pi_3=0} \Pi_3 + \ldots = 0 + g(\Pi_2)\Pi_3 + \ldots$$

where g is an unknown function. Thus, to first order, $\Pi_1 = g(\Pi_2)\Pi_3$.

- When the wind ceases, there is no (turbulent) evaporation:

$$V_T = 0 \text{ if } \Pi_2 = 0 \;\Rightarrow\; f(0, \Pi_3) = 0 \;\Leftrightarrow\; g(0) = 0.$$

- It is reasonable to postulate that function $g(\Pi_2)$ is homogeneous (of a certain degree γ)

$$g(k\Pi_2) = k^\gamma g(\Pi_2),$$

  which, in physical interpretation, implies a certain kind of similarity. But this is the equation in Theorem 2.1 and then, we get $g(x) = \beta x^\gamma$. Thus, the turbulent component of evaporation becomes

$$V_T = \beta U^\gamma (S^{1/2}/K)^{\gamma-1}(q_s - q).$$

■

3.3 A general type of equation

In this section we deal with equations of the form

$$H[f(x+y), f(x-y), f(x), f(y), x, y] = 0, \tag{3.3}$$

where H is a known function and f is the unknown function to be determined.

There are many possible methods to solve this equation. One of them consists of making $y = 0$ in (3.3) to get

$$H[f(x), f(x), f(x), f(0), x, 0] = 0, \tag{3.4}$$

and then solving this equation for $f(x)$.

Thus, we have the following theorem (see Aczél (1966), page 22).

Theorem 3.1 (Equation of the form 3.3). *If (3.4) can be solved for $f(x)$, then its solution is the only possible solution of the functional Equation (3.3). It contains, at most, one arbitrary constant.* ■

Example 3.3 (Cosine function). We know that the cosine function satisfies the functional equation

$$f(x+y) + \cos(x-y) = 2f(x)f(y); \quad x, y \in \mathbb{R}, \tag{3.5}$$

but, is the cosine the only function satisfying this equation?

Setting $x = 0$, we can write

$$f(y) + \cos(-y) = f(y) + \cos(y) = 2f(0)f(y) \;\Rightarrow\; f(y) = \frac{\cos(y)}{[2f(0)-1]},$$

where we have assumed $f(0) \neq 1/2$, because it leads to a contradiction.

Making $y = 0$, we finally obtain

$$f(0) = \frac{1}{[2f(0)-1]} \;\Rightarrow \begin{cases} f(0) = 1 & \Rightarrow \quad f(x) = \cos(x), \\ \text{or} & \\ f(0) = -\dfrac{1}{2} & \Rightarrow \quad f(x) = -\dfrac{1}{2}\cos(x), \end{cases}$$

but $f(x) = -\frac{\cos(x)}{2}$ is not a solution of (3.5). Thus, $f(x) = \cos(x)$ is its unique solution. ■

For another example related to the cosine function, see Example 2.2.

Another method for obtaining the solution of Equation (3.3) consists of taking $x = 0$ and $y = t$ and $y = -t$ to get

$$\begin{array}{rcl} H[f(t), f(-t), f(0), f(t), 0, t] & = & 0, \\ H[f(-t), f(t), f(0), f(-t), 0, -t] & = & 0. \end{array} \tag{3.6}$$

Thus, we have the following theorem.

Theorem 3.2 (Equation of the form 3.3). *If 0, t and $-t$ belong to the domain of Equation (3.3), and if (3.6) can be solved for $f(t)$, after elimination of $f(-t)$, then the function obtained from it is the only possible solution of the functional Equation (3.3). It contains at most one arbitrary constant.* ■

Example 3.4 (Replacing variables by constant values). Let us consider the functional equation

$$f(x+y) + 2f(x-y) - 3f(x) - y = 0; \quad x, y \in \mathbb{R}.$$

Setting $x = 0, y = t$ and $y = -t$, yields

$$\left. \begin{array}{l} f(t) + 2f(-t) - 3f(0) - t = 0 \\ 2f(t) + f(-t) - 3f(0) + t = 0 \end{array} \right\} \Rightarrow f(t) = K - t$$

where K is an arbitrary constant. ■

Other substitutions, such as the two above, lead to similar theorems which give general solutions of Equation (3.3). An illustrative example can be found in Chapter 2 (see Example 2.3).

3.4 Cauchy's equations

In this section we give the solutions for the following functional equations

$$\text{Type I}: f(x+y) = f(x) + f(y); \quad x, y \in \mathbb{R}. \tag{3.7}$$

$$\text{Type II}: f(x+y) = f(x)f(y); \quad x, y \in \mathbb{R} \text{ or } \mathbb{R}_{++}. \tag{3.8}$$

$$\text{Type III}: f(xy) = f(x) + f(y); \quad x, y \in \mathbb{R}_{++} \text{ or } \mathbb{R} \text{ or } \mathbb{R} - \{0\}. \tag{3.9}$$

$$\text{Type IV}: f(xy) = f(x)f(y); \quad x, y \in \mathbb{R}_{++} \text{ or } \mathbb{R} \text{ or } \mathbb{R} - \{0\}, \tag{3.10}$$

where f is a real function of a real variable. These equations are known as Cauchy's equations.

Equation $f(x+y) = f(x) + f(y)$ was already solved by Cauchy (1821). Its solution is given by the following theorem:

Theorem 3.3 (Cauchy's main equation). *If Equation (3.7) is satisfied for all real x, y, and if the function $f(x)$ is (a) continuous-at-a-point, or (b) nonnegative for small x, or (c) bounded in an interval or (d) integrable or (e) measurable, then*

$$f(x) = cx, \quad x \in \mathbb{R}, \tag{3.11}$$

where c is an arbitrary constant, for all real x. ∎

Proof: Here we prove this theorem under the assumption of continuity only. First we show that $f(nx) = nf(x)$ when n is a positive integer. We prove this by induction. It is obviously true for $n = 1$. If we assume that it is true for n, for $n + 1$ we get:

$$f[(n+1)x] = f(nx + x) = f(nx) + f(x) = nf(x) + f(x) = (n+1)f(x).$$

Now we can show that this is true for any positive rational x. Let m and n be positive integers and x and t positive rational numbers such that $nx = mt$. Then we have

$$f(nx) = f(mt) \quad \Rightarrow \quad nf(x) = mf(t) \quad \Rightarrow \quad f(x) = f\left(\frac{m}{n}t\right) = \frac{m}{n}f(t) \Rightarrow$$
$$\Rightarrow \quad f\left(\frac{m}{n}\right) = \frac{m}{n}f(1) = \frac{m}{n}c,$$

which shows that $f(x) = cx$ for any positive rational x. But from (3.7), making $x = y = 0, f(0) = 0$, and with $y = -x$, we obtain

$$f(-x) = -f(x) = -cx = c(-x) \quad \text{and,} \quad f(0) = 0.$$

Thus, $f(x) = cx$ for any rational x. Finally, by the assumed continuity of $f(x)$, which implies continuity everywhere, we obtain that $f(x) = cx$ for any real x.

For a complete demonstration of this theorem and other solutions of Cauchy's equation, see Aczél (1966) (pp. 31 and 35) and the references therein. ∎

Note that Cauchy's equation has already been used in Chapter 1 when the area of a rectangle and the simple interest examples were introduced (Examples 1.2.1 and 1.2.2, respectively).

Corollary 3.1 (Modified Cauchy equation). *If the following equation*

$$f(x + y) = f(x) + f(y) + K,$$

where K is a real constant, is satisfied for all real x and y and if the function $f(x)$ is (a) continuous-at-a-point, or (b) non smaller than K for small x, or (c) bounded in an interval, then

$$f(x) = cx - K,$$

where c is an arbitrary constant. ∎

Proof: Making $g(x) = f(x) + K$, the functional equation becomes

$$g(x+y) = g(x) + g(y),$$

which is Cauchy's Equation (3.7). Then, the result holds. ∎

Theorem 3.4 (Generalized Cauchy equation). *The general continuous solution of*

$$f(x_1 + x_2 + \ldots + x_n) = f(x_1) + f(x_2) + \ldots + f(x_n); \;\; x_i \in \mathbb{R} \;\; (i = 1, 2, \ldots, n) \tag{3.12}$$

is $f(x) = Cx$, where C is an arbitrary constant. ∎

Proof: Making $x_i = 0$ for all $i = 1, 2, \ldots, n$ we get $f(0) = 0$, and making $x_i = 0$ for $i = 3, 4, \ldots, n$ we get $f(x_1 + x_2) = f(x_1) + f(x_2)$, which is Cauchy's equation, whose solution satisfies (3.12). Thus, its general continuous solution is $f(x) = Cx$. ∎

The remaining Cauchy equations (3.8) to (3.10) can be easily solved by means of transformations. Thus Equation (3.8) has been solved in Theorem 2.2, whereas the general solution of Equation (3.9), which requires a combination of several methods to be obtained, is given in Theorem 2.4. A similar treatment leads to the following theorem:

Theorem 3.5 (Cauchy's Equation IV). *The most general solutions, which are continuous-at-a-point, of the functional equation*

$$f(xy) = f(x)f(y); \;\; x, y \in \mathbf{T} \tag{3.13}$$

are

$$\left.\begin{array}{l} f(x) = 1 \\ f(x) = \begin{cases} |x|^c & x \neq 0 \\ 0 & x = 0 \end{cases} \\ f(x) = \begin{cases} |x|^c sgn(x) & x \neq 0 \\ 0 & x = 0 \end{cases} \end{array}\right\} \; if \;\; \mathbf{T} = \mathbb{R}$$
$$\left.\begin{array}{l} f(x) = |x|^c \\ f(x) = |x|^c sgn(x) \end{array}\right\} \; if \;\; \mathbf{T} = \mathbb{R} - \{0\}$$
$$f(x) = x^c \;\; if \;\; \mathbf{T} = \mathbb{R}_{++} \tag{3.14}$$

where c is an arbitrary real number, together with

$$f(x) = 0 \;\; ; \;\; f(x) = \begin{cases} 0 & |x| \neq 1 \\ x & |x| = 1 \end{cases} \;\; ; \;\; f(x) = \begin{cases} 0 & |x| \neq 1 \\ 1 & |x| = 1 \end{cases}$$

which are common to the three domains. ∎

Example 3.5 (Characterization of the exponential distribution). Let us look for the continuous distribution functions which satisfy the no-aging property

$$P(X > s + t | X > t) = P(X > s) \text{ for all } s, t \geq 0,$$

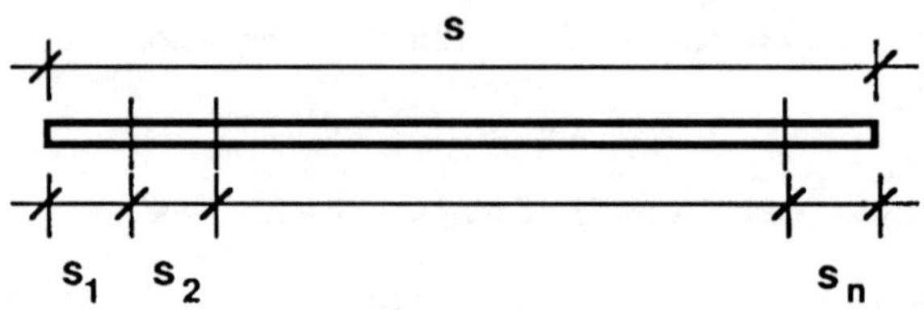

Figure 3.1: Longitudinal element and its constituting pieces.

that is, the survivor function does not change as time passes.

Then

$$1 - F_X(s) = P(X > s) = P(X > s + t/X > t) =$$
$$= \frac{P(X > s + t \text{ and } X > t)}{P(X > t)} = \frac{P(X > s + t)}{P(X > t)} = \frac{1 - F_X(s + t)}{1 - F_X(t)}$$

where $F_X(x)$ is the cumulative distribution function of X. Then, we get

$$1 - F_X(s + t) = [1 - F_X(s)][1 - F_X(t)] \quad \Rightarrow \quad G(s + t) = G(s)G(t)$$

where $G(x)$ is the survivor function.

According to Theorem 2.2

$$G(x) = \exp(cx) \;\Rightarrow\; F(x) = 1 - \exp(cx),$$

where $c < 0$ for $F(x)$ to be a cumulative distribution function. Hence, only the exponential distribution satisfies the above condition. ∎

Example 3.6 (Strength of longitudinal pieces). Let us assume a longitudinal element divided into non-overlapping and contiguous imaginary pieces (see Figure 3.1). Let $G(x, a)$ be the survivor function of the lifetime of a longitudinal piece of length "a" and assume that the strengths of all pieces are independent. Then, the reliability function $G(x, s)$ of the lifetime of an element of length s must satisfy the functional equation:

$$G(x, s) = G(x, y + z) = G(x, y)G(x, z) \; ; \; s = y + z$$

which is simply Equation (3.8) for every constant x. Thus, according to Theorem 2.2,

$$G(x, s) = \exp[c(x)s] = \{\exp[c(x)]\}^s = [g(x)]^s,$$

where $g(x)$ must be non-negative, non-increasing and such that $g(0) = 1$ and $\lim_{x\to\infty} g(x) = 0$, if it is to be a reliability function, but otherwise arbitrary. Note that this is Expression (5.53); that is, the solution obtained in Example 5.7. ∎

Example 3.7 (Characterization of the normal distribution). Let us now find the cumulative distribution function of the standardized (zero mean and unit variance) random variables X such that the family $\{aX/a \in \mathbb{R}\}$ is closed under sums of independent random variables.

Taking into account the two following properties of the characteristic function

$$\varphi_{\alpha X+\beta Y}(t) = \varphi_X(\alpha t)\varphi_Y(\beta t) \; ; \; \varphi_{\alpha X}(t) = \varphi_X(\alpha t),$$

where X and Y are independent random variables and α and β are real numbers, we can write

$$\varphi(\alpha t)\varphi(\beta t) = \varphi\left(\sqrt{\alpha^2+\beta^2}t\right),$$

where we have taken into consideration that the variance of $\alpha X+\beta Y$ is $\alpha^2+\beta^2$. Now, making the change of function $\varphi(\sqrt{t}) = f(t)\exp[ig(t)]$ and the change of variables $u = (\alpha t)^2$ and $v = (\beta t)^2$, the above expression becomes

$$\begin{aligned} f(u+v)\exp[ig(u+v)] &= f(u)\exp[ig(u)]f(v)\exp[ig(v)] = \\ &= f(u)f(v)\exp\{i[g(u)+g(v)]\}, \end{aligned}$$

where $f(x)$ and $g(x)$ are real functions.

Now, identifying modula and arguments we deduce

$$\begin{aligned} f(u+v) &= f(u)f(v), \\ g(u+v) &= g(u)+g(v)+2h(u,v)\pi, \end{aligned}$$

where $h(u,v)$ is an integer function. But the continuity of g implies $h(u,v) = k$, where k is a constant integer. Thus, we have (see Theorem 2.2 and Corollary 3.1):

$$f(t) = \exp(c_1 t) \; ; \; g(t) = c_2 t - 2k\pi \; ; \; \varphi(t) = \exp[(c_1 + c_2 i)t^2 - 2k\pi i].$$

Taking into account that the mean value and the variance must be, respectively, zero and one, this leads to

$$\varphi(t) = \exp\left[-\frac{t^2}{2}\right],$$

which is the standardized normal distribution. ∎

Example 3.8 (Composed Poisson distribution). Let us assume that $n_k(t)$ is the probability of occurrence of k events in a period of duration t. Assume also that:

1. The number of events occurring in non-overlapping time intervals are independent.
2. The probability $n_k(t)$ depends only on the length t of the time interval.
3. The function $n_0(t)$ is continuous

Then, we have

$$n_0(t+u) = n_0(t)n_0(u) \ ; \ t, u \geq 0,$$

which has the solution (see Theorem 2.2)

$$n_0(t) = \exp(-at),$$

where a is an arbitrary positive constant. ■

3.5 Jensen's equation

The equation

$$f\left(\frac{x+y}{2}\right) = \frac{f(x)+f(y)}{2} \ ; \ x, y \in \mathbb{R} \tag{3.15}$$

is known as Jensen's Equation (Jensen (1906)).

Making $x = 0, y = s+t$ and considering (3.15) again, we get

$$f\left(\frac{s+t}{2}\right) = \frac{f(0)+f(s+t)}{2} = \frac{f(s)+f(t)}{2} \Leftrightarrow f(s+t) = f(s)+f(t)-f(0) \tag{3.16}$$

which is the same equation as in Corollary 3.1.

Thus, we get the following theorem.

Theorem 3.6 (Jensen's equation). *The most general continuous solution of (3.15) in all* $\mathbb{R}$ *is*

$$f(x) = Cx + A, \tag{3.17}$$

where C *and* A *are arbitrary constants.* ■

Example 3.9 (Tax function). In some countries, the tax function applied to married couples is based on the splitting policy. This means that the total income of the couple is divided by two and then each member of the couple pays the tax associated with that amount. In this example we try to answer the following question : is there any tax function such that the splitting process becomes unnecessary? If this is true, the following functional equation must hold

$$2f\left(\frac{x+y}{2}\right) = f(x) + f(y)$$

which states that the same amount is paid by the couple either by using the splitting policy or by ignoring it. But this is Jensen's equation (3.15) and then the tax function must be of the form $f(x) = Cx + A$. ■

Theorem 3.7 (Generalized Jensen equation). *The most general continuous solution of*

$$f\left(\frac{x_1+x_2+\ldots+x_n}{n}\right) = \frac{f(x_1)+f(x_2)+\ldots+f(x_n)}{n}, \tag{3.18}$$

where $x_1, x_2, \ldots, x_n \in \mathbb{R}$, *is* $f(x) = Cx + A$, *where* C *and* A *are arbitrary constants.* ■

Proof: Put $x_1 = y_1 + y_2, x_2 = \ldots = x_n = 0$. Then, as in the case $n = 2$, we get

$$f\left(\frac{x_1 + \ldots + x_n}{n}\right) = f\left(\frac{y_1 + y_2}{n}\right) = \frac{1}{n}\left[f(y_1) + f(y_2) + (n-2)f(0)\right]$$

$$\frac{1}{n}\left[f(x_1) + \ldots + f(x_n)\right] = \frac{1}{n}\left[f(y_1 + y_2) + (n-1)f(0)\right]$$

hence

$$f(y_1 + y_2) = f(y_1) + f(y_2) - f(0)$$

and, by Corollary 3.1, the assertion follows. ∎

3.6 Generalizations of Cauchy's equations

A generalization of Cauchy's equations (3.7) and (3.8) is the functional equation

$$f(x + y) = F[f(x), f(y)], \tag{3.19}$$

where f is a real function of a real variable and F is a given function of two variables.

In Section 2.7 we presented a general method for solving (3.19).

Equations of form (3.19) are often called addition theorems. If F is a polynomial, rational or algebraic function we speak of a polynomial, rational or algebraic addition theorem. The following two theorems give some particular solutions for (3.19).

Theorem 3.8 (Polynomial addition theorem). *If $F(u, v)$ is a polynomial, the general solutions of Equation (3.19), which are continuous-at-a-point, are*

$$f(x) = Ax + C \quad \textit{and} \quad f(x) = \frac{\exp(Cx) - B}{A} \tag{3.20}$$

where $A \neq 0, B$ and C are arbitrary constants. The associated polynomials are

$$F(u, v) = u + v - C \quad \text{and} \quad F(u, v) = Auv + Bu + Bv + \frac{B^2 - B}{A} \tag{3.21}$$

∎

Theorem 3.9 (Rational addition theorem). *If $F(u, v)$ is a rational function, the general solutions of Equation (3.19), which are continuous-at-a-point, are*

$$f(x) = \frac{Mx + N}{Px + Q} \quad \text{and} \quad f(x) = \frac{M\exp(Kx) + N}{P\exp(Kx) + Q}, \tag{3.22}$$

where M, N, P, Q and K are arbitrary constants such that $PQ \neq 0$. The associated rational function F for the first solution in (3.22) is

$$F(u, v) = \frac{Auv + Bu + Cv + D}{Euv + Fu + Hv + J}, \tag{3.23}$$

where

$$
\begin{aligned}
A = 2MQ - PN \; ; \; B = C = \frac{-M^2Q}{P} \; ; \; J = \frac{M(2PN - MQ)}{P} \; ; \\
D = \frac{M^2N}{P} \; ; \; F = H = -PN \; ; \; E = PQ.
\end{aligned}
\tag{3.24}
$$

∎

Example 3.10 (Rational addition theorem). Find a real transformation, $y = f(x)$, such that it satisfies a rational addition theorem and such that $f(0) = 1/2, f(1) = 2/3, f(2) = 3/4$ and $f(3) = 4/5$.

According to Theorem 3.9 only two families of real transformations (see expressions in (3.22)) satisfy a rational addition theorem. Substitution of the above conditions into the first Equation (3.22) leads to the transformation

$$f(x) = \frac{x+1}{x+2}.$$

A different transformation can be obtained from the second equation in (3.22).

∎

Theorem 3.10 (Polynomial equation). *The general nonconstant continuous at-a-point solution of equation*

$$f(Ax + By + C) = Af(x) + Bf(y) + D \quad (AB \neq 0) \tag{3.25}$$

is

$$f(x) = \alpha x + \gamma, \tag{3.26}$$

where one of the three conditions must be satisfied:

- *(a)* $\alpha \neq 0$ *and* γ *are arbitrary constants if* $A + B = 1, C = D = 0$
- *(b)* $\alpha \neq 0$ *and* $\gamma = \dfrac{\alpha C - D}{A + B - 1}$ *if* $A + B \neq 1$
- *(c)* $\alpha = D/C$, *arbitrary* γ *if* $A + B = 1$ *and* $C \neq 0$.

∎

Finally, in Table 3.3 we give several equations of type (3.19) and their corresponding solutions. When dealing with domains, one has to be careful because the denominators in this table could be zero for certain pairs (x, y) and because of the double value of the square roots.

Example 3.11 (Interest formula). Assume that $f(x, t)$ is the capital you receive after a period of duration t when you deposit a capital x at the beginning of the period in a given savings account. Then, the function $f(x, t)$ must be increasing in both arguments and must satisfy the following conditions

$$
\left.
\begin{aligned}
f(x + y, t) &= f(x, t) + f(y, t) \\
f(x, t + u) &= f[f(x, t), u]
\end{aligned}
\right\} \; x, y, t, u \in \mathbb{R}_+
\tag{3.27}
$$

Table 3.3: Some functional equations and their corresponding general solutions on their natural domains.

EQUATION	SOLUTION
$f(x+y)=\dfrac{f(x)+f(y)}{1-f(x)f(y)}$	$f(x)=\tan(Ax)$
$f(x+y)=\dfrac{f(x)f(y)-1}{f(x)+f(y)}$	$f(x)=cot(Ax)$
$f(x+y)=\dfrac{f(x)+f(y)}{1+[f(x)f(y)/A^2]}$	$f(x)=A\tanh(Bx)$
$f(x+y)=\dfrac{f(x)f(y)}{f(x)+f(y)}$	$f(x)=\dfrac{C}{x}$
$f(x+y)=\dfrac{f(x)+f(y)-2f(x)f(y)}{1-2f(x)f(y)}$	$f(x)=\dfrac{1}{1+cot(Ax)}$
$f(x+y)=\dfrac{f(x)+f(y)-1}{2f(x)+2f(y)-2f(x)f(y)-1}$	$f(x)=\dfrac{1}{1+\tan(Ax)}$
$f(x+y)=\dfrac{f(x)+f(y)+2f(x)f(y)}{1-f(x)f(y)}$	$f(x)=\dfrac{Ax}{1-Ax}$
$f(x+y)=\dfrac{f(x)+f(y)-2f(x)f(y)}{1-f(x)f(y)}$	$f(x)=\dfrac{-Ax}{1-Ax}$
$f(x+y)=\dfrac{f(x)+f(y)-2f(x)f(y)\cos(a)}{1-f(x)f(y)}$	$f(x)=\dfrac{\sin(Ax)}{\sin(Ax+a)}$
$f(x+y)=\dfrac{f(x)+f(y)-2f(x)f(y)\cosh(a)}{1-f(x)f(y)}$	$f(x)=\dfrac{\sinh(Ax)}{\sinh(Ax+a)}$
$f(x+y)=\dfrac{f(x)+f(y)+2f(x)f(y)\cosh(a)}{1-f(x)f(y)}$	$f(x)=\dfrac{-\sinh(Ax)}{\sinh(Ax+a)}$
$f(x+y)=f(x)f(y)-\sqrt{1-f(x)^2}\sqrt{1-f(y)^2}$	$f(x)=\cos(Ax)$
$f(x+y)=f(x)f(y)+\sqrt{f(x)^2-1}\sqrt{f(y)^2-1}$	$f(x)=\cosh(Ax)$

The first equation states the fact that the final amount depends only on the total initial investment and not on the number of investments it can be divided into (in fact, it states this property for only two investments). The second equation states the fact that the final amount depends only on the total time $(t+u)$ the capital is deposited and not on the tentative period of deposit.

Of course, these two assumptions can be criticized but they are classical assumptions used everywhere.

It is interesting to point out the three following important facts:

- (i) if $f(x,t+u) < f[f(x,t),u]$ we get more if we cancel the actual deposit and initiate a new one by reinvestment.
- (ii) if $f(x+y,t) < f(x,t)+f(y,t)$ we are invited to make many small investments instead of a single one for the whole amount.
- (iii) if the inequalities in (i) and (ii) are reversed, the bank is offering more than necessary.

Thus, under this point of view, the optimal bank offer corresponds to the equalities in the above expressions.

The first equation in (3.27) is a family of Cauchy equations. Thus, assuming continuity with respect to x, its solution is given by

$$f(x,t) = c(t)x,$$

where $c(t)$ is an arbitrary increasing positive function of t.

Substitution into the second equation leads to

$$\left.\begin{array}{l} f(x,t+u) = c(t+u)x \\ f[f(x,t),u] = f[c(t)x,u] = c(u)c(t)x \end{array}\right\} \quad \Rightarrow \quad c(t+u) = c(u)c(t),$$

which is Equation (3.8). Thus, assuming continuity with respect to t, $c(t) = Q^t$, where Q is a constant larger than 1 but otherwise arbitrary. Consequently, the general continuous-at-a-point solution is

$$f(x,t) = xQ^t \quad \Leftrightarrow \quad f(x,t) = x(1+r)^t \ ; \ Q = 1+r \ ; \ r > 0,$$

which is the well known interest formula.

We wish to point out here that there are other interest formulas which do not satisfy the two assumptions (3.27) (see Chapter 12). ∎

Example 3.12 (Characterization of the normal distribution). (Aczél (1966), page 107) In this example we characterize the normal distribution by one important property. We try to find a probability density function of the form $f(x-a)$, where a is the parameter, such that the maximum likelihood estimate of "a" is the mean value of the sample. This implies that the maximum of the functions

$$L = \prod_{i=1}^{n} f(x_i - a) \quad \Leftrightarrow \quad V = \log L = \sum_{i=1}^{n} \log\left[f(x_i - a)\right]$$

must be attained at the sample mean. Thus, we must have

$$\sum_{i=1}^{n} h'\left(x_i - \frac{x_1 + x_2 + \ldots + x_n}{n}\right) = 0 \Leftrightarrow \sum_{i=1}^{n} h'(y_i) = 0 \ ; \ \sum_{i=1}^{n} y_i = 0,$$

where $h(x) = \log[f(x)]$.

Substituting $y_i = 0$ for all $i = 1, 2, \ldots, n$ we get $h'(0) = 0$, and substituting $y_i = 0$ for $i = 2, 3, \ldots, n-1$ we get $h'(y) = -h'(-y)$. Therefore, we have

$$\begin{aligned} h'(y_1 + y_2 + \ldots + y_{n-1}) &= h'(-y_n) = -h'(y_n) = \\ &= h'(y_1) + h'(y_2) + \ldots + h'(y_{n-1}), \end{aligned}$$

which is Equation (3.12). Thus, $h'(x) = cx$ and $h(x) = cx^2/2 + b$. Consequently, we have $f(x) = \exp(cx^2/2 + b)$. If we now impose the normalization condition

$$\int_{-\infty}^{\infty} f(x-a)dx = 1$$

we get

$$f(x-a) = \frac{1}{\sigma\sqrt{2\pi}} \exp\left[-\frac{(x-a)^2}{2\sigma^2}\right] \quad ; \quad -\infty < x < \infty,$$

which is a normal family of distributions. Note that $c = -1/\sigma^2$. ∎

3.7 D'Alembert's functional equation

Theorem 3.11 (D'Alembert's functional equation). *(D'Alembert (1750)) The functional equation*

$$f(x+y) + f(x-y) = 2f(x)f(y) \ ; \ x, y \in \mathbb{R} \tag{3.28}$$

has as general solutions the following continuous functions

$$f(x) = 1 \ ; \ f(x) = 0 \ ; \ f(x) = \cosh(Bx) \ ; \ f(x) = \cos(Bx), \tag{3.29}$$

where B is an arbitrary constant. ∎

Proof: See Theorem 2.3 in Section 2.10. ∎

3.8 Linear difference equations

Linear difference equations are equations of the form

$$\sum_{k=0}^{n} a_k(x) f(x-k) = h(x); \ \ a_0(x) = 1, \tag{3.30}$$

where $a_k(x), (k = 0, 1, \ldots, n)$ and $h(x)$ are given functions.

If all the $a_k(x)$ degenerate to constants we say that we have a *difference equation with constant coefficients* and if $h(x)$ is identically zero we say that we have a *homogeneous equation.*

The general solution of (3.30) can be written as the sum of a particular solution of (3.30) and the general solution of its associated homogeneous equation.

3.8.1 Solution of the homogeneous equation

In order to find the general solution of the constant coefficients homogeneous Equation (Equation (3.30) with $h(x) = 0$) we solve its associated *characteristic equation*

$$\sum_{k=0}^{n} a_k z^{n-k} = 0, \tag{3.31}$$

and we distinguish the following four cases:

- **Single real roots:**

  If z_1 is a single real root of (3.31), then its additive contribution to the general solution of (3.30) is $C_1 z_1^x$.

- **Multiple real roots:**

  If z_1 is a multiple root with multiplicity index p, then its additive contribution to the general solution is $(C_1 x^{p-1} + C_2 x^{p-2} + \ldots + C_p) z_1^x$.

- **Pairs of conjugate complex roots:**

  If $z_1 = \rho(\cos\alpha + i\sin\alpha)$ is a single complex root of (3.31), then the additive contribution of the pair of z_1 and its conjugate to the general solution of (3.30) is $\rho^x[A\cos(\alpha x) + B\sin(\alpha x)]$.

- **Pairs of multiple conjugate roots:**

  If $z_1 = \rho(\cos\alpha + i\sin\alpha)$ is a multiple complex root of (3.31) with multiplicity index p, then the additive contribution of the pair of z_1 and its conjugate to the general solution of (3.30) is

$$\begin{aligned} \rho^x[(A_1 x^{p-1} + A_2 x^{p-2} + \ldots + A_p)\cos(\alpha x) + \\ +(B_1 x^{p-1} + B_2 x^{p-2} + \ldots + B_p)\sin(\alpha x))]. \end{aligned} \tag{3.32}$$

3.8.2 Particular solution of the complete equation

To find a particular solution of Equation (3.30) we use the so called *method of variation of parameters*. We describe this method for second–order equations. Let us assume the second–order difference equation

$$f(x) + a_1(x)f(x-1) + a_2(x)f(x-2) = h(x), \tag{3.33}$$

and let $p(x)$ and $q(x)$ be two linearly independent solutions of the homogeneous difference equation

$$f(x) + a_1(x)f(x-1) + a_2(x)f(x-2) = 0. \tag{3.34}$$

We seek a solution of the form

$$f(x) = p(x)s(x) + q(x)t(x). \tag{3.35}$$

Substituting this into the complete Equation (3.33), we obtain

$$\begin{aligned}p(x)s(x)+q(x)t(x)+a_1(x)p(x-1)s(x-1)+a_1(x)q(x-1)t(x-1)+\\+a_2(x)p(x-2)s(x-2)+a_2(x)q(x-2)t(x-2)=h(x)\end{aligned} \tag{3.36}$$

and, taking into account that $p(x)$ and $q(x)$ satisfy (3.34), this equation becomes

$$\begin{aligned}-a_1(x)s(x)p(x-1)-a_2(x)s(x)p(x-2)-a_1(x)t(x)q(x-1)-\\-a_2(x)t(x)q(x-2)+a_1(x)p(x-1)s(x-1)+a_1(x)q(x-1)t(x-1)+\\+a_2(x)p(x-2)s(x-2)+a_2(x)q(x-2)t(x-2)=h(x)\end{aligned} \tag{3.37}$$

If we now choose $s(x)$ and $t(x)$ such that

$$p(x-1)[s(x)-s(x-1)]+q(x-1)[t(x)-t(x-1)]=0, \tag{3.38}$$

then, (3.37), taking into account (3.38) for x and $(x-1)$, shows that

$$-a_2(x)p(x-2)[s(x)-s(x-1)]-a_2(x)q(x-2)[t(x)-t(x-1)]=h(x). \tag{3.39}$$

Now we can solve the system (3.38)–(3.39) for $[s(x)-s(x-1)]$ and $[t(x)-t(x-1)]$ and we obtain

$$\begin{aligned}s(x)-s(x-1)&=+\frac{h(x)q(x-1)}{a_2(x)J(x)},\\ t(x)-t(x-1)&=-\frac{h(x)p(x-1)}{a_2(x)J(x)},\end{aligned} \tag{3.40}$$

where

$$J(x)=p(x-1)q(x-2)-q(x-1)p(x-2). \tag{3.41}$$

These are two inhomogeneous difference equations of first order with constant coefficients, whose solutions are

$$\begin{aligned}s(x+n)&=C_1+\sum_{j=r}^{n}\frac{h(x+j)q(x+j-1)}{J(x+j+1)},\\ t(x+n)&=C_2-\sum_{j=r}^{n}\frac{h(x+j)p(x+j-1)}{J(x+j+1)},\end{aligned} \tag{3.42}$$

where r is an arbitrary integer between 0 and n, C_1 and C_2 are arbitrary constants and we have taken into account that $J(x+1)=a_2(x)J(x)$.

Thus, the particular solution of (3.30) becomes

$$\begin{aligned}f(x+n)\;=\;&C_1p(x+n)+p(x+n)\sum_{j=r}^{n}\frac{h(x+j)q(x+j-1)}{J(x+j+1)}\\ &+C_2q(x+n)-q(x+n)\sum_{j=r}^{n}\frac{h(x+j)p(x+j-1)}{J(x+j+1)}.\end{aligned} \tag{3.43}$$

Example 3.13 (Sum of the internal angles of a polygon). In Section 1.2.3, we have obtained the function giving the sum of the internal angles of a polygon with n sides. Such a function f satisfies:

$$f(n+1) = f(n) + \alpha + \beta + \gamma = f(n) + f(3),$$

which is a difference equation. It can be solved by adding the homogeneous solution plus a particular solution. The characteristic equation now becomes $z - 1 = 0$, and then the homogeneous solution, according to all the above, becomes $f(x) = A$. A particular solution can be found by making $f(x) = Bx$ and then we get $B = f(3)$. Thus, its general solution is $f(x) = A + f(3)x$. The value of the constant A can be obtained by using the fact that $f(3) = \pi$ (see Example 1.2.3). Then, finally, we get the well known result $f(x) = \pi(x-2)$. ■

Example 3.14 (A three-parametric family of distributions to approximate the left tail). Sarabia and Castillo (1989) obtained a three parametric family of distributions to approximate the left tail in the following form. They base the work on a theorem given by Castillo et al. (1987) and Castillo and Ruiz-Cobo (1992) that says:

If F is the cumulative distribution function of a random variable and

$$\lim_{y \to 0} \frac{F^{-1}(y) - F^{-1}(2y)}{F^{-1}(2y) - F^{-1}(4y)} = 2^c = A > 0, \tag{3.44}$$

then $F(x)$ lies in the domain of attraction of $L_c(x)$ (Levy-Mises form). If $c > 0$ we have a Frechet type domain of attraction, if $c = 0$ we have a Gumbel type domain of attraction and if $c < 0$ we have a Weibull type domain of attraction.

Castillo and Sarabia force Equation (3.44) to be satisfied not only at the limit but in all range. Then, they obtain the functional equation

$$g(y) - (A+1)g(2y) + Ag(4y) = 0, \tag{3.45}$$

where $g(y) = F^{-1}(y)$ and $A = 2^c$.

Using now the following notation $a = \log 2, y = \exp(ax)$ and $f(x) = g[\exp(ax)]$, we get

$$Af(x+2) - (A+1)f(x+1) + f(x) = 0, \tag{3.46}$$

which is equivalent to

$$Af(x) - (A+1)f(x-1) + f(x-2) = 0. \tag{3.47}$$

The characteristic equation of (3.47) is

$$Az^2 - (A+1)z + 1 = 0. \tag{3.48}$$

There are two possible cases:

- Case 1: $A \neq 1$. Then, (3.48) has two real roots $z_1 = 1$ and $z_2 = 1/A$.
- Case 2: $A = 1$. Then, (3.48) has a double real root. $z_1 = 1$.

Thus, we have:

- Case 1:

$$f(x) = c_3 + c_1 A^{-x} \Rightarrow F^{-1}(x) = g(x) = c_3 + c_1 A^{-\frac{\log x}{\log 2}} \Rightarrow$$
$$\Rightarrow F(x) = \left(\frac{x - c_3}{c_1}\right)^{c_2}, \quad c_2 = -\frac{\log 2}{\log A}$$

- Case 2:

$$f(x) = c_3 + c_2 x \Rightarrow F^{-1}(x) = g(x) = c_3 + c_2 \frac{\log x}{\log 2} \Rightarrow$$
$$\Rightarrow F(x) = \exp\left(\frac{x - c_3}{c_1}\right), \quad c_1 = \frac{c_2}{\log 2}$$

Hence, if $F(x)$ is a cumulative distribution function we have the following three solutions:

- Case 1a ($A > 1$) :

$$F(x) = \begin{cases} \left(\dfrac{C_1}{C_3 - x}\right)^{c_2} & \text{if} \quad -\infty < x < C_3 - C_1 \\ 1 & \text{if} \quad x \geq C_3 - C_1 \end{cases}$$

for any arbitrary C_3.

- Case 1b ($A < 1$) :

$$F(x) = \begin{cases} 0 & \text{if} \quad x < C_3 \\ \left(\dfrac{x - C_3}{C_1}\right)^{c_2} & \text{if} \quad C_3 \leq x < C_1 + C_3 \\ 1 & \text{if} \quad x \geq C_1 + C_3 \end{cases}$$

for any arbitrary C_3.

- Case 2 (A=1) :

$$F(x) = \begin{cases} \exp\left(\dfrac{x - C_3}{C_1}\right) & \text{if} \quad x < C_3 \\ 1 & \text{if} \quad x \geq C_3 \end{cases}$$

for any arbitrary C_3.

where $C_1, C_2 > 0$.

■

Example 3.15 (Playing with scales). The functional equation

$$f(x) = af\left(\frac{x}{b}\right); \; a, b = \text{constant}; \; a > 0, b > 0 \tag{3.49}$$

can be solved by its reduction to a difference functional equation.

We shall distinguish two cases :

- First we assume $b \neq 1$. Then, making the change of function

$$h(x) = f(b^x) \;\; \Leftrightarrow \;\; f(x) = h\left(\frac{\log x}{\log b}\right),$$

  we have

$$h\left(\frac{\log x}{\log b}\right) = ah\left(\frac{\log x}{\log b} - 1\right),$$

  and calling now $u = \dfrac{\log x}{\log b}$ we get the difference equation

$$h(u) - ah(u-1) = 0 \;\; \Rightarrow \;\; h(u) = Ca^u,$$

  and then we finally obtain

$$f(x) = h\left(\frac{\log x}{\log b}\right) = Ca^{\frac{\log x}{\log b}} = Cx^{\frac{\log a}{\log b}}, \tag{3.50}$$

  which is the general solution of the functional Equation (3.49) for $b \neq 1$.

- If $b = 1$, (3.49) becomes $f(x) = af(x)$ and then the solution is $f(x) = 0$ if $a \neq 1$ and $f(x)$ arbitrary if $a = 1$.

■

Exercises

3.1 Show that the functional equation:

$$f(x+y) = f(x)\cos(y) + \sin(y)\cos(x)$$

characterizes the sine function.

3.2 Use Equation (3.19) to make a proposal for the tax function of a couple such that the tax of the couple could be determined as a function of the taxes of both partners. Use Theorem 3.8 to find the general solution when a polynomial relation is suggested, give a physical interpretation to it and choose a reasonable solution.

3.3 Use Equation (3.19) to derive a formula for the strength of a piece of size x. Assume that the strength of a piece of size $x + y$ is a function of the strengths of pieces of sizes x and y.

3.4 In Example 3.5 we have proved that the exponential distribution satisfies the functional equation

$$G(s+t) = G(s)G(t),$$

where $G(x)$ is the survivor function. An important bivariate extension is given by the functional equation

$$\overline{F}(s_1+t_1, s_2+t_2) = \overline{F}(s_1, s_2)\overline{F}(t_1, t_2) \quad \forall s_1, s_2, t_1, t_2 > 0,$$

where $F(s_1, s_2)$ is the cumulative distribution function. Prove that its general solution is

$$\overline{F}(s,t) = \exp\{-(\theta_1 s + \theta_2 t)\}; \quad \theta_1, \theta_2 > 0.$$

3.5 Try to give some physical or mathematic interpretations to the functional equations in Table 3.3.

Hint: Use their solutions for inspiration.

3.6 Write a functional equation to be satisfied by the tangent function based on some of their properties.

3.7 Discuss the solution of the functional equation

$$f(x+y) = F[f(x), f(y)],$$

such that

$$f(x_i) = \alpha_i; \quad i = 1, 2, 3, 4,$$

depending on the values of α_i; $i = 1, 2, 3, 4$

CHAPTER 4

Equations with several functions in one variable

4.1 Introduction

The last chapter was devoted to equations with one unknown function of one variable. The present chapter discusses the problem of equations with several unknown functions in one variable. The problem of several variables will be analyzed in Chapters 5 and 6. As indicated in Chapter 1, a single equation can, surprisingly, determine several unknown functions.

We look first, in Section 4.2, at Pexider's equations, which are natural generalizations of Cauchy's equations, treated in Chapter 3. Section 4.3 deals with an important type of functional equations, the sum of products equation. The key tool is Theorem 4.5 which is used in applications to finite elements and characterization of bivariate distributions by conditionals. This theorem will also be applied in Chapter 11 to solve some interesting problems arising in Computer Aided Geometric Design (CAGD), and in Chapter 13 to solve some problems in Probability Theory and Statistics. The section ends with a generalization of this theorem to n-dimensions. In Section 4.4, after solving other important generalizations, we present an interesting application to scale invariant equal sacrifice tax functions. Next, we give the general solution of a functional equation for complex functions of real variables, which is applied to characterize normal distributions. Finally, we give Theorems 4.11 and 4.12 which are useful in deriving solutions of functional equations if the solutions of other functional equations are known.

4.2 Pexider's equations

We start this chapter with the following direct generalizations of Cauchy's equations (3.7) to (3.10):

$$\begin{aligned}
\text{Type I}: &\ f(x+y) = g(x) + h(y); \quad x, y \in \mathbb{R} \text{ or } [a,b] \text{ with } a, b \in \mathbb{R} && (4.1)\\
\text{Type II}: &\ f(x+y) = g(x)h(y); \quad x, y \in \mathbb{R} && (4.2)\\
\text{Type III}: &\ f(xy) = g(x) + h(y); \quad x, y \in \mathbb{R} \text{ or } \mathbb{R}_{++} \text{ or } \mathbb{R} - \{0\} && (4.3)\\
\text{Type IV}: &\ f(xy) = g(x)h(y); \quad x, y \in \mathbb{R} \text{ or } \mathbb{R}_{++} \text{ or } \mathbb{R} - \{0\}, && (4.4)
\end{aligned}$$

where f, g and h are real functions of real variables. The general continuous-at-a-point solutions of the above equations are given in the following Theorems 4.1 to 4.4.

Theorem 4.1 (Pexider's equation I). *The most general system of solutions of (4.1) (Pexider (1903)) with f: (a) continuous-at-a-point, or (b) non-negative for small x, or (c) bounded in an interval, is*

$$f(x) = Ax + B + C \ ; \ g(x) = Ax + B \ ; \ h(x) = Ax + C, \tag{4.5}$$

where A, B and C are arbitrary constants. ■

Proof: Setting $x = 0$ in (4.1) we get

$$f(y) = g(0) + h(y) \quad \Rightarrow \quad h(y) = f(y) - a, \ \ a = g(0),$$

and making $y = 0$ we obtain

$$f(x) = g(x) + h(0) \quad \Rightarrow \quad g(x) = f(x) - b, \ \ b = h(0),$$

and substituting it back into (4.1) with $\alpha(x) = f(x) - a - b$ leads to

$$\alpha(x+y) = \alpha(x) + \alpha(y),$$

which is Cauchy's Equation (3.7). Thus, the theorem is proved. ■

Theorem 4.2 (Pexider's equation II). *The most general system of solutions of (4.2) with f continuous-at-a-point is*

$$f(x) = AB\exp(Cx); \ \ g(x) = A\exp(Cx); \ \ h(x) = B\exp(Cx), \tag{4.6}$$

where A, B and C are arbitrary non-zero constants, together with the trivial solutions

$$\left.\begin{array}{l} f(x) = 0; \ \ g(x) = 0; \ \ h(x) \ \textit{arbitrary} \\ f(x) = 0; \ \ g(x) \ \textit{arbitrary}; \ \ h(x) = 0. \end{array}\right. \tag{4.7}$$

■

Proof: Successively making $x = 0$ and $y = 0$ in (4.2), yields

$$f(y) = ah(y) \text{ with } a = g(0); \quad f(x) = g(x)b \text{ with } b = h(0).$$

If g(0) and h(0) are both different from zero, we have

$$f(x+y) = f(x)f(y)\frac{1}{ab},$$

which together with Theorem 2.2 leads to (4.6). If either g(0)=0 or $h(0) = 0$ then we get (4.7). ■

Theorem 4.3 (Pexider's equation III). *The most general system of solutions of (4.3) with f continuous-at-a-point is*

$$\left.\begin{array}{l} f(x) = A\log(BCx) \\ g(x) = A\log(Bx) \\ h(x) = A\log(Cx) \end{array}\right\} \text{ if (4.3) is valid for } x, y \in \mathbb{R}_{++},$$

$$\left.\begin{array}{l} f(x) = A\log(BC\,|x|) \\ g(x) = A\log(B\,|x|) \\ h(x) = A\log(C\,|x|) \end{array}\right\} \text{ if (4.3) is valid for } x, y \in \mathbb{R} - \{0\},$$

$$f(x) = A + B;\ g(x) = A;\ h(x) = B \ \text{ if } x, y \in \mathbb{R} \text{ or } \mathbb{R} - \{0\} \text{ or } \mathbb{R}_{++}. \tag{4.8}$$

■

Proof: Setting $x = 1$ and $y = 1$ in (4.3), it follows that

$$\begin{array}{lll} f(y) = g(1) + h(y) & \Rightarrow & h(y) = f(y) - a; \quad a = g(1) \\ f(x) = g(x) + h(1) & \Rightarrow & g(x) = f(x) - b; \quad b = h(1) \end{array}$$

and substituting it back into (4.3) and making $\alpha(x) = f(x) - a - b$ leads to

$$\alpha(xy) = \alpha(x) + \alpha(y),$$

which is Cauchy's Equation (2.35). Hence, according to Theorem 2.4, Expression (4.8) holds. ■

Theorem 4.4 (Pexider's equation IV). *The most general system of solutions of (4.4) with f continuous-at-a-point is*

$$f(x) = AB; g(x) = A; h(x) = B \ \text{ if } x, y \in \mathbb{R} \text{ or } \mathbb{R} - \{0\} \text{ or } \mathbb{R}_{++},$$

$$\left.\begin{array}{l} f(x) = ABx^C \\ g(x) = Ax^C \\ h(x) = Bx^C \end{array}\right\} \text{ if } x, y \in \mathbf{R}_{++},$$

$$\left.\begin{array}{l} f(x) = AB\,|x|^C \\ g(x) = A\,|x|^C \\ h(x) = B\,|x|^C \end{array}\right\} \text{ or } \left.\begin{array}{l} f(x) = AB\,|x|^C sgn(x) \\ g(x) = A\,|x|^C sgn(x) \\ h(x) = B\,|x|^C sgn(x) \end{array}\right\} \text{ if } x, y \in \mathbb{R} - \{0\}, \tag{4.9}$$

$$\left.\begin{array}{l} f(x) = \begin{cases} AB\,|x|^C & x \neq 0 \\ 0 & x = 0 \end{cases} \\ g(x) = \begin{cases} A\,|x|^C & x \neq 0 \\ 0 & x = 0 \end{cases} \\ h(x) = \begin{cases} B\,|x|^C & x \neq 0 \\ 0 & x = 0 \end{cases} \end{array}\right\} \quad or \tag{4.10}$$

$$\left.\begin{array}{l} f(x) = \begin{cases} AB\,|x|^C\,sgn(x) & x \neq 0 \\ 0 & x = 0 \end{cases} \\ g(x) = \begin{cases} A\,|x|^C\,sgn(x) & x \neq 0 \\ 0 & x = 0 \end{cases} \\ h(x) = \begin{cases} B\,|x|^C\,sgn(x) & x \neq 0 \\ 0 & x = 0 \end{cases} \end{array}\right\} \quad if\ x, y \in \mathbb{R}, \tag{4.11}$$

where A, B and C are arbitrary constants, together with the trivial solutions

$$\begin{array}{l} f(x) = 0;\ \ g(x) = 0;\ \ h(x)\ arbitrary, \\ f(x) = 0;\ \ g(x)\ arbitrary;\ \ h(x) = 0. \end{array} \tag{4.12}$$

■

Proof: If $g(1) \neq 0$ and $h(1) \neq 0$, then, setting $x = 1$ and $y = 1$ in (4.4), yields

$$\begin{array}{l} f(y) = g(1)h(y) \ \Rightarrow\ h(y) = f(y)/A,\ \ A = g(1), \\ f(x) = g(x)h(1) \ \Rightarrow\ g(x) = f(x)/B,\ \ B = h(1), \end{array}$$

and by substituting it back into (4.4) and making $\alpha(x) = f(x)/(AB)$ we obtain

$$\alpha(xy) = \alpha(x)\alpha(y),$$

which is Cauchy's Equation (3.10). Hence, according to Theorem 3.5, Expressions (4.9) hold. If, on the contrary, $g(1) = 0$ or $h(1) = 0$, then (4.12) holds. ■

Another alternative proof of this theorem for the $\mathbb{R}_{++}$ domain can be found in Section 4.4 (see Example 4.6).

4.3 The sum of products equation

In this section we deal with equations of the form $\sum_{k=1}^{n} f_k(x)g_k(y) = 0$, that appear very frequently in applications.

Theorem 4.5 (Sum of products equation). *All solutions of the equation*

$$\sum_{k=1}^{n} f_k(x)g_k(y) = 0 \tag{4.13}$$

can be written in the form

$$\begin{bmatrix} f_1(x) \\ f_2(x) \\ \cdots \\ f_n(x) \end{bmatrix} = \begin{bmatrix} a_{11} & a_{12} & \cdots & a_{1r} \\ a_{21} & a_{22} & \cdots & a_{2r} \\ \cdots & \cdots & \cdots & \cdots \\ a_{n1} & a_{n2} & \cdots & a_{nr} \end{bmatrix} \begin{bmatrix} \varphi_1(x) \\ \varphi_2(x) \\ \cdots \\ \varphi_r(x) \end{bmatrix},$$
$$\begin{bmatrix} g_1(y) \\ g_2(y) \\ \cdots \\ g_n(y) \end{bmatrix} = \begin{bmatrix} b_{1r+1} & b_{1r+2} & \cdots & b_{1n} \\ b_{2r+1} & b_{2r+2} & \cdots & b_{2n} \\ \cdots & \cdots & \cdots & \cdots \\ b_{nr+1} & b_{nr+2} & \cdots & b_{nn} \end{bmatrix} \begin{bmatrix} \psi_{r+1}(y) \\ \psi_{r+2}(y) \\ \cdots \\ \psi_n(y) \end{bmatrix}, \tag{4.14}$$

where r is an integer between 0 and n, and $\{\varphi_1(x), \varphi_2(x), \ldots, \varphi_r(x)\}$ on the one hand and $\{\psi_{r+1}(x), \psi_{r+2}(x), \ldots, \psi_n(x)\}$ on the other are arbitrary systems of mutually linearly independent functions and the constants a_{ij} and b_{ij} satisfy

$$\begin{bmatrix} a_{11} & a_{21} & \cdots & a_{n1} \\ a_{12} & a_{22} & \cdots & a_{n2} \\ \cdots & \cdots & \cdots & \cdots \\ a_{1r} & a_{2r} & \cdots & a_{nr} \end{bmatrix} \begin{bmatrix} b_{1r+1} & b_{1r+2} & \cdots & b_{1n} \\ b_{2r+1} & b_{2r+2} & \cdots & b_{2n} \\ \cdots & \cdots & \cdots & \cdots \\ b_{nr+1} & b_{nr+2} & \cdots & b_{nn} \end{bmatrix} = \mathbf{0}. \tag{4.15}$$

■

Example 4.1 (Cover with polynomial cross sections). Assume that we have been asked to design a cover for a sports area in such a way that its associated construction process is easy. To this end we look for a cover with polynomial cross sections. In other words, we look for the most general function $Z = z(x, y)$ such that all sections with planes parallel to the axes are second degree polynomials, i.e., such that:

$$\begin{aligned} z(x,y) &= a(y)x^2 + b(y)x + c(y), \\ z(x,y) &= d(x)y^2 + e(x)y + f(x). \end{aligned} \tag{4.16}$$

Expressions (4.16) imply

$$a(y)x^2 + b(y)x + c(y) - d(x)y^2 - e(x)y - f(x) = 0, \tag{4.17}$$

which is a functional equation of the form (4.13).

According to Theorem 4.5, and because the sets of functions $\{x^2, x, 1\}$ and $\{y^2, y, 1\}$ are linearly independent, then $r = 3$ and (4.14) becomes

$$\begin{bmatrix} x^2 \\ x \\ 1 \\ d(x) \\ e(x) \\ f(x) \end{bmatrix} = \begin{bmatrix} 1 & 0 & 0 \\ 0 & 1 & 0 \\ 0 & 0 & 1 \\ a_{41} & a_{42} & a_{43} \\ a_{51} & a_{52} & a_{53} \\ a_{61} & a_{62} & a_{63} \end{bmatrix} \begin{bmatrix} x^2 \\ x \\ 1 \end{bmatrix}, \tag{4.18}$$

$$\begin{bmatrix} a(y) \\ b(y) \\ c(y) \\ -y^2 \\ -y \\ -1 \end{bmatrix} = \begin{bmatrix} b_{14} & b_{15} & b_{16} \\ b_{24} & b_{25} & b_{26} \\ b_{34} & b_{35} & b_{36} \\ -1 & 0 & 0 \\ 0 & -1 & 0 \\ 0 & 0 & -1 \end{bmatrix} \begin{bmatrix} y^2 \\ y \\ 1 \end{bmatrix}, \tag{4.19}$$

and (4.15) leads to

$$\begin{bmatrix} 1 & 0 & 0 & a_{41} & a_{51} & a_{61} \\ 0 & 1 & 0 & a_{42} & a_{52} & a_{62} \\ 0 & 0 & 1 & a_{43} & a_{53} & a_{63} \end{bmatrix} \begin{bmatrix} b_{14} & b_{15} & b_{16} \\ b_{24} & b_{25} & b_{26} \\ b_{34} & b_{35} & b_{36} \\ -1 & 0 & 0 \\ 0 & -1 & 0 \\ 0 & 0 & -1 \end{bmatrix} = \mathbf{0}, \tag{4.20}$$

which is equivalent to the system

$$\begin{array}{lllll} b_{14} = a_{41}; & b_{15} = a_{51}; & b_{16} = a_{61}; & b_{24} = a_{42}; & b_{25} = a_{52}; \\ b_{26} = a_{62}; & b_{34} = a_{43}; & b_{35} = a_{53}; & b_{36} = a_{63}. & \end{array} \tag{4.21}$$

Thus, with a new reparameterization we can write

$$\begin{array}{l} a(y) = A + By + Cy^2; \; b(y) = D + Ey + Fy^2; \; c(y) = G + Hy + Iy^2; \\ d(x) = I + Fx + Cx^2; \; e(x) = H + Ex + Bx^2; \; f(x) = G + Dx + Ax^2, \end{array} \tag{4.22}$$

and then we finally obtain

$$z(x,y) = Cx^2y^2 + Bx^2y + Fxy^2 + Ax^2 + Exy + Iy^2 + Dx + Hy + G, \tag{4.23}$$

where A, B, C, D, E, F, G, H and I are arbitrary constants, which is the desired solution.

Now a careful selection of the arbitrary constants leads to the cover in Figure 4.1.

■

Theorem 4.5 was generalized by Losonczi (1963), who gave the following theorem.

Theorem 4.6 (Extension of the sums of product equation). *For $k \geq 2$ the general solutions of the functional equation*

$$\sum_{i=1}^{n} f_i^1(x_1) f_i^2(x_2) \cdots f_i^k(x_k) = 0 \tag{4.24}$$

is

$$f_i^j(x_j) = \sum_{s=1}^{n_j} c_{is}^j F_s^j(x_j) \quad (i = 1, 2, \ldots, n; \; j = 1, 2, \ldots, k) \tag{4.25}$$

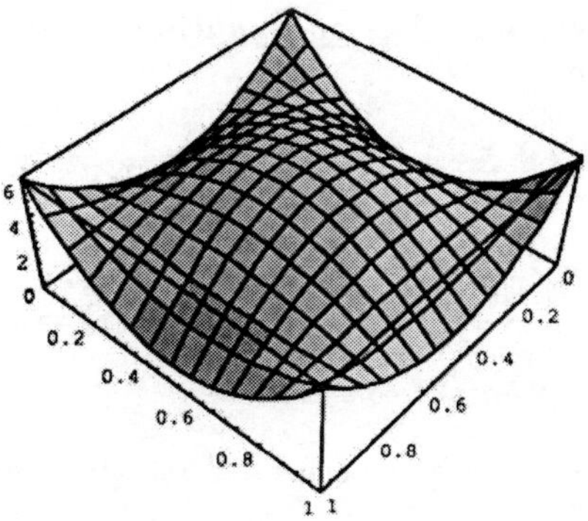

Figure 4.1: Cover with polynomial cross sections.

where $n_1, n_2, \ldots, n_k$ satisfy the following

$$0 \le n_j \le n \quad (j = 1, 2, \ldots, k) \tag{4.26}$$

$$\sum_{j=1}^{k} n_j = n(k-1) \tag{4.27}$$

(defining $\sum_1^0 = 0$).

The components of the functional system

$$\begin{array}{llll} F_1^1(x_1), & F_2^1(x_1), & \ldots & F_{n_1}^1(x_1) \\ F_1^2(x_2), & F_2^2(x_2), & \ldots & F_{n_2}^2(x_2) \\ \vdots & & & \\ F_1^k(x_k), & F_2^k(x_k), & \ldots & F_{n_k}^k(x_k) \end{array} \tag{4.28}$$

are grouped into linearly independent sets $\{F_s^j(x_j)\}$ $s = 1, 2, \ldots, n_j$ for $j = 1, 2, \ldots, k$, and are otherwise arbitrary.

The c_{is}^j are arbitrary, for j with $n_j = 0$, and otherwise must satisfy

$$\sum_{i=1}^{n} c_{is_1}^1 c_{is_2}^2 \cdots c_{is_k}^k = 0 \quad (s_j = 1, 2, \ldots, n_j;\ j = 1, 2, \ldots, k) \tag{4.29}$$

(Because of (4.26) and (4.27), there will be no more than one $n_j = 0$.) ∎

4.4 Other generalizations

This section includes another important sets of generalizations of functional equations.

The following theorem gives the general continuous solution of the generalized Jensen equation.

Theorem 4.7 (Generalized Jensen equation). *The most general continuous solution of equation*

$$f\left(\frac{x+y}{2}\right) = \frac{g(x)+h(y)}{2}; \quad x, y \in \mathbb{R} \tag{4.30}$$

is

$$f(x) = Ax + \frac{B+C}{2}; \quad g(x) = Ax + B; \quad h(x) = Ax + C, \tag{4.31}$$

where A, B and C are arbitrary constants. ■

Proof: This equation can be transformed into Pexider's Equation (4.1) by making

$$\alpha(x) = f\left(\frac{x}{2}\right), \quad \beta(x) = \frac{g(x)}{2}, \quad \gamma(x) = \frac{h(x)}{2},$$

and then, using Theorem 4.1, Expression (4.31) is obtained. ■

Theorem 4.8 (Other equation). *The most general systems of continuous solutions of equation*

$$f(x+y) = f(x)g(y) + h(y); \quad x, y \in \mathbb{R} \tag{4.32}$$

are

$$\begin{array}{l} f(x) = Ax + B; \quad g(x) = 1; \quad h(x) = Ax \\ f(x) = A\exp(Cx) + B; \quad g(x) = \exp(Cx); \quad h(x) = B[1 - \exp(Cx)], \end{array} \tag{4.33}$$

and the trivial solutions

$$f(x) = A; \quad g(x) = \textit{arbitrary continuous}; \quad h(x) = A[1 - g(x)], \tag{4.34}$$

where A, B and C are arbitrary constants. ■

Proof: Setting $x = 0$ in (4.32), leads to

$$f(y) = f(0)g(y) + h(y),$$

and, by subtracting this from the previous equation and the symmetry in x and y, we obtain

$$v(x+y) = v(x)g(y) + v(y) = v(y)g(x) + v(x); \quad v(x) = f(x) - f(0).$$

To solve this equation we consider two cases:

- $f(x)$ is not constant:
  - if $g(x) = 1$: Then, we have Cauchy's Equation (3.7) and we find the general continuous-at-a-point solution

$$v(x) = Ax \Rightarrow f(x) = Ax + B; \quad h(x) = Ax.$$

– if $g(x) \neq 1$: Then, there exists one y_0 such that $g(y_0) \neq 1$ and then

$$v(x) = a[g(x) - 1]; \quad a = \frac{v(y_0)}{g(y_0) - 1}.$$

Since f, and thus v, is not constant, $a \neq 0$ and the functional equation becomes

$$a[g(x+y) - 1] = a[g(x) - 1]g(y) + a[g(y) - 1] \Rightarrow g(x+y) = g(x)g(y),$$

which is Cauchy's Equation (3.8). Thus, we have (see Theorem 2.2):

$$g(x) = \exp(Cx); \quad f(x) = A\exp(Cx) + B; \quad h(x) = B[1 - \exp(Cx)].$$

- $f(x)$ constant: In this case (4.34) holds.

■

Theorem 4.9 (Another equation). *The most general non-constant and continuous solutions of equation*

$$f(xy) = f(x)g(y) + h(y); \quad x, y \in \mathbb{R}_{++} \tag{4.35}$$

are

$$\begin{array}{lll} f(x) = A\log(x) + B; & g(x) = 1; & h(x) = A\log(x); \\ f(x) = Ax^C + B; & g(x) = x^C; & h(x) = B[1 - x^C]; \\ f(x) = A; & g(x) = \textit{arbit. cont.}; & h(x) = A[1 - g(x)], \end{array} \tag{4.36}$$

where A, B and C are constants such that $A \neq 0$ and $C \neq 0$ but otherwise arbitrary. ■

Proof: Making the following change of variables and functions

$$\begin{array}{l} x = \exp(u); \quad y = \exp(v); \\ f^*(x) = f[\exp(x)]; \quad g^*(x) = g[\exp(x)]; \quad h^*(x) = h[\exp(x)]; \end{array}$$

the functional Equation (4.35) becomes

$$f^*(u+v) = f^*(u)g^*(v) + h^*(v),$$

which is Equation (4.32). Thus, using (4.33) and (4.34) we get (4.36). ■

Example 4.2 (Scale-invariant equal sacrifice taxation). (Aczél (1987b) page 22) The "equal sacrifice principle" in taxation states that taxes should be organized in such a manner that everybody's sacrifice (loss of utility) after taxes becomes the same. It is not the amount, x, of income that counts, but its utility $u(x)$. Under the equal sacrifice principle, $u(x) - u(y)$, i.e., the difference between the utilities of gross, x, and net, y, incomes should be a constant.

We postulate that the equality of utility sacrifices is scale-invariant, that is

$$u(x) - u(y) = u(x') - u(y') \Rightarrow u(rx) - u(ry) = u(rx') - u(ry'),$$

which is equivalent to

$$u(rx) - u(ry) = F[u(x) - u(y), r].$$

Thus, we can write

$$u(ry) - u(rz) = F[u(y) - u(z), r],$$

and then

$$u(rx) - u(rz) = F[t, r] + F[s, r] = F[t + s, r],$$

where we have called $t = u(x) - u(y)$ and $s = u(y) - u(z)$.

But this is Cauchy's Equation (3.7) for r held constant and then (see Example 2.14)

$$\begin{aligned} F[t, r] &= \alpha(r)t, \\ u(rx) - u(ry) &= \alpha(r)[u(x) - u(y)], \end{aligned}$$

which for $y = y_0$ becomes

$$u(rx) = \alpha(r)u(x) + \beta(r); \quad \beta(r) = u(ry_0) - \alpha(r)u(y_0).$$

But this is Equation (4.35). Thus, we conclude that the general continuous nonconstant utility functions leading to scale-invariant equal sacrifices are

$$u(x) = A\log(x) + B; \quad u(x) = Ax^C + B,$$

where A, B and C are constants such that $A \neq 0$ and $C \neq 0$ but otherwise arbitrary.

In addition, we can obtain the tax function associated with the above utility functions. In fact, if we consider that the utility loss must be constant, we obtain

$$u(x) - u[x - g(x)] = d,$$

where $g(x)$ is the tax function and d is the constant sacrifice. Thus, we can write

$$g(x) = x - u^{-1}[u(x) - d],$$

and, using the above utility functions, we finally get

$$g(x) = x\left[1 - exp\left(-\frac{d}{A}\right)\right], \quad g(x) = x\left[1 - \left(1 - \frac{d}{Ax^C}\right)^{\frac{1}{C}}\right].$$

Note that the first tax function implies constant tax rate. On the contrary, the second tax function is progressive for $A < 0$ and $C < 0$. ■

Theorem 4.10 (A measurable solution). *The general measurable solutions of the functional equation*

$$f(x)g(y) = h(ax + by)k(cx + dy), \tag{4.37}$$

where f, g, h and k are complex functions of real variables and a, b, c and d are fixed non-zero real numbers with $\Delta = ad - bc \neq 0$ are

$$\begin{aligned} f(x) &= \alpha_1 \exp(a_1 x + b_1 x^2) \; ; \; g(x) = \alpha_2 \exp\left(a_2 x - \frac{bd}{ac} b_1 x^2\right), \\ h(x) &= \beta_1 \alpha_1 \alpha_2 \exp\left(\frac{a_1 d - a_2 c}{\Delta} x + \frac{d}{\Delta a} b_1 x^2\right), \\ k(x) &= \tfrac{1}{\beta_1} \exp\left(\frac{a_2 a - a_1 b}{\Delta} x - \frac{b}{\Delta c} b_1 x^2\right), \end{aligned} \tag{4.38}$$

where a_1, a_2, b_1 are real arbitrary constants and $\alpha_1, \alpha_2, \beta_1$ are non-zero real arbitrary constants (see Baker (1974) and Lajkó (1973)). ■

Theorem 4.11 (Obtaining solutions from other equations). *Let A be a set and D a subset of A. Let $f : A \to A$ and $\wedge : A \times A \to A$ be a function and an operation in A, respectively. If the general invertible solution of equation*

$$f(x \wedge y) = f(x) \wedge f(y); \;\; \forall x, y \in D \subset A$$

is $f(x)$, then the general invertible solution of the functional equation

$$g(x * y) = g(x) * g(y); \;\; \forall x, y \in h^{-1}(D) \subset A,$$

where

$$x * y = h^{-1}[h(x) \wedge h(y)],$$

$g : A \to A, * : A \times A \to A$ and $h : A \to A$ is invertible, assuming that the obvious compatibility of domains and ranges of the implied functions holds, is the invertible function $g(x) = h^{-1} f h(x)$. ■

Proof: In fact we have

$$\left.\begin{array}{r} g(x * y) = g(x) * g(y) \\ x * y = h^{-1}\left[h(x) \wedge h(y)\right] \end{array}\right\} \Rightarrow h^{-1}\left[hg(x) \wedge hg(y)\right] = gh^{-1}\left[h(x) \wedge h(y)\right]$$
$$\Rightarrow hg(x) \wedge hg(y) = hgh^{-1}\left[h(x) \wedge h(y)\right],$$

and then

$$f(x) = hgh^{-1}(x) \;\; \Leftrightarrow \;\; g(x) = h^{-1} f h(x),$$

but, because we know the general invertible solution of $f(x)$ and function h, we know the general invertible solution of $g(x * y) = g(x) * g(y)$. ■

This theorem has many interesting applications. In particular, the general invertible solutions of Cauchy's equations (3.9), (3.8) and (3.10) can be obtained from the general invertible solution of Cauchy's Equation (3.7).

Example 4.3 (Using Cauchy's Equation I). Consider the functional equation

$$g\left(\frac{x}{y}\right) = \frac{g(x)}{g(y)}; \quad x, y \in (0, \infty). \tag{4.39}$$

Since we have

$$x * y = \frac{x}{y} = h^{-1}[h(x) \wedge h(y)] \text{ with } h(x) = \log(x),$$

and we know that the general invertible solution of the functional equation

$$f(x - y) = f(x) - f(y); \quad \forall x, y \in \mathbb{R}$$

is $f(x) = cx$ (see Theorem 3.3), then $x \wedge y = x - y$ and

$$g(x) = h^{-1} f h(x) = \exp[c \log(x)] = x^c; \quad x \in (0, \infty),$$

which gives the solution of (4.39). ■

Example 4.4 (Using Cauchy's Equation I). Similarly, the functional equation

$$g\left[\left(\sqrt{x} + \sqrt{y}\right)^2\right] = \left(\sqrt{g(x)} + \sqrt{g(y)}\right)^2; \quad x, y \in (0, \infty) \tag{4.40}$$

can be solved by considering

$$x * y = \left(\sqrt{x} + \sqrt{y}\right)^2 = h^{-1}[h(x) \wedge h(y)] \; ; \; h(x) = \sqrt{x}.$$

and that the general invertible solution of the functional equation

$$f(x + y) = f(x) + f(y); \quad \forall x, y \in (0, \infty)$$

is $f(x) = cx$ (see Theorem 3.3). Then, $x \wedge y = x + y$ and

$$g(x) = h^{-1} f h(x) = \left(c\sqrt{x}\right)^2 = c^2 x; \quad \forall x \in (0, \infty),$$

which solves (4.40). ■

Example 4.5 (Using Cauchy's Equation IV). For the functional equation

$$g[x^{\log(y)}] = g(x)^{\log[g(y)]}; \quad x, y \in (1, \infty) \tag{4.41}$$

we have

$$x * y = x^{\log(y)} = h^{-1}[h(x) \wedge h(y)]; \quad h(x) = \log(x),$$

and since the general invertible solution of the functional equation

$$f(xy) = f(x) f(y); \quad x, y \in (0, \infty)$$

is $f(x) = x^c$ (see Theorem 3.5), then $x \wedge y = xy$ and we finally obtain

$$g(x) = h^{-1} f h(x) = \exp\left[\log(x)\right]^c; \quad x \in (1, \infty),$$

which is the solution of (4.41). ■

Note that the previous Examples 4.3 to 4.5 illustrate one of the methods described in Chapter 2 to solve functional equations, namely, the use of a more general Equation (see Section 2.5).

Theorem 4.11 can be easily generalized in the following way.

Theorem 4.12 (Obtaining solutions from other equations). *Let A be a set and D a subset of A. Let $f_i : A \to A, (i = 1,2,3), \wedge : A \times A \to A$ and $\vee : A \times A \to A$ three functions and two operations in A, respectively. If the general invertible solution of equation*

$$f_1(x \wedge y) = f_2(x) \vee f_3(y); \quad x, y \in D \subset A$$

is $\{f_1(x), f_2(x), f_3(x)\}$, then the general invertible solution of the functional equation

$$g_1(x * y) = g_2(x)\Delta g_3(y); \quad x, y \in p^{-1}(D) \subset A, \tag{4.42}$$

where $g_i : A \to A, (i = 1,2,3), * : A \times A \to A, \Delta : A \times A \to A$,

$$x * y = p^{-1}[p(x) \wedge p(y)], \quad x\Delta y = q^{-1}[q(x) \vee q(y)],$$

and $p : A \to A$ and $q : A \to A$ are invertible, assuming that the obvious compatibility of domains and ranges of the implied functions holds, is invertible and $g_i(x) = q^{-1} f_i p(x)$, $(i = 1,2,3)$. ■

Proof: From (4.42) with $x * y = p^{-1}[p(x) \wedge p(y)]$ and $x\Delta y = q^{-1}[q(x) \vee q(y)]$ we have

$$\begin{aligned} g_1 p^{-1}[p(x) \wedge p(y)] &= q^{-1}[qg_2(x) \vee qg_3(y)], \\ qg_1 p^{-1}[p(x) \wedge p(y)] &= qg_2(x) \vee qg_3(y), \end{aligned}$$

which implies

$$qg_1 p^{-1}(x \wedge y) = qg_2 p^{-1}(x) \vee qg_3 p^{-1}(y),$$

and then

$$[\, f_i(x) = qg_i p^{-1}(x) \;\Leftrightarrow\; g_i(x) = q^{-1} f_i p(x) \,]; \quad i = 1,2,3.$$

■

Example 4.6 (Pexider's equation). The general invertible solution of the functional equation

$$g_1(xy) = g_2(x) g_3(y); \quad x, y \in (0, \infty)$$

can be obtained in the following manner. Here we have

$$x * y = x\Delta y = xy \quad \Rightarrow \quad p(x) = q(x) = \log(x).$$

But the general invertible solution of equation

$$f_1(x+y) = f_2(x) + f_3(y); \quad \forall x, y \in \mathbb{R}$$

is

$$f_1(x) = Ax + B + C; \quad f_2(x) = Ax + B; \quad f_3(x) = Ax + C$$

(see Theorem 4.1). Thus, we obtain

$$x \wedge y = x \vee y = x + y,$$

and then, we finally get

$$\begin{aligned} g_1(x) &= q^{-1} f_1 p(x) = \exp[A \log(x) + B + C] = B^* C^* x^A \; ; x \in \mathbb{R}_{++}, \\ g_2(x) &= q^{-1} f_2 p(x) = \exp[A \log(x) + B] = B^* x^A \; ; x \in \mathbb{R}_{++}, \\ g_3(x) &= q^{-1} f_3 p(x) = \exp[A \log(x) + C] = C^* x^A \; ; x \in \mathbb{R}_{++}, \end{aligned}$$

which is the invertible solution given by Theorem 4.4 . ∎

Corollary 4.1 *If $\{g_1, g_2, g_3\}$ is a invertible solution of (4.42) and $* = \Delta$, then $\{g_1^{-1}, g_2^{-1}, g_3^{-1}\}$ is also a solution of (4.42).*

Proof: Expression (4.42) implies

$$x * y = g_1^{-1}\left[g_2(x) * g_3(y)\right],$$

and making $x = g_2^{-1}(x)$ and $y = g_3^{-1}(y)$ we obtain

$$g_2^{-1}(x) * g_3^{-1}(y) = g_1^{-1}(x * y).$$

∎

Exercises

4.1 Characterize all bivariate distributions such that one family of conditionals is gamma and the other is normal. Using Equation (4.13) characterize all independent subfamilies.

4.2 Characterize all bivariate distributions with Pareto conditionals, i.e., with conditional probability density functions of the form

$$f(x; \alpha, \sigma) = \frac{\alpha}{\sigma}\left(1 + \frac{x}{\sigma}\right)^{-(\alpha+1)}; \quad x > 0; \quad \alpha, \sigma > 0, \tag{4.43}$$

where we assume α=constant.

4.3 Assume that $x_1, x_2, \ldots, x_n$ is a sample of n observations coming from a m-parameter exponential family, i.e., with likelihood

$$f(\underline{x}; \underline{\theta}) = r_n(\underline{x}) \exp\left\{\sum_{i=1}^{m} \theta_i s_i(\underline{x}) + n\lambda(\underline{\theta})\right\}. \tag{4.44}$$

Find the most general t-parameter exponential family of priors for $\underline{\theta}$ conjugate with respect to (4.44).

4.4 Use Theorem 4.11 to solve the functional equation

$$f(\sqrt[k]{x^k + y^k}) = \sqrt[k]{f^k(x) + f^k(y)}; \quad x, y \in \mathbb{R}_{++}; \quad k > 0,$$

in the class of continuous-at-a-point functions.

4.5 Use Theorem 4.12 to solve the functional equation

$$f(\sqrt[k]{x^k + y^k}) = \sqrt[k]{g^k(x) + h^k(y)} \quad ; \quad x, y \in \mathbb{R}_{++} \quad ; \quad k > 0,$$

in the class of continuous-at-a-point functions.

4.6 Let $(X_1, X_2, \ldots, X_n)$ be the order statistics of an independent and identically distributed sample of size n coming from a given population. Show that if $(X_2 - X_1)$ and X_1 are independent, then the population is either exponential or geometric.

(**Hint**: The joint probability density function, $g(u, v)$, of X_1 and $X_2 - X_1$ satisfies

$$g(u, v) \propto f(u)f(u + v)[1 - F(u + v)]^{n-2},$$

where $f(x)$ and $F(x)$ are the probability density and the cumulative distribution functions of the parent population, respectively.)

4.7 Solve the functional equation

$$f(x)e^y + g(x)\log y = xp(y) + x^2q(y)$$

4.8 Solve the functional equation

$$f(x)yz + xg(y)h(z) + x^2yq(z) = 0$$

CHAPTER 5

Equation for one function of several variables

5.1 Introduction

In this chapter we study equations for a single function of several variables. Some particular examples, such as the characterization of the area of a rectangle or the interest formula, were dealt with in Chapter 1. Most of these equations are direct generalizations of equations studied in previous chapters. The case of complex variables arises as a simple particular case.

We look first, in Section 5.2, at some generalizations of Cauchy's and Jensen's equations, which are applied to derive the formula for the area of a trapezoid and to design a fatigue model. In Sections 5.3 and 5.4, we study the equation for homogeneous functions and the translation equation and we discuss its application to iterative methods. Finally, in Section 5.5 we illustrate the usefulness of these equations presenting several examples of applications from the mathematical, physical and engineering sciences as the characterizations of the complex, the scalar and the dot and vector products of vectors.

5.2 Generalized Cauchy and Jensen equations

In this section we give the solutions of some generalizations of Cauchy's equations (3.7) to (3.10). We include some cases of complex domains and ranges.

Theorem 5.1 (Generalized Cauchy equation). *The general continuous non-vanishing solution of the equation*

$$f(x_1 + y_1, \ldots, x_n + y_n) = f(x_1, \ldots, x_n) + f(y_1, \ldots, y_n); \quad x_i, y_i \in \mathbb{R} \text{ or } \mathbb{R}_+, \tag{5.1}$$

where f is a real function of real variables is

$$f(x_1, x_2, \ldots, x_n) = a_1 x_1 + a_2 x_2 + \ldots + a_n x_n, \tag{5.2}$$

where $a_i, i = 1, 2, \ldots, n$, *are arbitrary constants.* ∎

Proof: We shall prove this by induction.

- The theorem is true for $n = 1$: For $k = 1$, (5.1) becomes

$$f(x_1 + y_1) = f(x_1) + f(y_1); \quad x_1, y_1 \in \mathbb{R} \text{ or } \mathbb{R}_+,$$

which is Cauchy's equation, and then its solution is (see Theorem 3.3):

$$f(x_1) = a_1 x_1.$$

- We assume the theorem true for $n = k - 1$, that is, the solution of

$$f(x_1 + y_1, \ldots, x_{k-1} + y_{k-1}) = f(x_1, \ldots, x_{k-1}) + f(y_1, \ldots, y_{k-1})$$

is

$$f(x_1, \ldots, x_{k-1}) = a_1 x_1 + \ldots + a_{k-1} x_{k-1}.$$

- Finally, we show that the theorem is true for $n = k$: In this case the functional equation becomes

$$f(x_1+y_1, x_2+y_2, \ldots, x_k+y_k) = f(x_1, x_2, \ldots, x_k)+f(y_1, y_2, \ldots, y_k), \quad (5.3)$$

from which by first letting $x_k = 0$ and then $y_i = 0$; $i = 1, \ldots, k - 1$ and $y_k = x_k$, we get

$$f(x_1, x_2, \ldots, x_k) = f(x_1, x_2, \ldots, x_{k-1}, 0) + f(0, 0, \ldots, 0, x_k). \quad (5.4)$$

Next, with $x_k = y_k = 0$ we obtain

$$f(x_1 + y_1, \ldots, x_{k-1} + y_{k-1}, 0) = f(x_1, \ldots, x_{k-1}, 0) + f(y_1, \ldots, y_{k-1}, 0), \quad (5.5)$$

and, finally, letting $x_i = y_i = 0$; $i = 1, \ldots, k - 1$, we get:

$$f(0, 0, \ldots, 0, x_k + y_k) = f(0, 0, \ldots, 0, x_k) + f(0, 0, \ldots, 0, y_k). \quad (5.6)$$

Equations (5.4) to (5.6) prove that $f(x_1, \ldots, x_{k-1}, 0)$ and $f(0, \ldots, 0, x_k)$ satisfy Equation (5.1) for $n = k - 1$ and $n = 1$, respectively. Thus, we have:

$$f(x_1, \ldots, x_{k-1}, 0) = a_1 x_1 + \ldots + a_{k-1} x_{k-1}$$

and

$$f(0, 0, \ldots, 0, x_k) = a_k x_k,$$

which together with (5.4) implies (5.2).

∎

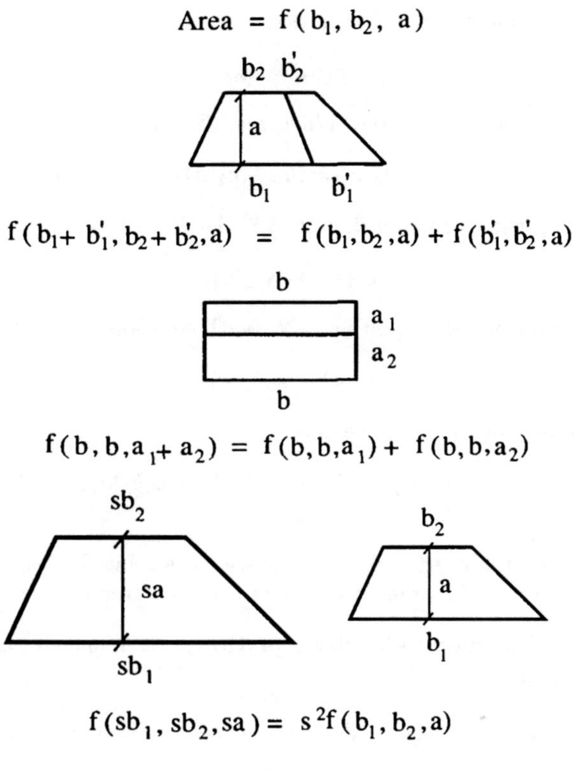

Figure 5.1: Basic trapezoids.

Example 5.1 (Area of a trapezoid). Assume that the area of a trapezoid is unknown but given by $f(b_1, b_2, a)$, where b_1 and b_2 are the bases, a is the height and f is a non-negative function to be determined.

According to Figure 5.1, we have

$$\begin{aligned} f(b_1 + b_1', b_2 + b_2', a) &= f(b_1, b_2, a) + f(b_1', b_2', a), \\ f(b, b, a_1 + a_2) &= f(b, b, a_1) + f(b, b, a_2), \\ f(sb_1, sb_2, sa) &= s^2 f(b_1, b_2, a). \end{aligned} \tag{5.7}$$

The solution of the first equation is (see Theorem 5.1):

$$f(b_1, b_2, a) = c_1(a)b_1 + c_2(a)b_2, \tag{5.8}$$

and by substituting this into the second equation of (5.7), we get

$$c_1(a_1 + a_2)b + c_2(a_1 + a_2)b = c_1(a_1)b + c_2(a_1)b + c_1(a_2)b + c_2(a_2)b. \tag{5.9}$$

If we now call $u(a) = c_1(a) + c_2(a)$, Equation (5.9) can be written as

$$u(a_1 + a_2) = u(a_1) + u(a_2), \tag{5.10}$$

which is Cauchy's equation, with solution

$$u(a) = ka, \tag{5.11}$$

where k is an arbitrary constant. With this, (5.8) leads to:

$$\text{Area} = f(b_1, b_2, a) = (ka - c_2(a))b_1 + c_2(a)b_2. \tag{5.12}$$

Replacing this in the third equation of (5.7), leads to

$$c_2(sa) = sc_2(a), \tag{5.13}$$

which is the homogeneous Equation (2.2), with solution

$$c_2(a) = k_2 c_2(a). \tag{5.14}$$

Finally, replacing this in (5.12) we get

$$\text{Area} = f(b_1, b_2, a) = (k_1 b_1 + k_2 b_2)a, \tag{5.15}$$

where $k_1 = k - k_2$.

The interpretation of the arbitrary constants is related to the units used to measure b_1, b_2, a and the area, as in the case of the area of a rectangle. ■

Theorem 5.2 (Extended Cauchy equation). *The general continuous solution of the equation*

$$f(x_1+y_1, x_2+y_2, \ldots, x_n+y_n) = f(x_1, x_2, \ldots, x_n) + f(y_1, y_2, \ldots, y_n) + a, \tag{5.16}$$

where f is a real function of a real variable and a is a constant is

$$f(x_1, x_2, \ldots, x_n) = c_1 x_1 + c_2 x_2 + \ldots + c_n x_n - a. \tag{5.17}$$

■

Proof: With $g(x_1, x_2, \ldots, x_n) = f(x_1, x_2, \ldots, x_n) + a$, we get

$$g(x_1 + y_1, x_2 + y_2, \ldots, x_n + y_n) = g(x_1, x_2, \ldots, x_n) + g(y_1, y_2, \ldots, y_n),$$

which is Equation (5.1). Thus, we get (5.17) as the general continuous solution of (5.16). ■

Theorem 5.3 (Complex Cauchy equation). *The general continuous complex solution of the equation*

$$f(x + y) = f(x) + f(y); \quad x, y \in \mathbb{C}, \tag{5.18}$$

where f is a complex function of complex variable, is

$$f(x) = ax + b\bar{x}, \tag{5.19}$$

where a and b are arbitrary complex constants. ■

Proof: Calling

$$x = x_1 + ix_2; \;\; y = y_1 + iy_2; \;\; f(x) = f_1(x_1, x_2) + if_2(x_1, x_2),$$

Equation (5.18) becomes

$$\begin{aligned} f_1(x_1 + y_1, x_2 + y_2) &= f_1(x_1, x_2) + f_1(y_1, y_2), \\ f_2(x_1 + y_1, x_2 + y_2) &= f_2(x_1, x_2) + f_2(y_1, y_2), \end{aligned}$$

which are Cauchy's equations of real functions of the type (5.1) for $n = 2$. Thus, the most general continuous solution of (5.18) is

$$f(x) = f_1(x_1, x_2) + if_2(x_1, x_2) = a_1x_1 + a_2x_2 + i(a_3x_1 + a_4x_2) = ax + b\bar{x},$$

where a and b are complex constants that satisfy

$$a = \frac{1}{2}\,[a_1 + a_4 + i(a_3 - a_2)]; \;\; b = \frac{1}{2}\,[a_1 - a_4 + i(a_3 + a_2)],$$

and, since $a_1, \ldots, a_4$ are arbitrary, so are a and b. ■

Theorem 5.4 (Generalized Complex Cauchy equation). *The general continuous solution of the equation*

$$f(x_1 + y_1, \ldots, x_n + y_n) = f(x_1, \ldots, x_n) + f(y_1, \ldots, y_n); \;\; x_i, y_i \in \mathbb{C}, \tag{5.20}$$

where f is a complex function of complex variables, is

$$f(x_1, \ldots, x_n) = a_1x_1 + \ldots + a_nx_n + b_1\bar{x}_1 + \ldots + b_n\bar{x}_n, \tag{5.21}$$

where $a_i, b_i, i = 1, \ldots, n$, are arbitrary complex constants. ■

The proof of this theorem is identical to that of Theorem 5.1.

Theorem 5.5 (Generalized Cauchy equation II). *The general continuous solution of the equation*

$$f(x_1 + y_1, \ldots, x_n + y_n) = f(x_1, \ldots, x_n)f(y_1, \ldots, y_n), \tag{5.22}$$

where f is a real function of n real variables is

$$f(x_1, \ldots, x_n) = \exp(c_1x_1 + \ldots + c_nx_n), \tag{5.23}$$

where $c_i, i = 1, \ldots, n$ are arbitrary constants. ■

Theorem 5.6 (Complex Cauchy's equation II). *The general continuous non-vanishing solution of the equation*

$$f(x + y) = f(x)f(y); \;\; x, y \in \mathbb{C}, \tag{5.24}$$

where f is a complex function of complex variables is

$$f(x) = \exp(ax + b\bar{x}), \tag{5.25}$$

where a and b are arbitrary complex constants. ■

Theorem 5.7 (Jensen's equation). *The general continuous solution of Jensen's equation*

$$f\left(\frac{x_1+y_1}{2}, \frac{x_2+y_2}{2}\right) = \frac{f(x_1,x_2)+f(y_1,y_2)}{2}; \quad x_i, y_i \in \mathbb{R} \ (i=1,2), \tag{5.26}$$

where f is a real function of real variables is

$$f(x,y) = ax + by + c, \tag{5.27}$$

where a, b and c are arbitrary constants. ∎

Proof: Making $x_1 = 0, x_2 = 0, y_1 = u_1 + v_1, y_2 = u_2 + v_2$ in (5.26) we get

$$f\left(\frac{u_1+v_1}{2}, \frac{u_2+v_2}{2}\right) = \frac{f(0,0)+f(u_1+v_1, u_2+v_2)}{2} = \frac{f(u_1,u_2)+f(v_1,v_2)}{2},$$

but this is Equation (5.16) for $n = 2$, and then the general solution of (5.26) is (5.27). ∎

Theorem 5.8 (Sincov's equation). *The general solution of the equation*

$$f(x,y) + f(y,z) = f(x,z) \tag{5.28}$$

is

$$f(x,y) = g(y) - g(x), \tag{5.29}$$

where g is an arbitrary function. ∎

Proof: Making $z = 0$ in (5.28) and calling $g(x) = f(x, 0)$, we get (5.29). ∎

Example 5.2 (Fatigue model). (Castillo et al. (1990a)) One important model for fatigue analysis states the following functional form for the relation between the survivor functions of two longitudinal elements of lengths x and y:

$$G(t,x) = G(t,y)^{k(x,y)}, \tag{5.30}$$

where t is the fatigue lifetime and $k(x,y)$ is a positive function.

If we apply the above expression to three elements of lengths x, y and z, in all its possible combinations, we get

$$\left.\begin{array}{l} G(t,x) = G(t,y)^{k(x,y)} \\ G(t,x) = G(t,z)^{k(x,z)} \\ G(t,y) = G(t,z)^{k(y,z)} \end{array}\right\} \Rightarrow k(x,z) = k(x,y)k(y,z). \tag{5.31}$$

Taking logarithms and calling $f(x,y) = \log k(x,y)$, we get

$$f(x,z) = f(x,y) + f(y,z). \tag{5.32}$$

Hence, according to Theorem 5.8, its general solution is

$$f(x,y) = g(y) - g(x) \;\Rightarrow\; k(x,y) = \frac{\exp[g(y)]}{\exp[g(x)]} = \frac{m(y)}{m(x)}, \tag{5.33}$$

where m is an arbitrary positive function.
With this, the model (5.30) becomes

$$G(t,x) = G(t,y)^{\frac{m(y)}{m(x)}}. \tag{5.34}$$

One remarkable conclusion is that the function $K(x,y)$ cannot be arbitrarily chosen as one might think at a first glance looking at (5.30). ∎

5.3 Other equations

In this section we study other important functional equations.

Theorem 5.9 (Homogeneous functions). *The general solution of the equation*

$$f(xz,yz) = z^k f(x,y); \;\; x,y \in \mathbb{R}_+, \;\; z \in \mathbb{R}_{++}, \tag{5.35}$$

where f is a real function of a real variable is

$$f(x,y) = \begin{cases} (\sqrt{xy})^k h\left(\dfrac{x}{y}\right) \;\; or \;\; x^k g\left(\dfrac{x}{y}\right) & if \;\; x \neq 0, \; y \neq 0, \\ ax^k & if \;\; y = 0, x \neq 0, \\ by^k & if \;\; x = 0, y \neq 0, \\ c(ck = 0) & if \;\; x = y = 0, \end{cases} \tag{5.36}$$

where h and g are arbitrary functions. ∎

For the proof of this theorem see Aczél (1966) page 229.

Let us now look at the so called translation equation. Its general continuous solution is given by the following theorem.

Theorem 5.10 (Translation equation). *The general continuous solution of the translation equation*

$$F[F(x,u),v] = F(x,u+v); \;\; x, F(x,u) \in (a,b); \;\; u,v \in (-\infty,\infty) \tag{5.37}$$

is

$$F(x,y) = f[f^{-1}(x) + y], \tag{5.38}$$

where f is an arbitrary function which is continuous and strictly monotonic in $(-\infty,\infty)$, if one of the following conditions holds:

- *(a) $F(x,u)$ is strictly monotonic for each value of x with respect to u and for uncountably many values of u with respect to x.*

- *(b) $F(x,u)$ is continuous for each value of u with respect to x and for $x = x_0$ with respect to u and nonconstant for every fixed value of x.*

The interval of definition with respect to x can only be open. ∎

For the proof of this theorem see Example 2.7.

Example 5.3 (Interest formula). In Example 3.11 we have characterized the compound interest formula by means of Cauchy's equation. In this example we remind the reader that one of the required conditions was simply the translation equation

$$f(x, t+u) = f[f(x,t), u]; \quad x, y, t, u \in \mathbb{R}_+.$$

According to (5.38) one solution of this equation is

$$f(x,t) = g[g^{-1}(x) + t].$$

For the case of the solution in Example 3.11, we have $g(x) = (1+r)^x$. ∎

Corollary 5.1 *Any solution of Equation (5.37) can also be written as*

$$F(x,y) = g[g^{-1}(x) * y],$$

where $G(x,y) = x * y$ is any solution of the translation Equation (5.37) and g is a continuous and strictly monotonic function.

Proof: If $F(x,y)$ and $G(x,y)$ are solutions of Equation (5.37), then, according to Theorem 5.10, we have

$$\begin{aligned} F(x,y) &= f[f^{-1}(x) + y], \\ x * y &= G(x,y) = h[h^{-1}(x) + y], \end{aligned}$$

where f and h are continuous and strictly monotonic functions.
Thus, we can write

$$\begin{aligned} &h^{-1}(x * y) = h^{-1}(x) + y \;\Rightarrow\; h^{-1}[h(u) * v] = u + v \;\Rightarrow \\ &\Rightarrow F(x,y) = f[f^{-1}(x) + y] = f\{h^{-1}[h(f^{-1}(x)) * y]\} \end{aligned}$$

and calling

$$g(x) = fh^{-1}(x) \;\Leftrightarrow\; g^{-1}(x) = hf^{-1}(x)$$

we get

$$F(x,y) = g[g^{-1}(x) * y],$$

where the continuity and invertibility of g comes from those of f and h. ∎

Example 5.4 (Stress-strain law). Let us assume that we try to find a stress-strain law of the form $\varepsilon = f(\varepsilon_0, \Delta\sigma)$, where ε_0 is the initial strain, $\Delta\sigma$ is the stress increment and f is such that the strain, ε, is independent of the way the stress level, σ, has been reached. This condition can be formulated as the translation equation

$$\varepsilon = f[f(\varepsilon_0, \Delta\sigma_1), \Delta\sigma_2] = f(\varepsilon_0, \Delta\sigma_1 + \Delta\sigma_2), \tag{5.39}$$

which implies

$$\varepsilon = f(\varepsilon_0, \Delta\sigma) = h[h^{-1}(\varepsilon_0) + \Delta\sigma],$$

where

$$\Delta\sigma = \Delta\sigma_1 + \Delta\sigma_2,$$

and calling $\sigma_0 = h^{-1}(\varepsilon_0)$ and $\sigma = \sigma_0 + \Delta\sigma$ we trivially get

$$\varepsilon = h(\sigma).$$

If we force the strain to be of the form

$$\varepsilon = f(\varepsilon_0, \Delta\sigma) = \varepsilon_0 + g(\Delta\sigma),$$

its substitution into (5.39) leads to

$$g(\Delta\sigma_1 + \Delta\sigma_2) = g(\Delta\sigma_1) + g(\Delta\sigma_2) \;\Rightarrow\; g(\Delta\sigma) = K\Delta\sigma \;\Rightarrow\; \varepsilon = \varepsilon_0 + K\Delta\sigma,$$

and then, the only possible solution is the linear law. ■

5.4 Application to iterative methods

The translation equation has many applications to the theory of continuous iteration of functions. Calling $F(x, n) = f^{(n)}(x)$ to the n-th f-iterate, it is clear that

$$F(x, m + n) = f^{(n+m)}(x) = f^{(n)}(f^{(m)}(x)) = F(F(x, m), n), \tag{5.40}$$

which proves that $F(x, n)$ satisfies the translation equation, the solution of which, as given by Theorem 5.10, is

$$f^{(n)}(x) = F(x, n) = g^{-1}[g(x) + n]. \tag{5.41}$$

This implies

$$f(x) = F(x, 1) = g^{-1}[g(x) + 1] \Leftrightarrow g(f(x)) = g(x) + 1. \tag{5.42}$$

Thus, the problem of finding the n-th iterate is merely a question of solving the equivalent functional equation

$$g(f(x)) = f(x + 1), \tag{5.43}$$

or

$$f^{-1}(g(x)) = f^{-1}(x) + 1, \tag{5.44}$$

which is a particular case of the Abel equation (see Example 2.17). For an extensive study of the Abel equation see Kuczma et al. (1990).

According to the above, there exists a correspondence between g and f. Given an invertible function $g(x)$, we can obtain $f(x)$ using (5.42) and its n-th iterate by (5.41). However, the problem of obtaining $g(x)$ when $f(x)$ is known is much more complicated.

Example 5.5 (Application to iterative methods). Let us assume that we are looking for the n-th iterate of the transformation $g(x) = Ax + B$. Then, the functional Equation (5.43) becomes the difference equation

$$Af(x) + B = f(x+1),$$

which has as its general continuous solution

$$f(x) = CA^x + \frac{B}{1-A},$$

and then we obtain

$$g_n(x) = F(x,n) = f[f^{-1}(x) + n] =$$

$$= CA^{\frac{\log(x - \frac{B}{1-A}) - \log(C)}{\log(A)} + n} + \frac{B}{1-A} = xA^n + \frac{B(1-A^n)}{1-A},$$

which is the desired n-th iterate. ∎

Example 5.6 (Application to iterative methods). Let us assume that we are looking for the n-th iterate of the transformation

$$g(x) = 2x(1-x).$$

Then, the functional Equation (5.43) becomes the difference equation

$$g[f(x)] = 2f(x)[1 - f(x)] = f(x+1),$$

with continuous solution

$$f(x) = \frac{[1 - (1-2a)^{2^{x-1}}]}{2} \Leftrightarrow f^{-1}(x) = \frac{\log\left[\dfrac{\log(1-2x)}{\log(1-2a)}\right]}{\log(2)} + 1$$

and then we obtain

$$g_n(x) = F(x,n) = f[f^{-1}(x) + n] =$$

$$= \frac{1 - (1-2a)^{2^{\frac{\log\left[\frac{\log(1-2x)}{\log(1-2a)}\right]}{\log(2)} + n}}}{2} = \frac{1 - (1-2x)^{2^n}}{2}.$$

∎

Finally, we analyze the uniqueness of solutions for the Expression (5.43).

Theorem 5.11 (Translation Functional Equation (uniqueness)). *If Expression (5.43) can be written in terms of two different functions $g(x)$ and $h(x)$, i.e, if*

$$g^{-1}[g(x)+1] = h^{-1}[h(x)+1], \tag{5.45}$$

then, with c being an arbitrary constant, we obtain

$$g(u) = h(u) + c. \tag{5.46}$$

■

Proof: From (5.45) we can write

$$hg^{-1}[g(x)+1] = h(x)+1; \ \ \forall x, \tag{5.47}$$

and making $x = h^{-1}(x)$ we get

$$hg^{-1}[g(h^{-1}(x))+1] = h(h^{-1}(x))+1 = x+1; \ \ \forall x, \tag{5.48}$$

that is, calling $s(x) = g(h^{-1}(x))$,

$$s(x)+1 = s(x+1), \tag{5.49}$$

which is a difference equation with general solution

$$s(x) = x+c \Leftrightarrow g(h^{-1}(x)) = x+c \Leftrightarrow h^{-1}(x) = g^{-1}(x+c), \tag{5.50}$$

which implies $g(u) = h(u) + c$. ■

5.5 Some examples

In this section we include other examples of functional equations involving functions of several variables.

Example 5.7 (Fatigue model). Let us assume that $G(x,s)$ is the reliability or survivor function, under fatigue, of a longitudinal element of length s, i.e., that

$$P[X > x] = G(x,s), \tag{5.51}$$

where P means probability and X is the fatigue lifetime of the element.

Because of the weakest link principle, the fatigue life of an element of length ks is the minimum of the fatigue life of its k constituting pieces of length s (see Figure 5.2).

Then, if independence holds we must have

$$G(x,ks) = G(x,s)^k \ \ \Rightarrow \ \ H(x,ks) = kH(x,s), \tag{5.52}$$

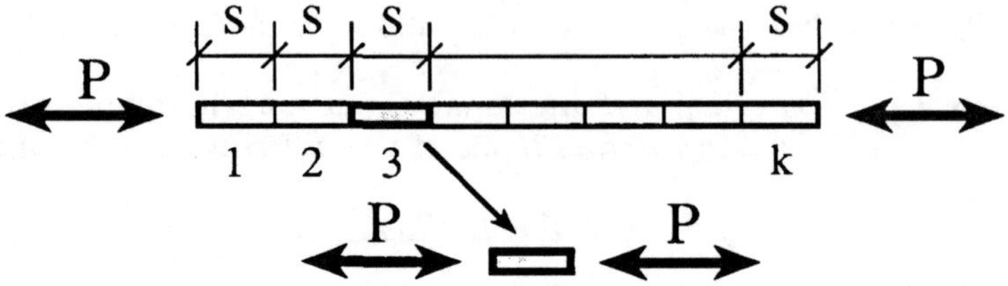

Figure 5.2: Longitudinal element and its constituting pieces.

where $H(x,s) = \log[G(x,s)]$ and k can be generalized by allowing it to be any positive real number.

But, the equation on the right hand side of (5.52) is, for any fixed x, an equation of the form (3.1). Thus, its general solution is

$$H(x,s) = C(x)s \quad \Rightarrow \quad G(x,s) = \exp[C(x)s] \quad \Rightarrow \quad [g(x)]^s, \tag{5.53}$$

where $g(x)$ is an arbitrary positive function.

If $G(x,s)$ is to be a reliability function, then $g(x)$ must be a non-increasing function, such that $g(0) = 1$ and $\lim_{x\to\infty} g(x) = 0$. ■

A more detailed description about the application of functional equations to fatigue problems will be given in Chapter 10.

Example 5.8 (Characterization of the product of two complex numbers). (Castillo et al. (1990e)) In this example we characterize the product of two complex numbers by means of the following conditions:

- (i) A rotation of an angle φ of the two factors causes a rotation of the product of an angle 2φ and no change in the modulus.
- (ii) Multiplication of the modulus of any one of the factors by a factor K leads to a multiplication of the modulus of the product by the same factor K and no change in the argument.
- (iii) The product remains unchanged if one of the factors is rotated through an angle φ and the other factor is rotated through an angle $-\varphi$.
- (iv) The product of two unit factors leads to a unit factor.
- (v) The product of any complex number by its conjugate gives a real number.

In order to prove that the above five conditions characterize the product of two complex numbers, we denote by $M(m_1, m_2, \alpha_1, \alpha_2)$ and $A(m_1, m_2, \alpha_1, \alpha_2)$ two functions that give the modulus and argument, respectively, of the product of the two complex numbers with modulus m_1 and m_2 and arguments α_1 and α_2, respectively.

Condition (i) yields

$$\begin{aligned} M(m_1, m_2, \alpha_1, \alpha_2) &= M(m_1, m_2, \alpha_1 - \alpha_2 + \alpha_2, 0 + \alpha_2) \\ &\overset{(i)}{=} M(m_1, m_2, \alpha_1 - \alpha_2, 0) \\ &= M_1(m_1, m_2, \alpha_1 - \alpha_2) \end{aligned}$$

and

$$\begin{aligned} A(m_1, m_2, \alpha_1, \alpha_2) &= A(m_1, m_2, \alpha_1 - \alpha_2 + \alpha_2, 0 + \alpha_2) \\ &\overset{(i)}{=} 2\alpha_2 + A(m_1, m_2, \alpha_1 - \alpha_2, 0) \\ &= 2\alpha_2 + A_1(m_1, m_2, \alpha_1 - \alpha_2), \end{aligned}$$

where the meaning of $M_1(m_1, m_2, \alpha)$ and $A_1(m_1, m_2, \alpha)$ is obvious, and above the equal signs we have included the numbers of the conditions that allow the corresponding equalities to be written.

Taking into account condition (ii) and Theorem 2.1, we get

$$M_1(Km_1, m_2, \alpha) \overset{(ii)}{=} KM_1(m_1, m_2, \alpha) \Rightarrow M_1(m_1, m_2, \alpha) = m_1 c(m_2, \alpha),$$

where c is an arbitrary function, and

$$\begin{aligned} M_1(m_1, Km_2, \alpha) &\overset{(ii)}{=} KM_1(m_1, m_2, \alpha) \\ &\Downarrow \\ m_1 c(Km_2, \alpha) &= Km_1 c(m_2, \alpha) \\ &\Downarrow \\ c(m_2, \alpha) &= m_2 d(\alpha), \end{aligned}$$

where d is an arbitrary function. Then, we can write

$$M_1(m_1, m_2, \alpha) = m_1 m_2 d(\alpha) \quad \Rightarrow \quad M(m_1, m_2, \alpha_1, \alpha_2) = m_1 m_2 d(\alpha_1 - \alpha_2)$$

Condition (ii) leads to

$$\begin{aligned} A(Km_1, m_2, \alpha_1, \alpha_2) &= 2\alpha_2 + A_1(Km_1, m_2, \alpha_1 - \alpha_2) \\ &= A(m_1, m_2, \alpha_1, \alpha_2) \\ &= 2\alpha_2 + A_1(m_1, m_2, \alpha_1 - \alpha_2) \\ &\Downarrow \\ A_1(m_1, m_2, \alpha_1 - \alpha_2) &= f(m_2, \alpha_1 - \alpha_2), \end{aligned}$$

where f is an arbitrary function, and

$$\begin{aligned} A(m_1, Km_2, \alpha_1, \alpha_2) &= A(m_1, m_2, \alpha_1, \alpha_2) \\ &\Downarrow \\ f(Km_2, \alpha_1 - \alpha_2) &= f(m_2, \alpha_1 - \alpha_2) \\ &\Downarrow \\ f(m_2, \alpha_1 - \alpha_2) &= g(\alpha_1 - \alpha_2), \end{aligned}$$

where g is an arbitrary function. This implies

$$A(m_1, m_2, \alpha_1, \alpha_2) = 2\alpha_2 + g(\alpha_1 - \alpha_2).$$

Condition (iii) implies

$$d(\alpha_1 - \alpha_2 + 2\phi) = d(\alpha_1 - \alpha_2) \Rightarrow d(\alpha_1 - \alpha_2) = D \Rightarrow$$
$$\Rightarrow M(m_1, m_2, \alpha_1, \alpha_2) = Dm_1m_2,$$

where D is an arbitrary constant, and

$$2(\alpha_2 - \phi) + g(\alpha_1 - \alpha_2 + 2\phi) = 2\alpha_2 + g(\alpha_1 - \alpha_2) \Rightarrow g(x + y) - y = g(x)$$

With $x = 0$ we obtain

$$g(x) = x + K \quad \Rightarrow \quad A(m_1, m_2, \alpha_1, \alpha_2) = \alpha_1 + \alpha_2 + K,$$

where K is an arbitrary constant.

Finally, from conditions (iv) and (v) we get

$$M(1, 1, \alpha_1, \alpha_2) = D = 1 \quad \Rightarrow \quad M(m_1, m_2, \alpha_1, \alpha_2) = m_1m_2,$$

$$A(m_1, m_1, \alpha, -\alpha) = \alpha - \alpha + K = 0 \quad \Rightarrow \quad A(m_1, m_2, \alpha_1, \alpha_2) = \alpha_1 + \alpha_2,$$

which are the well known rules used for the product of two complex numbers.

The following three conditions also characterize the same product:

- (i) If the argument of one of the factors is modified by an angle φ and the other factor is kept fixed, the argument of the product becomes modified by the same angle but the modulus remains unchanged.
- (ii) If the modulus of one of the factors is multiplied by a positive constant K, the modulus of the product becomes multiplied by the same factor but the argument remains unchanged.
- (iii) The complex number 1 is a unit element for the product of complex numbers.

Proof: Using the above conditions we get.

$$\begin{aligned} M(m_1, m_2, \alpha_1, \alpha_2) &\overset{(ii)}{=} m_1m_2M(1, 1, \alpha_1, \alpha_2) \\ &\overset{(i)}{=} m_1m_2M(1, 1, 0, 0) \\ &\overset{(iii)}{=} m_1m_2, \end{aligned}$$

$$\begin{aligned} A(m_1, m_2, \alpha_1, \alpha_2) &\overset{(ii)}{=} A(1, 1, \alpha_1, \alpha_2) \\ &\overset{(i)}{=} \alpha_1 + \alpha_2 + A(1, 1, 0, 0) \\ &\overset{(iii)}{=} \alpha_1 + \alpha_2, \end{aligned}$$

where again, on the equality signs, we have indicated the properties being used. ∎

In the following two examples the dot and vector products are characterized.

Example 5.9 (Vector dot product). In this example we characterize the vector dot product by means of the following set of properties:

- (i) The dot product is invariant when the two vectors undergo the same rotation.
- (ii)
$$(c\mathbf{a}) \cdot \mathbf{b} = c(\mathbf{a} \cdot \mathbf{b}) = \mathbf{a} \cdot (c\mathbf{b}),$$
where $\mathbf{a}$ and $\mathbf{b}$ are vectors and c is a scalar.
- (iii)
$$(\mathbf{e}_1 + \mathbf{e}_2) \cdot \mathbf{e}_3 = \mathbf{e}_1 \cdot \mathbf{e}_3 + \mathbf{e}_2 \cdot \mathbf{e}_3,$$
where $\mathbf{e}_1, \mathbf{e}_2$ and $\mathbf{e}_3$ are unit and coplanar vectors.

Proof: Because of the property (i), the dot product can be written as

$$\mathbf{a} \cdot \mathbf{b} = h(\alpha, a, b),$$

where $h(\alpha, a, b)$ is a function to be determined; that is, the dot product of two vectors $\mathbf{a}$ and $\mathbf{b}$ depends on the modula a and b of the vectors $\mathbf{a}$ and $\mathbf{b}$, respectively, and their angle α.

Condition (ii) gives

$$h(\alpha, a, b) = ah(\alpha, 1, b) = abh(\alpha, 1, 1) = abf(\alpha),$$

where $f(\alpha)$ is a function to be determined.

From condition (iii), taking into account that the vector sum of two unit and coplanar vectors making an angle α is a vector with modulus $2\cos(\alpha/2)$, we can write

$$2\cos\left(\frac{u-v}{2}\right) f\left(\frac{u+v}{2}\right) = f(u) + f(v),$$

where, as indicated in Figure 5.3, u and v are the angles associated with the pairs $(\mathbf{e}_1, \mathbf{e}_3)$ and $(\mathbf{e}_2, \mathbf{e}_3)$, respectively.

Carrying out the change of variable

$$u = x + y; \quad v = x - y$$

leads to

$$2\cos(y) f(x) = f(x+y) + f(x-y),$$

which is the equation in Example 2.3. Thus, its general solution is

$$f(t) = M\cos(t) + N\sin(t).$$

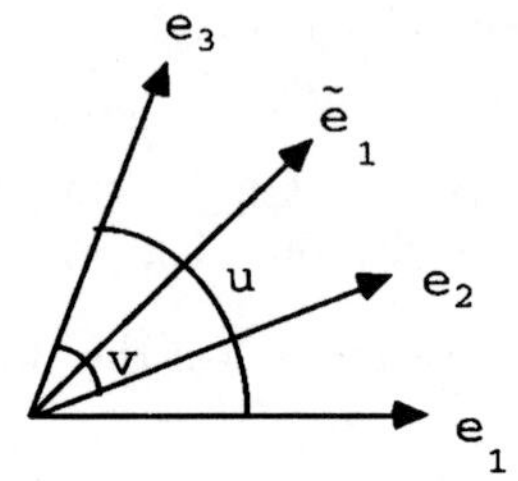

Figure 5.3: Illustration of the meaning of angles u and v.

If we perform a rotation of an angle π around the axis $\mathbf{e_2}$ we transform the pair $(\mathbf{e_1}, \mathbf{e_2})$ into $(\tilde{\mathbf{e}}_1, \mathbf{e_2})$ and, because of property (i), we have

$$\begin{aligned} \mathbf{e}_1 \cdot \mathbf{e}_2 &= f(\alpha) = \tilde{\mathbf{e}}_1 \cdot \mathbf{e}_2 = f(-\alpha) \\ &\Downarrow \\ M\cos(t) + N\sin(t) &= M\cos(-t) + N\sin(-t) \\ &\Downarrow \\ N &= 0 \end{aligned}$$

Consequently, we get

$$\mathbf{a} \cdot \mathbf{b} = Mab\cos(\alpha),$$

where M is a constant.

Remark: If we add the following property: the dot product of two co-linear vectors with the same direction is the product of their modula, we obtain $M = 1$. ■

Example 5.10 (Vector product). In this example we characterize the formula for the modulus of the vector product by means of the following set of properties:

- (i) If we perform a rotation of the factor vectors, then the vector product undergoes the same rotation.
- (ii)
$$(c\mathbf{a}) \times \mathbf{b} = c(\mathbf{a} \times \mathbf{b}) = \mathbf{a} \times (c\mathbf{b}),$$
where $\mathbf{a}$ and $\mathbf{b}$ are vectors and c is a scalar.
- (iii)
$$(\mathbf{e}_1 + \mathbf{e}_2) \times \mathbf{e}_3 = \mathbf{e}_1 \times \mathbf{e}_3 + \mathbf{e}_2 \times \mathbf{e}_3,$$
where $\mathbf{e}_1$, $\mathbf{e}_2$ and $\mathbf{e}_3$ are unit and coplanar vectors.

Proof: Because of the property (i), the modulus of the vector product can be written as

$$|\mathbf{a} \times \mathbf{b}| = h(\alpha, a, b),$$

where $h(\alpha, a, b)$ is a function to be determined; that is, the modulus of the vector product of two vectors **a** and **b** depends on the modula a and b of the vectors **a** and **b**, respectively, and their angle α.

Conditions (ii) and (iii) have the same structure as in the case of Example 5.9, and thus, following the same steps, we have

$$h(\alpha, a, b) = ah(\alpha, 1, b) = abh(\alpha, 1, 1) = abf(\alpha),$$

where

$$f(\alpha) = M\cos(\alpha) + N\sin(\alpha).$$

But, because of property (i), the vector product $\mathbf{e} \times \mathbf{e}$ can only be a vector in the direction of $\mathbf{e}$. Thus, $\mathbf{e} \times \mathbf{e} = f(0)\mathbf{e}$.

However, since $\mathbf{e}$ is transformed into $-\mathbf{e}$ under a rotation of π, then, (i) and (ii) lead to

$$\mathbf{e} \times \mathbf{e} = f(0)\mathbf{e} = (-\mathbf{e}) \times (-\mathbf{e}) = -f(0)\mathbf{e} \;\;\Rightarrow\;\; f(0) = 0 \;\;\Rightarrow\;\; M = 0.$$

Consequently, we conclude

$$|\mathbf{a} \times \mathbf{b}| = Nab\sin(\alpha),$$

where N is a constant. ■

Exercises

5.1 Using geometric considerations, show that the formula giving the area of a circular ring satisfies the Sincov equation.

5.2 Using Equation (5.35), derive a general formula for the area of families of plane surfaces depending on two length parameters, as ellipses, rectangles, etc.

5.3 Using Equation (5.35), derive a general formula for the volume of families of volumes depending on two length parameters, as cylinders, cones, etc.

5.4 A bivariate random variable (X, Y) is said to have the loss of memory property if

$$\overline{F}(s_1 + t, s_2 + t) = \overline{F}(s_1, s_2)\overline{F}(t, t), \quad \forall s_1, s_2, t \geq 0 \tag{5.54}$$

where $\overline{F}(s_1, t) = Prob(X > s, Y > t)$.

Show that the general solution of (5.54) is

$$\overline{F}(x, y) = \begin{cases} e^{-\theta y}\overline{F}_1(x - y); & x \geq y \\ \\ e^{-\theta x}\overline{F}_2(y - x); & y \geq x \end{cases} \tag{5.55}$$

where $F_1(x)$ and $F_2(y)$ are the marginal distributions $F(x,0)$ and $F(0,y)$, respectively, and find the most general bivariate distribution that satisfies (5.54) and has exponential marginals.

5.5 Obtain the n-iterate of the function

$$f(x) = \log(1 + \exp(x))$$

using (5.41) and (5.42).

5.6 Assuming that the area of a triangle is $f(b,h)$, where b and h are its basis and its height, respectively, write a system of functional equations that characterize such formula.

5.7 Using Theorem 5.8 that gives the general solution of the Sincov equation, show that there exists a function $v(x)$ such that

$$\int_x^y h(u)du = v(y) - v(x)$$

and give an expression for $v(x)$.

5.8 Assume a geometric figure such that can be defined by two length parameters a and b. Derive the general formula for its area based on the homogeneous functional equation. Apply it to the particular case in Figure 5.4.

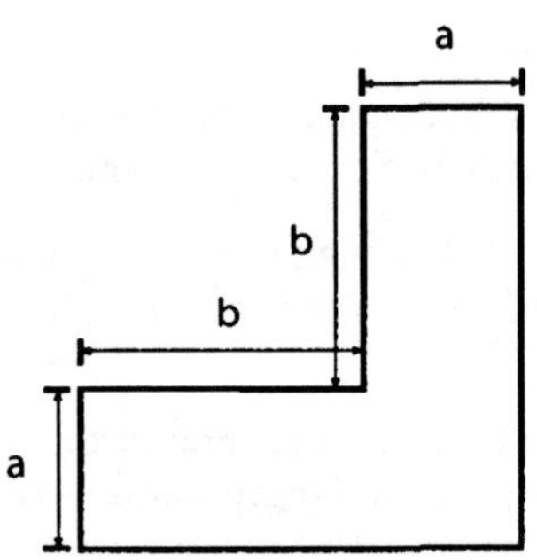

Figure 5.4: Two-parameter geometric figure.

5.9 Solve the functional equation

$$f(x_1 + y_1, x_2 + y_2, \ldots, x_n + y_n) = f(x_1, x_2, \ldots, x_n)f(y_1, y_2, \ldots, y_n)$$

CHAPTER 6

Equations with functions of several variables

6.1 Introduction

In this chapter we study equations with several unknown functions of several variables. We start with some generalizations of Pexider's, Jensen's and Sincov's equations in Sections 6.2 and 6.3, respectively, and we apply them to some problems of taxation functions and fatigue life of longitudinal elements. In Section 6.4, we solve a general equation which is used: (a) to derive the general solutions of special cases of important equations in later sections and (b) to characterize the random variables that can be obtained by sums of independent random variables of given families, some probability based models in expert systems and the three-variate distributions generated from their bivariate marginals. In Section 6.5 we study the associativity equation and we apply it to the problem of synthesis of judgments. Finally, Sections 6.6, 6.7 and 6.8 are devoted to the transitivity, bisymmetry and transformation equations, respectively.

6.2 Generalized Pexider and Jensen equations

In this section we give the solutions of a generalization of Pexider's Equation (4.1).

Theorem 6.1 (Generalized Pexider equation). *The general continuous system of solutions of the functional equation*

$$F(x_1 + y_1, \ldots, x_n + y_n) = G(x_1, \ldots, x_n) + H(y_1, \ldots, y_n)$$
$$x_i, y_i \in \mathbb{R} \;\; or \;\; \mathbb{R}_+ \; ; \; i = 1, \ldots, n \tag{6.1}$$

is

$$\begin{aligned} F(x_1,\ldots,x_n) &= C_1x_1+\ldots+C_nx_n+a+b,\\ G(x_1,\ldots,x_n) &= C_1x_1+\ldots+C_nx_n+a,\\ H(x_1,\ldots,x_n) &= C_1x_1+\ldots+C_nx_n+b, \end{aligned} \tag{6.2}$$

where a, b and C_i $(i=1,\ldots,n)$ are arbitrary constants. ■

Proof: Making first $x_1=x_2=\ldots=x_n=0$ and then $y_1=y_2=\ldots=y_n=0$ in (6.1) leads to

$$\begin{aligned} F(y_1,\ldots,y_n) &= a+H(y_1,\ldots,y_n);\ a=G(0,\ldots,0),\\ F(x_1,\ldots,x_n) &= b+G(x_1,\ldots,x_n);\ b=H(0,\ldots,0), \end{aligned} \tag{6.3}$$

which after substitution into (6.1), gives

$$F(x_1+y_1,\ldots,x_n+y_n)=F(x_1,\ldots,x_n)+F(y_1,\ldots,y_n)-a-b, \tag{6.4}$$

and calling

$$U(x_1,\ldots,x_n)=F(x_1,\ldots,x_n)-a-b, \tag{6.5}$$

we get

$$U(x_1+y_1,\ldots,x_n+y_n)=U(x_1,\ldots,x_n)+U(y_1,\ldots,y_n),$$

which is Equation (5.1), with a general continuous solution

$$U(x_1,\ldots,x_n)=C_1x_1+\ldots+C_nx_n. \tag{6.6}$$

Substitution of (6.6) into (6.5) and (6.3) leads to (6.2). ■

Example 6.1 (Taxation function). Let us assume that due taxes depend on the salary and capital incomes x and y, respectively. If we want a married couple to pay the same amount independently of whether it is a separate or joint tax return we must have

$$F(x_1+x_2,y_1+y_2)=G(x_1,y_1)+H(x_2,y_2),\quad x_i,y_i\in\mathbb{R}_+,\quad (i=1,2),$$

where we have initially assumed a different tax function for wife (G), husband (H) and couples (F).
According to (6.2), we have

$$\begin{aligned} F(x,y)&=C_1x+C_2y+a+b,\\ G(x,y)&=C_1x+C_2y+a,\\ H(x,y)&=C_1x+C_2y+b. \end{aligned}$$

If we add the condition of the same treatment for wife and husband, we get $a=b$. Finally, a zero tax amount for zero income implies $a=b=0$. ■

Theorem 6.2 (Generalized Jensen equation). *The general continuous system of solutions of the functional equation*

$$F\left(\frac{x_1+y_1}{2},\ldots,\frac{x_n+y_n}{2}\right)=\frac{G(x_1,\ldots,x_n)+H(y_1,\ldots,y_n)}{2};\qquad x_i,y_i\in\mathbb{R}\ \ or\ \ \mathbb{R}_+,\ \ i=1,2,\ldots,n \tag{6.7}$$

is

$$\begin{aligned} F(x_1,\ldots,x_n) &= C_1x_1 + \ldots + C_nx_n + \frac{a+b}{2}, \\ G(x_1,\ldots,x_n) &= C_1x_1 + \ldots + C_nx_n + a, \\ H(x_1,\ldots,x_n) &= C_1x_1 + \ldots + C_nx_n + b, \end{aligned} \tag{6.8}$$

where a, b and C_i $(i = 1, 2, \ldots, n)$ are arbitrary constants. ■

Proof: Calling

$$U(x_1,\ldots,x_n) = 2F\left(\frac{x_1}{2},\ldots,\frac{x_n}{2}\right), \tag{6.9}$$

i.e.,

$$F(x_1,\ldots,x_n) = \frac{U(2x_1,\ldots,2x_n)}{2}, \tag{6.10}$$

we have

$$U(x_1+y_1,\ldots,x_n+y_n) = G(x_1,\ldots,x_n) + H(y_1,\ldots,y_n),$$

that is Equation (6.1), with the general continuous solution

$$\begin{aligned} U(x_1,\ldots,x_n) &= C_1x_1 + \ldots + C_nx_n + a + b, \\ G(x_1,\ldots,x_n) &= C_1x_1 + \ldots + C_nx_n + a, \\ H(x_1\ldots,x_n) &= C_1x_1 + \ldots + C_nx_n + b. \end{aligned} \tag{6.11}$$

Finally, by substituting function U into (6.9), we obtain

$$F(x_1,\ldots,x_n) = C_1x_1 + \ldots + C_nx_n + \frac{a+b}{2}. \tag{6.12}$$

■

6.3 Generalized Sincov equation

Theorem 6.3 (Generalized Sincov equation). *The general system of solutions of the functional equation*

$$F(x,z) = G(x,y) + H(y,z) \tag{6.13}$$

is

$$F(x,z) = h(z) - f(x); \quad G(x,y) = g(y) - f(x); \quad H(y,z) = h(z) - g(y), \tag{6.14}$$

where f, g and h are arbitrary functions. ■

Proof: Making $y = a$ in (6.13) and calling $h(z) = H(a,z)$ and $f(x) = -G(x,a)$ we get

$$F(x,z) = G(x,a) + H(a,z) = h(z) - f(x),$$

and with $r(y) = H(y,b)$ we have

$$\begin{aligned} G(x,y) &= F(x,b) - H(y,b) = h(b) - f(x) - r(y),, \\ H(y,z) &= F(c,z) - G(c,y) = h(z) - h(b) + r(y), \end{aligned}$$

and calling $g(y) = h(b) - r(y)$ Expression (6.14) is obtained. ■

Example 6.2 (Fatigue strength of a longitudinal element). Bogdanoff and Kozin (1987), basing their work on some experimental results of Picciotto (1970), suggest the following model for the survivor function

$$F(x,z) = F(y,z)^{N(x,y)}$$

where $F(x,z)$ and $F(y,z)$ are the survivor functions associated with two elements of lengths x and y, respectively, and $N(x,y)$ is an unknown positive function.

For $0 < F < 1$ we can take logarithms twice in the previous equation and get the functional equation

$$\log[-\log F(x,z)] = \log N(x,y) + \log[-\log F(y,z)],$$

which, taking into account (6.14), leads to

$$\begin{array}{rcl} \log[-\log F(x,z)] & = & h(z) - f(x), \\ \log N(x,y) & = & g(y) - f(x), \\ \log[-\log F(y,z)] & = & h(z) - g(y) = h(z) - f(y), \end{array}$$

which implies $g(y) = f(y)$. Thus, we finally get

$$\begin{array}{c} F(x,z) = p(z)^{q(x)}; \quad N(x,y) = \dfrac{q(x)}{q(y)}, \\ p(z) = \exp\{-\exp[h(z)]\}; \quad q(x) = \exp[-f(x)], \end{array}$$

where $q(x)$ is an arbitrary positive function, and $p(x)$ is a survivor function, which can be associated with a reference length r_0, i.e., the value of r_0 such that $q(r_0) = 1$. ■

The following lemma analyzes the problem of uniqueness of representation of functions of the form $f[g(x)+h(y)]$, which will be used in the following sections.

Lemma 6.1 (Uniqueness of representation of $F(x,y) = f[g(x) + h(y)]$). *If the function $F(x,y)$ has the two following representations*

$$\begin{array}{c} F(x,y) = f_1[g_1(x) + h_1(y)] = f_2[g_2(x) + h_2(y)]; \\ x,y \in \mathbb{R} \text{ or } [\alpha,\beta] \text{ with } \alpha,\beta \in \mathbb{R}, \end{array} \tag{6.15}$$

where the functions f_i, g_i and h_i (i=1,2) are continuous and strictly monotonic functions, then

$$f_2(x) = f_1\left(\frac{x-a-b}{c}\right); \quad g_2(x) = cg_1(x) + a; \quad h_2(y) = ch_1(y) + b, \tag{6.16}$$

where a, b and c are arbitrary constants. We assume that the obvious restrictions for the domains and ranges of f_i, g_i and h_i $(i = 1,2)$ hold in order (6.15) to make sense. ■

Proof: Making $u = g_1(x)$ and $v = h_1(y)$ in (6.15), yields

$$f_1(u+v) = f_2[g_2 g_1^{-1}(u) + h_2 h_1^{-1}(v)] \;\Rightarrow\; f_2^*(u+v) = g_2^*(u) + h_2^*(v), \tag{6.17}$$

where

$$f_2^*(u) = f_2^{-1} f_1(u);\;\; g_2^*(u) = g_2 g_1^{-1}(u);\;\; h_2^*(u) = h_2 h_1^{-1}(u). \tag{6.18}$$

But the right equation in (6.17) is a Pexider's Equation (4.1) with solution

$$f_2^*(u) = cu + a + b;\;\; g_2^*(u) = cu + a;\;\; h_2^*(u) = cu + b, \tag{6.19}$$

which, taking into account (6.18), leads to (6.16). ∎

6.4 A general equation

In this section we solve a very general equation which later will lead to many important results.

Theorem 6.4 (Generalized bisymmetry equation). *The general solution continuous on a real rectangle of the functional equation*

$$F[G(x,y), H(u,v)] = K[M(x,u), N(y,v)], \tag{6.20}$$

with G invertible in both variables, F and M invertible in the first variable for a fixed value of the second variable, and H, K and N invertible in the second variable for a fixed value of the first, is

$$\begin{array}{rclrcl} F(x,y) &=& k[f(x)+g(y)]; & G(x,y) &=& f^{-1}[p(x)+q(y)],\\ K(x,y) &=& k[l(x)+m(y)]; & H(x,y) &=& g^{-1}[r(x)+s(y)],\\ M(x,y) &=& l^{-1}[p(x)+r(y)]; & N(x,y) &=& m^{-1}[q(x)+s(y)], \end{array} \tag{6.21}$$

where f, g, k, l, m, p, q and s are arbitrary continuous and strictly monotonic functions and r is an arbitrary continuous function. The lower case functions in (6.21) are not uniquely determined. If there are two sets of f, g, k, l, m, p, q, s (indicated by subscripts 1 and 2), then they must be connected by the relations:

$$\begin{array}{rcll} f_2(x) &=& cf_1(x)+a, & l_2(x) = cl_1(x)+a+e-d,\\ k_2(x) &=& k_1\left(\dfrac{x-a-b}{c}\right), & g_2(x) = cg_1(x)+b,\\ m_2(x) &=& cm_1(x)+b+d-e, & p_2(x) = cp_1(x)+a-d,\\ q_2(x) &=& cq_1(x)+d, & r_2(x) = cr_1(x)+e.\\ s_2(x) &=& cs_1(x)+b-e. & \end{array} \tag{6.22}$$

The two sides of (6.20)can be written as

$$k[p(x)+q(y)+r(u)+s(v)]. \tag{6.23}$$

∎

Proof: For the proof of (6.21) we refer the reader to Aczél (1966), page 314. The uniqueness relation (6.22) is demonstrated in the following paragraphs.

Let us assume that we have two different representations of the functions in (6.21). Then we must have

$$\begin{aligned} k_1[f_1(x)+g_1(y)] &= k_2[f_2(x)+g_2(y)], \\ f_1^{-1}[p_1(x)+q_1(y)] &= f_2^{-1}[p_2(x)+q_2(y)], \\ k_1[l_1(x)+m_1(y)] &= k_2[l_2(x)+m_2(y)], \\ g_1^{-1}[r_1(x)+s_1(y)] &= g_2^{-1}[r_2(x)+s_2(y)], \\ l_1^{-1}[p_1(x)+r_1(y)] &= l_2^{-1}[p_2(x)+r_2(y)], \\ m_1^{-1}[q_1(x)+s_1(y)] &= m_2^{-1}[q_2(x)+s_2(y)]. \end{aligned} \tag{6.24}$$

Applying Lemma 6.1 to all equations in (6.24), we obtain

$$\begin{array}{lll} k_2(x) = k_1\left(\dfrac{x-a-b}{c}\right); & f_2(x) = cf_1(x)+a, \\ g_2(x) = cg_1(x)+b, & f_2(x) = c_1\ f_1(x)+a_1+b_1, \\ p_2(x) = c_1\ p_1(x)+a_1, & q_2(x) = c_1\ q_1(x)+b_1, \\ g_2(x) = c_2\ g_1(x)+a_2+b_2, & r_2(x) = c_2\ r_1(x)+a_2, \\ s_2(x) = c_2\ s_1(x)+b_2, & l_2(x) = c_3\ l_1(x)+a_3+b_3, \\ p_2(x) = c_3\ p_1(x)+a_3, & r_2(x) = c_3\ r_1(x)+b_3, \\ m_2(x) = c_4\ m_1(x)+a_4+b_4, & q_2(x) = c_4\ q_1(x)+a_4, \\ s_2(x) = c_4\ s_1(x)+b_4, & l_2(x) = cl_1(x)+a_5, \\ m_2(x) = cm_1(x)+b_5 & \end{array} \tag{6.25}$$

which implies

$$\begin{array}{l} c=c_1=c_2=c_3=c_4;\quad a_1=a_3=a-b_1;\quad b_5=b_1+b-a_2; \\ a_5\ =a-b_1+a_2;\quad b_3=a_2;\quad b_2=b_4=b-a_2;\quad a_4=b_1, \end{array} \tag{6.26}$$

and calling $d=b_1$ and $e=a_2$ and substituting (6.26) into (6.25) we finally get (6.22).

Note that because of (6.23) and the associativity of the sum operation, expressions of the form (6.20) can be evaluated by parallel computation. ∎

Corollary 6.1 (Generalized associativity equation). *The general solution continuous on a real rectangle of the functional equation*

$$F[G(x,y),z] = K[x,N(y,z)], \tag{6.27}$$

with G invertible in both variables, F invertible in the first variable for a fixed value of the second variable and K and N invertible in the second variable for a fixed value of the first, is

$$\begin{array}{ll} F(x,y) = k[f(x)+g(y)]; & G(x,y) = f^{-1}[p(x)+q(y)], \\ K(x,y) = k[p(x)+n(y)]; & N(x,y) = n^{-1}[q(x)+g(y)], \end{array} \tag{6.28}$$

where f, g, k, n, p and q are arbitrary continuous and strictly monotonic functions which are determined up to the following relations

$$\begin{array}{ll} f_2(x) = cf_1(x) + a; & g_2(x) = cg_1(x) + b, \\ n_2(x) = cn_1(x) + b + d; & p_2(x) = cp_1(x) + a - d, \\ k_2(x) = k_1\left(\dfrac{x-a-b}{c}\right); & q_2(x) = cq_1(x) + d. \end{array} \tag{6.29}$$

■

Proof: Equation (6.27) is a particular case of Equation (6.20), with

$$\begin{array}{l} H(u,z) = z \;\Rightarrow\; g(x) = s(x) + a \;\text{ and }\; r(x) = a, \\ M(x,u) = x \;\Rightarrow\; l(x) = p(x) + a \;\text{ and }\; r(x) = a, \end{array}$$

which, with (6.21), leads to

$$\begin{array}{rcl} F(x,y) & = & k[f(x) + g(y)], \\ G(x,y) & = & f^{-1}[p(x) + q(y)], \\ K(x,y) & = & k[p(x) + m(y) + a] = k[p(x) + n(y)], \\ N(x,y) & = & m^{-1}[q(x) + g(y) - a] = n^{-1}[q(x) + g(y)], \end{array} \tag{6.30}$$

where we have made $m(x) = n(x) - a$.
To prove relations (6.29), we assume that we have two different representations of the functions in (6.28). Then, we must have

$$\begin{array}{rcl} k_1[f_1(x) + g_1(y)] & = & k_2[f_2(x) + g_2(y)], \\ f_1^{-1}[p_1(x) + q_1(y)] & = & f_2^{-1}[p_2(x) + q_2(y), \\ k_1[p_1(x) + n_1(y)] & = & k_2[p_2(x) + n_2(y)], \\ n_1^{-1}[q_1(x) + g_1(y)] & = & n_2^{-1}[q_2(x) + g_2(y)]. \end{array} \tag{6.31}$$

If we now apply Lemma 6.1 to all equations in (6.31) we get

$$\begin{array}{rcll} k_2(x) & = & k_1\left(\dfrac{x-a-b}{c}\right); & f_2(x) = cf_1(x) + a, \\ g_2(x) & = & cg_1(x) + b; & n_2(x) = c_1n_1(x) + a_1 + b_1, \\ q_2(x) & = & c_1q_1(x) + a_1; & g_2(x) = c_1g_1(x) + b_1, \\ f_2(x) & = & c_2f_1(x) + a_2 + b_2; & p_2(x) = c_2p_1(x) + a, \\ q_2(x) & = & c_2q_1(x) + b_2; & p_2(x) = cp_1(x) + a_3, \\ n_2(x) & = & cn_1(x) + b_3 & \end{array} \tag{6.32}$$

from which we obtain

$$c = c_1 = c_2;\;\; a_2 = a_3 = a - a_1;\;\; b_1 = b;\;\; b_2 = a_1;\;\; b_3 = a_1 + b, \tag{6.33}$$

and calling $d = a_1$ and substitution back into (6.32) leads to (6.29). ■

The two sides of (6.27) can be written as

$$k[p(x) + q(y) + g(z)],$$

whose representation is not unique, but given by the following corollary.

Corollary 6.2 (Generalized associativity Equation (uniqueness)). *Assume there are two sets of functions* $\{k_1, p_1, q_1, r_1\}$ *and* $\{k_2, p_2, q_2, r_2\}$ *such that*

$$k_1[p_1(x) + q_1(y) + r_1(z)] = k_2[p_2(x) + q_2(y) + r_2(z)]. \tag{6.34}$$

Then, we must have

$$\begin{aligned} k_2(u) &= k_1\left(\frac{u-b-c-d}{a}\right), \\ p_2(x) &= ap_1(x) + b, \\ q_2(y) &= aq_1(y) + c, \\ r_2(z) &= ar_1(z) + d, \end{aligned} \tag{6.35}$$

where a, b, c *and* d *are arbitrary constants.* ■

Example 6.3 (Application to expert systems). (Castillo et al. (1990c)) Corollary 6.1 can be applied to derive some probability models to be used in expert systems. When dealing with probabilistic type expert systems for medical diagnosis, the normal procedure consists of starting with some "a priori" probability for a given disease and, as soon as new symptoms become known, this probability is updated; the process is repeated until the resulting value is high or low enough to make it possible to either confirm or discard the disease. The main problem with these kind of models is the huge number of parameters involved if a full dependence model, i.e., a model taking care of all dependencies, is considered. To make the model feasible some important simplifications must be carried out.

To simplify the example, we deal here with a population of patients in which we consider only three continuous symptoms X, Y and Z, with ranges $[\alpha_X, \omega_X]$, $[\alpha_Y, \omega_Y]$ and $[\alpha_Z, \omega_Z]$, respectively. Let us assume that the probability density function of the random variable (X, Y, Z) in the given population is $Q(x, y, z)$. In addition, we shall make the assumption that the information given by a set of symptoms about $Q(x, y, z)$ can be summarized into a function of the values of the symptoms in the set. More precisely, we assume that functions F, G, H, K, L and M exist so that we have

$$Q(x, y, z) = F[G(x, y), z] = K[x, N(y, z)] = L[y, M(x, z)].$$

Thus, the probability density function, Q, of the random variable (X, Y, Z) can be calculated by means of a function F of $G(x, y)$, a summary of the influences of symptoms X and Y, and the value associated with symptom Z. The same argument is valid for any permutation of the X, Y and Z symptoms. In other words, if at a given time, we know the values associated with symptoms X and Y, we can calculate the influence of these two symptoms on Q by means of $G(x, y)$, and later incorporate the influence of symptom Z by means of function F.

In this case the regularity conditions in Corollary 6.1 hold. If the symptoms are relevant for the disease, it is reasonable to assume that functions

F, K, L, G, N and M are invertible (strictly monotonic) with respect to both variables. This means that the higher (lower) level of one symptom, the higher (lower) probability of the disease.

According to (6.28), the general continuous solution of the functional equation above is

$$\begin{array}{rclrcl} F(x,y) & = & k[f(x)+g(y)]; & G(x,y) & = & f^{-1}[p(x)+q(y)], \\ K(x,y) & = & k[p(x)+n(y)]; & N(x,y) & = & n^{-1}[q(x)+g(y)], \\ L(x,y) & = & k[q(x)+m(y)]; & M(x,y) & = & m^{-1}[p(x)+g(y)], \end{array}$$

which leads to

$$Q(x,y,z) = k[p(x)+q(y)+g(z)].$$

For the function $Q(x,y,z)$ to be a probability density function it must be integrable and non-negative. To this end, we can replace $k(x)$ by $\exp[k(x)]$. Then, our general model becomes

$$Q(x,y,z) = \exp\{k[p(x)+q(y)+g(z)]\},$$

where we can now choose parametric families for k, p, q and g if we so wish, the only condition being that the associated joint density be an integrable function in $[\alpha_X, \omega_X] \times [\alpha_Y, \omega_Y] \times [\alpha_Z, \omega_Z]$. ■

Example 6.4 (Three-variate distributions generated from their bivariate marginals). (Castillo et al. (1990d)) If in Example 6.3 $Q(x,y,z)$ is the cumulative distribution function instead of the probability density function, we arrive at the same solutions but with a different meaning. Now we can assume without loss of generality that k is increasing.

For the function $Q(x,y,z)$ to be a cumulative distribution function, functions p, q and g must be increasing and we must have

$$\begin{array}{lcl} Q(\alpha_X,y,z) & = & k[p(\alpha_X)+q(y)+g(z)] = 0 \Rightarrow p(\alpha_X) = -\infty; k(-\infty) = 0, \\ Q(x,\alpha_Y,z) & = & k[p(x)+q(\alpha_Y)+g(z)] = 0 \Rightarrow q(\alpha_Y) = -\infty; k(-\infty) = 0, \\ Q(x,y,\alpha_Z) & = & k[p(x)+q(y)+g(\alpha_Z)] = 0 \Rightarrow g(\alpha_Z) = -\infty; k(-\infty) = 0, \\ Q(\omega_X,\omega_Y,\omega_Z) & = & k[p(\omega_X)+q(\omega_Y)+g(\omega_Z)] = 1, \end{array}$$

and calling $a = p(\omega_X), b = q(\omega_Y)$ and $c = g(\omega_Z)$ we get $k(a+b+c) = 1$.

Figure 6.1 shows a graphical example of functions p, q, g and k.

If we now oblige functions G, N and M to be marginal cumulative distribution functions we get

$$\begin{array}{rcl} G(x,y) & = & Q(x,y,\omega_Z) = f^{-1}[p(x)+q(y)] = k[p(x)+q(y)+c], \\ N(y,z) & = & Q(\omega_X,y,z) = n^{-1}[q(y)+g(z)] = k[a+q(y)+g(z)], \\ M(x,z) & = & Q(x,\omega_Y,z) = m^{-1}[p(x)+g(z)] = k[p(x)+b+g(z)], \end{array}$$

from which we obtain

$$k(x) = f^{-1}(x-c) = n^{-1}(x-a) = m^{-1}(x-b)$$

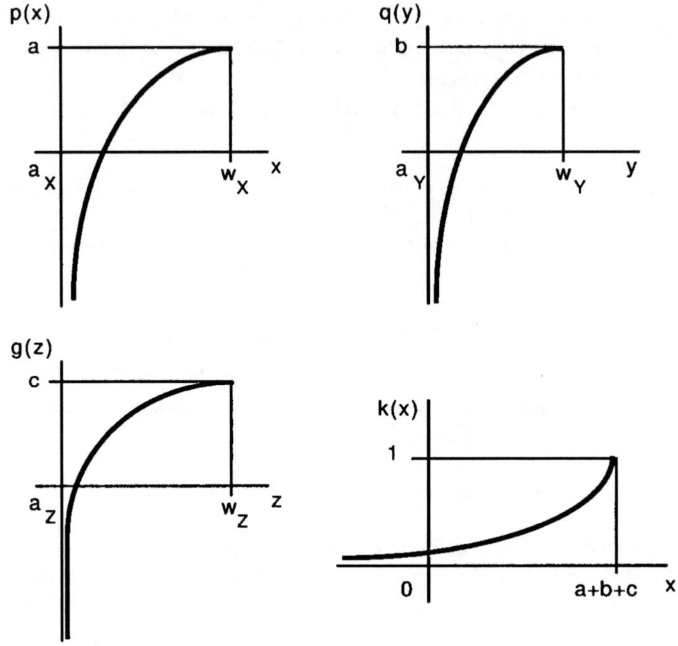

Figure 6.1: Example of functions p, q, g and k.

and

$$\begin{aligned} F(x,y) &= k[k^{-1}(x)+g(y)-c], \; K(x,y) = k[p(x)+k^{-1}(y)-a], \\ L(x,y) &= k[q(x)+k^{-1}(y)-b]. \end{aligned}$$

Note that the regularity conditions in Corollary 6.1 hold if we assume the cumulative distribution function Q to be increasing in all its arguments. Thus, we have obtained all three-variate distributions such that its cumulative distribution function can be obtained from any of its bivariate marginals and the other variable. This is

$$Q(x,y,z) = \exp\{k[p(x)+q(y)+g(z)]\}.$$

■

Theorem 6.5 (Generalized associativity equation). *The general solution continuous on a real rectangle of the functional equation*

$$G(x,y) = K[M(x,z),N(y,z)], \tag{6.36}$$

with N invertible in the first argument for any value of the second, M invertible in both arguments, K invertible in the first argument for some fixed value of the *second, and G invertible in the second for a fixed value of the first, is*

$$\begin{array}{rclrcl} G(x,y) &=& f^{-1}[p(x)+q(y)]; & K(x,y) &=& f^{-1}[l(x)+n(y)], \\ M(x,y) &=& l^{-1}[p(x)+r(y)]; & N(x,y) &=& n^{-1}[q(x)-r(y)], \end{array} \quad (6.37)$$

where f, l, n, p, q and r are arbitrary continuous and strictly monotonic functions which are determined up to the following relations

$$\begin{array}{rclrcl} f_2(x) &=& c_1 f_1(x)+a_1+b_1; & p_2(x) &=& c_1 p_1(x)+a_1, \\ q_2(x) &=& c_1 q_1(x)+b_1; & l_2(x) &=& c_1 l_1(x)+a_4, \\ n_2(x) &=& c_1 n_1(x)+b_1+a_1-a_4; & r_2(x) &=& c_1 r_1(x)+a_4-a_1, \end{array} \quad (6.38)$$

where the a's and b's are arbitrary constants. ∎

Proof: If $N(y,z)$ is invertible with respect to its first argument for any z, we have

$$N(y,z)=\omega \Leftrightarrow y=L(z,\omega),$$

and substituting this into (6.36) we can write

$$K[M(x,z),\omega]=G[x,L(z,\omega)],$$

which is an equation of the form (6.27). Thus, from (6.28) we get (6.37).

To prove relations (6.38), we assume that we have two different representations of the functions in (6.37), i.e.,

$$\begin{array}{rcl} f_1^{-1}[p_1(x)+q_1(y)] &=& f_2^{-1}[p_2(x)+q_2(y)], \\ f_1^{-1}[l_1(x)+n_1(y)] &=& f_2^{-1}[l_2(x)+n_2(y)], \\ l_1^{-1}[p_1(x)+r_1(y)] &=& l_2^{-1}[p_2(x)+r_2(y)], \\ n_1^{-1}[q_1(x)-r_1(y)] &=& n_2^{-1}[q_2(x)-r_2(y)]. \end{array} \quad (6.39)$$

Applying now Lemma 6.1 to all equations in (6.39) we get

$$\begin{array}{rclrcl} f_2(x) &=& c_1 f_1(x)+a_1+b_1; & p_2(x) &=& c_1 p_1(x)+a_1; \\ q_2(x) &=& c_1 q_1(x)+b_1; & l_2(x) &=& c_2 l_1(x)+a_2+b_2; \\ p_2(x) &=& c_2 p_1(x)+a_2; & r_2(x) &=& c_2 r_1(x)+b_2; \\ n_2(x) &=& c_3 n_1(x)+a_3+b_3; & q_2(x) &=& c_3 q_1(x)+a_3; \\ -r_2(x) &=& -c_3 r_1(x)+b_3; & l_2(x) &=& c_1 l_1(x)+a_4; \\ n_2(x) &=& c_1 n_1(x)+b_4, && & \end{array} \quad (6.40)$$

from which we obtain

$$a_2=a_1; \quad a_3=b_1; \quad b_2=-b_3=a_4-a_1; \quad c_1=c_2=c_3; \quad b_4=a_1+b_1-a_4, \quad (6.41)$$

and back substitution into (6.40) leads to (6.38). ∎

The two sides of (6.36) can be written as

$$f^{-1}[p(x)+q(y)].$$

6.5 The associativity equation

The following theorem gives the solution of the associativity equation and analyzes the uniqueness of its representation.

Theorem 6.6 (Associativity equation). *The general solution continuous and invertible in both variables on a real rectangle of the functional equation*

$$F[F(x,y),z] = F[x,F(y,z)] \tag{6.42}$$

is

$$F(x,y) = f^{-1}[f(x)+f(y)], \tag{6.43}$$

where f is one arbitrary continuous and strictly monotonic function, which can be replaced only by $cf(x)$, where c is an arbitrary constant. ∎

Proof: Equation (6.42) is a particular case of Equation (6.27) with $G = F = K = N$.
Applying Lemma 6.1 to the equality of F and K in (6.28), yields

$$\begin{aligned} &k(x) = k\left(\frac{x-a_1-b_1}{c_1}\right) \ \Rightarrow\ c_1 = 1,\ a_1 = -b_1,\\ &p(x) = f(x)+a_1,\\ &n(y) = g(y)-a_1, \end{aligned} \tag{6.44}$$

and doing the same for the equality $F = N$ in (6.28) we can write

$$\begin{aligned} g(y) &= c_2 g(y) + b_2 \ \Rightarrow\ c_2 = 1,\ b_2 = 0,\\ q(x) &= f(x)+a_2,\\ n^{-1}(x) &= k(x-a_2). \end{aligned} \tag{6.45}$$

Substituting now (6.44) and (6.45) into the G function of (6.28), and making $A = a_1 + a_2$ we get

$$\begin{aligned} G(x,y) &= f^{-1}[p(x)+q(y)]\\ &= f^{-1}[f(x)+a_1+f(y)+a_2]\\ &= f^{-1}[f(x)+f(y)+A], \end{aligned} \tag{6.46}$$

which satisfies Equation (6.42).
Defining $f^*(x) = f(x) + A$, we obtain

$$G(x,y) = f^{*-1}[f^*(x)+f^*(y)],$$

which is Equation (6.43).

To study the uniqueness of representation of the form (6.43) we assume that the following two representations exist

$$f^{-1}[f(x)+f(y)] = g^{-1}[g(x)+g(y)]. \tag{6.47}$$

According to Lemma 6.1 we have

$$g^{-1}(x) = f^{-1}\left(\frac{x-a-b}{c}\right); \; g(x) = cf(x) + a; \; g(y) = cf(y) + b \tag{6.48}$$

from which we get

$$g(x) = cf(x) + a = cf(x) + b = cf(x) + a + b \Rightarrow \begin{cases} b = 0, \\ a = 0, \end{cases} \tag{6.49}$$

and then we obtain

$$g(x) = cf(x). \tag{6.50}$$

■

The two sides of (6.42) can be written as

$$f^{-1}[f(x) + f(y) + f(z)].$$

Corollary 6.3 (Alternative representation of associative operations). *Any solution continuous and invertible in both variables of Equation (6.42) can also be written as*

$$F(x, y) = g^{-1}[g(x) * g(y)],$$

*where * is used to denote any associative continuous and cancelable operation and g is a continuous and strictly monotonic function.* ■

Proof: If $F(x, y)$ is one solution of Equation (6.42) and * is associative, then, according to Theorem 6.6, we have

$$\begin{aligned} F(x, y) &= f^{-1}[f(x) + f(y)], \\ x * y &= G(x, y) = h^{-1}[h(x) + h(y)], \end{aligned}$$

where f and h are continuous and strictly monotonic functions.
Thus, we obtain

$$\begin{aligned} &h(x * y) = h(x) + h(y) \;\Rightarrow\; h[h^{-1}(u) * h^{-1}(v)] = u + v \\ &\quad\Rightarrow F(x, y) = f^{-1}[f(x) + f(y)] = f^{-1}\{h[h^{-1}f(x) * h^{-1}f(y)]\}, \end{aligned}$$

and calling

$$g(x) = h^{-1}f(x) \;\Leftrightarrow\; g^{-1}(x) = f^{-1}h(x),$$

we get

$$F(x, y) = g^{-1}[g(x) * g(y)],$$

where the continuity and invertibility of g comes from those of f and h. ■

Example 6.5 (Synthesis of judgments). (Aczél (1987b), pp. 122-125) Let us assume that we have n quantifiable judgments $x_1, x_2, \ldots, x_n$ which we want to synthesize into a "consensus" judgment $f(x_1, x_2, \ldots, x_n)$. We assume that the consensus function f is separable, that is, it satisfies the condition

$$f(x_1, x_2, \ldots, x_n) = g_1(x_1)\Delta g_2(x_2)\Delta \ldots \Delta g_n(x_n)$$

where Δ is associative, commutative and cancelable.

The associative and commutative property of the operation Δ is imposed to avoid any possible type of manipulation of the final judgment. If Δ is not associative the order of the judges can be adequately replaced to get desired results. Otherwise, the final result is invariant with respect to the evaluation order.

The cancelable property ensures that all judges have influence on the consensual result.

Theorem 6.6 implies that Δ can be written as

$$y_1 \Delta y_2 = \varphi^{-1}[\varphi(y_1) + \varphi(y_2)],$$

and then, we have

$$f(x_1, x_2, \ldots, x_n) = \varphi^{-1}\left\{\sum_{i=1}^{n} \varphi[g_i(x_i)]\right\}.$$

We can add extra conditions to be satisfied by function f such as *'unanimity'*, which is

$$f(x, x, \ldots, x) = x \;\Rightarrow\; \varphi^{-1}\left\{\sum_{i=1}^{n} \varphi[g_i(x)]\right\} = x.$$

If, in addition, all judges have the same weight, then $g_i(x) = g(x)$ for all $i = 1, 2, \ldots, n$ and we get

$$f(x, x, \ldots, x) = x \;\Rightarrow\; \varphi^{-1}\{n\varphi[g(x)]\} = x \;\Rightarrow\; g(x) = \varphi^{-1}\left[\frac{\varphi(x)}{n}\right],$$

and, finally, f becomes

$$f(x_1, x_2, \ldots, x_n) = \varphi^{-1}\left\{\frac{\sum_{i=1}^{n} \varphi(x_i)}{n}\right\}.$$

Some interesting particular examples are:

1. The arithmetic mean: $\varphi(x) = x$.

2. The geometric mean: $\varphi(x) = \log x$.

3. The L_p mean: $\varphi(x) = x^p$.

■

6.6 The transitivity equation

The following theorem gives the solution of the transitivity equation and analyzes the uniqueness of its representation.

Theorem 6.7 (Transitivity equation). *The general solution continuous and invertible in both variables on a real rectangle of the functional equation*

$$S(x,y) = S[S(x,z), S(y,z)] \tag{6.51}$$

is

$$S(x,y) = f^{-1}[f(x) - f(y)], \tag{6.52}$$

where f is one arbitrary continuous and strictly monotonic function which can be replaced only by $g(x) = cf(x)$, where c is an arbitrary constant. ∎

Proof: Equation (6.51) is a particular case of Equation (6.36) with $S = G = K = M = N$.
Applying Lemma 6.1 to the equality of K and N in (6.37), we deduce

$$\begin{array}{rcl} n^{-1}(x) & = & f^{-1}\left(\dfrac{x - a_1 - b_1}{c_1}\right), \\ q(x) & = & c_1 l(x) + a_1, \\ r(y) & = & -c_1 n(y) - b_1, \end{array} \tag{6.53}$$

and doing the same for the equality $M = G$ in (6.37), we obtain

$$\begin{array}{rcl} p(x) & = & c_2 p(x) + a_2 \Rightarrow c_2 = 1, \; a_2 = 0, \\ l^{-1}(x) & = & f^{-1}\left(\dfrac{x - a_2 - b_2}{c_2}\right) = f^{-1}(x - b_2), \\ r(y) & = & q(y) + b_2. \end{array} \tag{6.54}$$

Substituting now the expressions (6.53) and (6.54) into the N function in (6.37), we get

$$\begin{array}{rcl} S(x,y) & = & N(x,y) = n^{-1}[q(x) - r(y)] = n^{-1}[r(x) - b_2 - r(y)] = \\ & = & n^{-1}\{c_1[n(y) - n(x)] - b_2\} = n^{-1}\{c_1[n(y) - n(x)] + A\}, \end{array} \tag{6.55}$$

where we have made $A = -b_2$. Substitution now into Equation (6.51) leads to

$$c_1^2 + c_1 = 0 \Rightarrow \left\{ \begin{array}{l} c_1 = 0, \\ c_1 = -1, \end{array} \right. \tag{6.56}$$

which, for non-constant $S(x,y)$, gives

$$S(x,y) = n^{-1}[n(x) - n(y) + A]. \tag{6.57}$$

Defining $n^*(x) = n(x) - A$, Expression (6.57) becomes

$$S(x,y) = n^{*-1}[n^*(x) - n^*(y)].$$

To study the uniqueness of representation of the form (6.52) we assume that the two following representations exist

$$f^{-1}[f(x) - f(y)] = g^{-1}[g(x) - g(y)]. \tag{6.58}$$

Lemma 6.1 yields

$$g^{-1}(x) = f^{-1}\left(\frac{x-a-b}{c}\right); \quad g(y) = cf(y)+a; \quad B-g(x) = c[-f(x)]+b, \tag{6.59}$$

from which we deduce that

$$g(x) = cf(x) + a = cf(x) - b = cf(x) + a + b \Rightarrow \begin{cases} b = 0, \\ a = 0, \end{cases} \tag{6.60}$$

and then

$$g(x) = cf(x). \tag{6.61}$$

■

Corollary 6.4 (Alternative representation of the solutions of the transitivity equation). *Any solution of Equation (6.51) can also be written as*

$$S(x,y) = g^{-1}[g(x) * g(y)],$$

*where * is any transitive continuous and cancelable operation; that is, any operation $x * y = G(x,y)$ such that $G(x,y)$ satisfies Equation (6.51), and g is a continuous and strictly monotonic function.* ■

Proof: If $S(x,y)$ is one solution of Equation (6.51) and * is transitive, then, from Theorem 6.7, it follows that

$$\begin{aligned} S(x,y) &= n^{-1}[n(x) - n(y)], \\ x * y &= G(x,y) = h^{-1}[h(x) - h(y)], \end{aligned}$$

where n and h are continuous and strictly monotonic functions.

This leads to

$$\begin{aligned} &h(x * y) = h(x) - h(y) \;\Rightarrow\; h[h^{-1}(u) * h^{-1}(v)] = u - v \;\Rightarrow \\ &\Rightarrow S(x,y) = n^{-1}[n(x) - n(y)] = n^{-1}\{h[h^{-1}n(x) * h^{-1}n(y)]\} \end{aligned}$$

and calling

$$g(x) = h^{-1}n(x) \;\Leftrightarrow\; g^{-1}(x) = n^{-1}h(x),$$

yields

$$S(x,y) = g^{-1}[g(x) * g(y)],$$

where the continuity and monotonicity of g come from those of n and h. ■

6.7 The bisymmetry equation

The following theorem gives the solution of the bisymmetry equation and analyzes the uniqueness of its representation.

Theorem 6.8 (Bisymmetry equation). *The general solution continuous and invertible in both variables on a real rectangle of the functional equation*

$$S[S(x,y),S(u,z)] = S[S(x,u),S(y,z)] \tag{6.62}$$

is

$$S(x,y) = g^{-1}[Ag(x)+Bg(y)+C], \tag{6.63}$$

where g is one arbitrary continuous and strictly monotonic function which can be replaced only by $g_1(x) = cg(x)+d$, where c and d are arbitrary constants. ∎

Proof: Equation (6.62) is a particular case of Equation (6.20) with $S = F = G = H = K = M = N$.

Lemma 6.1 and the equality of F and H in (6.21), lead to

$$\begin{array}{rcl} g^{-1}(x) &=& k\left(\dfrac{x-a_1-b_1}{c_1}\right), \\ r(x) &=& c_1 f(x)+a_1, \\ s(y) &=& c_1 g(y)+b_1, \end{array} \tag{6.64}$$

and doing the same for the equality $F = M$ in (6.21), we obtain

$$\begin{array}{rcl} l^{-1}(x) &=& k\left(\dfrac{x-a_2-b_2}{c_2}\right), \\ p(x) &=& c_2 f(x)+a_2, \\ r(y) &=& c_2 g(y)+b_2. \end{array} \tag{6.65}$$

If we now substitute Expressions (6.64) and (6.65) into the H function in (6.21), and let $A = c_2$, $B = c_1$, $C = b_1+b_2$, we get

$$\begin{array}{rcl} S(x,y) &=& H(x,y) = g^{-1}[r(x)+s(y)] \\ &=& g^{-1}[c_2 g(x)+c_1 g(y)+b_1+b_2] \\ &=& g^{-1}[Ag(x)+Bg(y)+C] \end{array} \tag{6.66}$$

which satisfies Equation (6.62).

To study the uniqueness of representation of the form (6.66) we assume that the following two representations exist

$$g^{-1}[A_1 g(y)+B_1 g(x)+C_1] = g_1^{-1}[A_2 g_1(y)+B_2 g_1(x)+C_2]. \tag{6.67}$$

According to Lemma 6.1 we have

$$\begin{array}{rcl} g_1^{-1}(x) &=& g^{-1}\left(\dfrac{x-a-b}{c}\right), \\ A_2 g_1(y) &=& cA_1 g(y)+a, \\ B_2 g_1(x)+C_2 &=& c[B_1 g(x)+C_1]+b, \end{array} \tag{6.68}$$

from which we get

$$g_1(x) = \frac{cA_1g(x) + a}{A_2} = \frac{c[B_1g(x) + C_1] + b - C_2}{B_2} = cg(x) + a + b, \tag{6.69}$$

and then

$$\frac{A_1}{A_2} = 1; \ \frac{a}{A_2} = a + b; \ \frac{B_1}{B_2} = 1; \ \frac{cC_1 + b - C_2}{B_2} = a + b, \tag{6.70}$$

which implies

$$A_1 = A_2; \ B_1 = B_2; \ \begin{cases} a + b = \dfrac{cC_1 - C_2}{A_1 + B_1 - 1} & \text{if} \quad A_1 + B_1 \neq 1 \\ cC_1 - C_2 = 0; \ \dfrac{a}{b} = \dfrac{A_2}{1 - A_2} & \text{if} \quad A_1 + B_1 = 1 \end{cases} \tag{6.71}$$

and then

$$g_1(x) = cg(x) + \frac{cC_1 - C_2}{A_1 + B_1 - 1}; \ A_1 = A_2; \ B_1 = B_2. \tag{6.72}$$

Thus, we finally obtain

$$g_1(x) = cg(x) + d. \tag{6.73}$$

■

The two sides of (6.62) can be written as

$$g^{-1}\{A^2g(x) + AB[g(y) + g(u)] + B^2g(z)\}.$$

6.8 The transformation equation

The following theorem gives the solution of the transformation equation and analyzes the uniqueness of its representation.

Theorem 6.9 (Transformation equation). *The general solution continuous and invertible in both variables on a real rectangle of the functional equation*

$$S[S(x,y),z] = S[x, N(y,z)] \tag{6.74}$$

is

$$S(x,y) = k[k^{-1}(x) + n(y)], \ N(x,y) = n^{-1}[n(x) + n(y)], \tag{6.75}$$

where k and n are arbitrary continuous and strictly monotonic functions which can be replaced only by $k^*(x) = k[(x-a)/c]$ and $n^*(x) = cn(x)$, where a and c are arbitrary constants. ■

Proof: Equation (6.74) is a particular case of Equation (6.27) with $S = F = G = K$.

Lemma 6.1 and the equality of K and G in (6.28), yield

$$\left.\begin{array}{lll} f^{-1}(x) & = & k\left(\dfrac{x-a_1-b_1}{c_1}\right) \\ p(x) & = & = c_1 p(x) + a_1 \Rightarrow c_1 = 1,\ a_1 = 0 \\ q(y) & = & = c_1 n(y) + b_1 = n(y) + b_1 \end{array}\right\} \Rightarrow f^{-1}(x) = k(x-b_1), \tag{6.76}$$

and doing the same for the equality $F = K$ in (6.28), we obtain

$$\begin{array}{lllll} k(x) & = & k\left(\dfrac{x-a_2-b_2}{c_2}\right) & \Rightarrow & c_2 = 1,\ a_2 = -b_2, \\ p(x) & = & c_2 f(x) + a_2 & = & f(x) - b_2, \\ n(y) & = & c_2 g(y) + b_2 & = & g(y) + b_2. \end{array} \tag{6.77}$$

Substituting the expressions (6.76) and (6.77) into the F and N functions in (6.28) leads to

$$\begin{array}{lllll} S(x,y) & = & k[k^{-1}(x) + n(y) + b_1 - b_2] & = & k[k^{-1}(x) + n(y) + A], \\ N(x,y) & = & n^{-1}[n(x) + n(y) + b_1 - b_2] & = & n^{-1}[n(x) + n(y) + A], \end{array} \tag{6.78}$$

which satisfies Equation (6.74). Now, making $n^*(x) = n(x) + A$, we finally obtain (6.75).

To study the uniqueness of representation of the form (6.75) we assume that the two following representations exist

$$\begin{array}{lll} k_1[k_1^{-1}(x) + n_1(y)] & = & k_2[k_2^{-1}(x) + n_2(y)], \\ n_1^{-1}[n_1(x) + n_1(y)] & = & n_2^{-1}[n_2(x) + n_2(y)]. \end{array} \tag{6.79}$$

According to Lemma 6.1, we have

$$\begin{array}{llllll} k_2(x) & = & k_1\left(\dfrac{x-a_3-b_3}{c_3}\right); & n_2^{-1}(x) & = & n_1^{-1}\left(\dfrac{x-a_4-b_4}{c_4}\right); \\ k_2^{-1}(x) & = & c_3 k_1^{-1}(x) + a_3; & n_2(y) & = & c_4 n_1(y) + b_4; \\ n_2(y) & = & c_3 n_1(y) + b_3; & n_2(x) & = & c_4 n_1(x) + a_4, \end{array} \tag{6.80}$$

from which we obtain

$$\begin{array}{lll} n_2(x) & = & c_4 n_1(x) + b_4 \\ & = & c_4 n_1(x) + a_4 \\ & = & c_3 n_1(x) + b_3 \\ & = & c_4 n_1(x) + a_4 + b_4 k_2^{-1}(x) \\ & = & c_3 k_1^{-1}(x) + a_3 \\ & = & c_3 k_1^{-1}(x) + a_3 + b_3, \end{array} \tag{6.81}$$

and then

$$b_3 = b_4 = 0;\ c_3 = c_4;\ a_4 = 0, \tag{6.82}$$

which implies

$$k_2(x) = k_1\left(\frac{x-a_3}{c_3}\right);\ n_2(x) = c_3 n_1(x). \tag{6.83}$$

■

The two sides of (6.74) can be written as

$$k[k^{-1}(x) + n(y) + n(z)].$$

Exercises

6.1 Find the general solution continuous on a real rectangle of the functional equation

$$F(x + y, uv) = K[M(x, u), N(y, v)],$$

with F and M invertible in the first variable for a fixed value of the second variable and K and N invertible in the second variable for a fixed value of the first.

6.2 Find the general solution continuous on a real rectangle of the functional equation

$$F(x + y, z) = K(x, y + z),$$

with F invertible in the first variable for a fixed value of the second variable and K invertible in the second variable for a fixed value of the first. Give this functional equation a physical interpretation.

6.3 Find the general solution continuous on a real rectangle of the functional equation

$$G(x, y) = K(x + z, y - z),$$

with K invertible in the first argument for some fixed value of the second and G invertible in the second for a fixed value of the first.

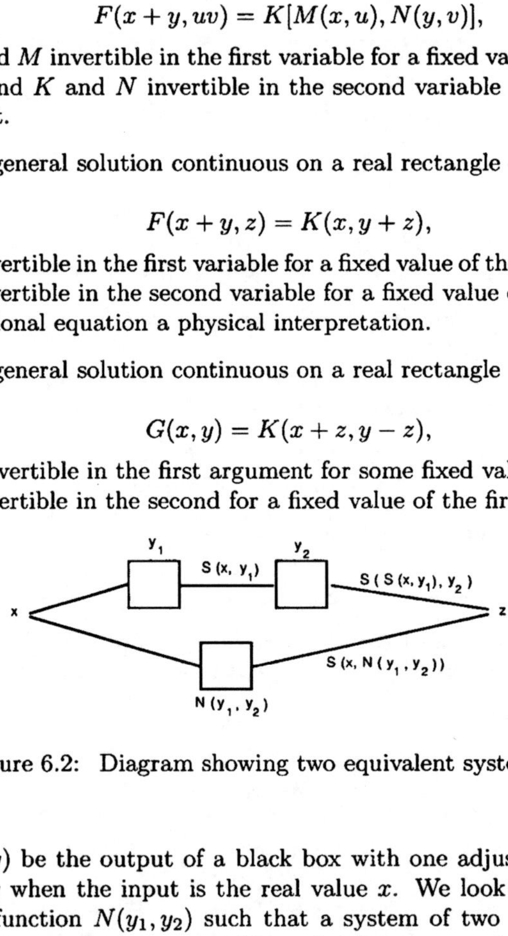

Figure 6.2: Diagram showing two equivalent systems.

6.4 Let $S(x, y)$ be the output of a black box with one adjustable parameter of value y when the input is the real value x. We look for the function S and a function $N(y_1, y_2)$ such that a system of two such boxes with parameters y_1 and y_2, connected in series, becomes equivalent to a single box with parameter $N(y_1, y_2)$ (see Figure 6.2). Find the $S(x, y)$ function if

$$N(u, v) = \frac{\ln(2^y + 2^x)}{\ln 2}$$

CHAPTER 7

Functional equations and differential equations

7.1 Introduction

The methods developed in previous chapters for solving functional equations or systems can be called direct methods. However, functional equations and systems can also be solved by their reduction to differential equations, i.e., by solving their corresponding equivalent differential equations or systems. The term equivalent is used in the sense of Definition 1.6. Thus, when we use this term here, it must be understood that we refer to the class of differentiable functions. The main shortcoming of this method is that we need to assume differentiability up to a certain order of the functions involved, a property not required in the use of direct methods. The existence of equivalent systems of differential and functional equations, in the sense of having the same sets of solutions, allows us not only to use differential equations for solving functional equations but also to use functional equations for solving differential equations. In this chapter we analyze the problem of equivalence of differential, functional and difference equations and give methods to move between them. Figure 7.1 illustrates these relationships.

The present chapter is organized in the following manner. Section 7.2 introduces a motivating example: a mass supported by two springs and a viscous damper is used to illustrate the concept of equivalence of differential, difference and functional equations. Section 7.3 deals with the problem of reduction of functional equations to equivalent differential equations. We shall start, in Section 7.3.1, with the case of equations with functions of one variable and then we get ordinary differential equations. In this section we also show how functional equations can be used to obtain numerical solutions to differential equations. It is interesting to point out that there exist functional equations which are exact reproductions of some sets of differential equations. In particular, exact associated difference equations, in the sense of having the same solutions at the

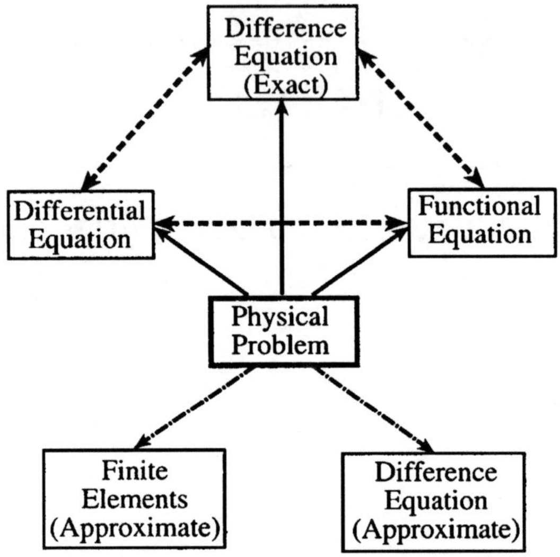

Figure 7.1: Illustration of the relation between differential equations, functional equations and their associated numerical methods.

grid points, are obtained. In Section 7.3.2 we analyze equations with functions of several variables and then partial differential equations will result. In particular, a generalized auto-distributivity equation is solved. Section 7.4 shows how to go from difference to differential equations. Then, Section 7.5 deals with the problem of reduction of differential equations to equivalent functional equations. Section 7.6 shows how to obtain a difference equation equivalent to a given functional equation. Finally, Section 7.7 presents an alternative approach to differential equations for solving physical and engineering problems, based on considering discrete pieces and establishing the equilibrium or balance for such pieces, obtaining the corresponding mathematical model in terms of functional equations. As will be shown, this approach makes it possible to use new numerical and exact methods that can be more intuitive and efficient than those associated with the differential equation scheme.

7.2 A motivating example

Consider the system in Figure 7.2 consisting of a mass m supported by two identical springs and a viscous damper or dashpot (see Richart et al. (1970)). The spring constants $k/2$ are defined as the change in force per unit spring length change. The force in the dashpot is directly proportional with a constant m to velocity $z'(t)$.

The differential equation of motion of the system in Figure 7.2 may be ob-

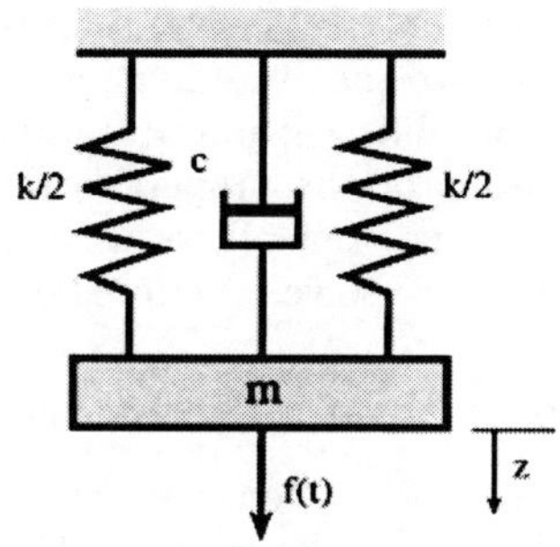

Figure 7.2: One degree of freedom system with springs and viscous damping.

tained by making use of Newton's second law. Measuring displacement from the rest position, the equilibrium of vertical forces at position $z(t)$ leads to the differential equation

$$mz''(t) + cz'(t) + kz(t) = f(t). \tag{7.1}$$

As will be shown, in the case of regular damping ($c^2 < 4km$), the differential Equation (7.1) is equivalent (in the sense of having the same solutions) to the functional equation

$$z(t+u_2) = \alpha_0(u_1,u_2)z(t) + \alpha_1(u_1,u_2)z(t+u_1) + \delta(t;u_1,u_2), \tag{7.2}$$

where

$$\begin{aligned} \alpha_0(u_1,u_2) &= \frac{exp(au_2)(\cos(b\,u_2)\sin(b\,u_1) - \cos(b\,u_1)\sin(b\,u_2))}{\sin(b\,u_1)}, \\ \alpha_1(u_1,u_2) &= \frac{\sin(b\,u_2)exp(a(u_2-u_1))}{\sin(b\,u_1)}, \end{aligned} \tag{7.3}$$

with a and b arbitrary constants, and $\delta(t;u_1,u_2)$ is a function associated with a particular solution.

Similarly, the differential Equation (7.1) is equivalent to the difference equation

$$z(t+2u) = \alpha_0(u)z(t) + \alpha_1(u)z(t+u) + \delta(t,u), \tag{7.4}$$

where $\alpha_0(u)$ and $\alpha_1(u)$ are functions of u (constants if u is assumed constant) and function $\delta(t)$ is associated with a particular solution.

What is important here is that Equations (7.2) and (7.4) are exact in the sense that they gives exact values of the solution at any point or the grid points $(t, t+u, t+2u, \ldots t+nu, \ldots)$, respectively.

7.3 From functional to differential equations

In this section we deal with two important problems: first, we show how a functional equation can be transformed into an equivalent differential equation and second, we solve it to obtain the general solution to the initial equation. It is worthwhile mentioning here that the first problem is of interest in itself, because, on one hand we get a different physical interpretation of the problem under consideration, by means of the properties revealed by the differential equation, and, on the other hand, this allows us to use some techniques from the differential equations world to solve functional equations.

7.3.1 Reduction to ordinary differential equations

The general solution to a functional equation with functions of one variable depends on one or several arbitrary constants. Thus, if we differentiate the general solution one or several times we can eliminate all the constants and obtain an ordinary differential equation. Similarly, differentiating one or several times both sides of a functional equation leads to the same result.

To clarify this methodology we apply it to several of the equations encountered in previous chapters.

Example 7.1 (The equation for homogeneous functions of one variable). Let us start with the following Equation (already studied in Section 3.2)

$$f(yx) = y^k f(x); \quad x, y \in \mathbb{R}_+. \tag{7.5}$$

By differentiating, first with respect to x and then with respect to y, we get

$$yf'(yx) = y^k f'(x); \quad xf'(yx) = ky^{k-1} f(x), \tag{7.6}$$

and assuming $x \neq 0, y \neq 0$ and eliminating $f'(yx)$ we obtain the following differential equation with associated solution

$$xf'(x) - kf(x) = 0 \Rightarrow f(x) = cx^k; \quad x \in \mathbb{R}_+. \tag{7.7}$$

Because of the continuity of f in $\mathbb{R}_+$ we get $f(0) = 0$ if $k \neq 0$ and $f(0) = c$ if $k = 0$. Thus, (7.7) is the general differentiable solution of the functional Equation (7.5), which was already given in (3.1). Thus, the one parameter (k) family of functional equations (7.5) is equivalent to the one parameter family of differential equations (7.7).

It is interesting to point out here that both Equation (7.5) and the left equation in (7.7) lead to $f(0) = 0$ if $k \neq 0$, that is, an extra condition is implied by the initial equations. ∎

Example 7.2 (Cauchy's equations). In this section we apply the above technique to the four Cauchy equations (3.7) to (3.10). Let us start with Cauchy's equation I

$$f(x + y) = f(x) + f(y); \quad x, y \in \mathbb{R}. \tag{7.8}$$

Note that this equation also implies $f(0) = 0$.

By differentiating, with respect to y, we get

$$f'(x + y) = f'(y), \tag{7.9}$$

and substitution of $y = 0$ leads to the following differential equation with its associated general differentiable solution

$$f'(x) = C \Rightarrow f(x) = Cx + b. \tag{7.10}$$

But substitution into (7.8), or taking into account that $f(0) = 0$, leads to $b = 0$ and then, the general differentiable solution of the initial functional equation is $f(x) = Cx$. Thus, here the functional equation (7.8) is equivalent to the one-parameter (C) family of differential equations $\{f'(x) = C\}$ plus the extra condition $f(0) = 0$.

Similarly, for Cauchy's equation II

$$f(x+y) = f(x)f(y); \quad x, y \in \mathbb{R}, \tag{7.11}$$

differentiating with respect to x and making $y = 0$ leads to

$$f'(x) = Cf(x) \Rightarrow f(x) = k\exp(Cx). \tag{7.12}$$

Substitution of (7.12) into (7.11), or taking into account that (7.11) implies either $f(0) = 0$ or $f(0) = 1$, leads to $f(x) = 0$ or $f(x) = \exp(cx)$. Consequently, (7.11) is equivalent to (7.12) with the extra condition $\{f(0) = 0 \text{ or } f(0) = 1\}$.

Cauchy's equation III

$$f(xy) = f(x) + f(y); \quad x, y \in \mathbb{R}_{++}, \tag{7.13}$$

can be solved by the same method; i.e., by differentiating we get

$$\left.\begin{array}{l} yf'(xy) = f'(x) \\ xf'(xy) = f'(y) \end{array}\right\} \Rightarrow xf'(x) = yf'(y) = C \Rightarrow f(x) = C\log x + b, \tag{7.14}$$

where we have used the conditions $x \neq 0$ and $y \neq 0$. However, Equation (7.13) implies $f(1) = 0$, and then $b = 0$. Thus, we conclude that (7.13) is equivalent to the one-parameter family of differential equations $\{xf'(x) = C\}$ plus $f(1) = 0$.

Finally, Cauchy's equation IV

$$f(xy) = f(x)f(y); \quad x, y \in \mathbb{R}_{++} \tag{7.15}$$

gives

$$\left.\begin{array}{l} yf'(xy) = f(y)f'(x) \\ xf'(xy) = f(x)f'(y) \end{array}\right\} \Rightarrow \frac{xf'(x)}{f(x)} = \frac{yf'(y)}{f(y)} = C \Rightarrow f(x) = kx^C, \tag{7.16}$$

where we have used the conditions $f(x) \neq 0$ and $x \neq 0$. Substituting this now into (7.15), leads to $f(x) = 0$ or $f(x) = x^C$. Thus, (7.15) is equivalent to $\{xf'(x) = Cf(x)\}$ plus $\{f(1) = 0 \text{ or } f(1) = 1\}$. ∎

Example 7.3 (Jensen's equation). By differentiating with respect to x and y in Jensen's equation

$$f\left(\frac{x+y}{2}\right) = \frac{f(x)+f(y)}{2}; \quad x, y \in \mathbb{R} \tag{7.17}$$

we get

$$\left.\begin{array}{l}\frac{1}{2}f'\left(\frac{x+y}{2}\right)=\frac{f'(x)}{2}\\ \frac{1}{2}f'\left(\frac{x+y}{2}\right)=\frac{f'(y)}{2}\end{array}\right\}\Rightarrow f'(x)=f'(y)=C\Rightarrow f(x)=Cx+b, \tag{7.18}$$

which is the general differentiable solution. Note that this general differentiable solution depends on two arbitrary constants even though the functional equation contains a single unknown.

In this case, (7.17) is equivalent to (7.18) with no extra condition. ∎

Example 7.4 (D'Alembert's functional equation). D'Alembert's equation

$$f(x+y)+f(x-y)=2f(x)f(y);\quad x,y\in\mathbb{R} \tag{7.19}$$

can be solved by initially setting $y=0$ and then $x=0$ to obtain

$$f(0)=1\quad\text{or}\quad f(x)=0\quad\text{and}\quad f(y)=f(-y). \tag{7.20}$$

Then, differentiating twice with respect to y and setting $y=0$ we get

$$f''(x)=kf(x)\Rightarrow f(x)=\begin{cases}a\cosh(\sqrt{k}x)+b\sinh(\sqrt{k}x) & \text{if}\quad k>0,\\ a+bx & \text{if}\quad k=0,\\ a\cos(\sqrt{-k}x)+b\sin(\sqrt{-k}x) & \text{if}\quad k<0,\end{cases} \tag{7.21}$$

where we have made $k=f''(0)$. If we now use (7.20) we get $a=1$ and $b=0$. Thus, the general differentiable solution of (7.21) becomes

$$f(x)=\begin{cases}\cosh(\sqrt{k}x) & \text{if}\quad k>0,\\ 1 & \text{if}\quad k=0,\\ \cos(\sqrt{-k}x) & \text{if}\quad k<0,\end{cases} \tag{7.22}$$

This proves the equivalence of (7.19) and (7.21)(see Theorem 3.11). ∎

Example 7.5 (Pexider's equations). Pexider's equation I

$$f(x+y)=g(x)+h(y);\quad x,y\in\mathbb{R} \tag{7.23}$$

leads to

$$\left.\begin{array}{l}f'(x+y)=g'(x)\\ f'(x+y)=h'(y)\end{array}\right\}\Rightarrow g'(x)=h'(y)=A\Rightarrow\begin{cases}f(x)=Ax+D,\\ g(x)=Ax+B,\\ h(x)=Ax+C.\end{cases} \tag{7.24}$$

Thus, Equation (7.23) is equivalent to (7.24) with the extra condition $f(0)=g(0)+h(0)$.

Pexider's equation II

$$f(x+y)=g(x)h(y);\quad x,y\in\mathbb{R},\quad g(x),h(x)\neq 0 \tag{7.25}$$

gives

$$\left.\begin{array}{l} f'(x+y) = g'(x)h(y) \\ f'(x+y) = g(x)h'(y) \end{array}\right\} \Rightarrow \frac{g'(x)}{g(x)} = \frac{h'(y)}{h(y)} = C \Rightarrow$$

$$\Rightarrow \left\{\begin{array}{l} f(x) = D\exp(Cx), \\ g(x) = A\exp(Cx), \\ h(x) = B\exp(Cx). \end{array}\right. \tag{7.26}$$

In this case, the extra condition is $f(0) = g(0)h(0)$.
Pexider's equation III

$$f(xy) = g(x) + h(y); \quad x, y \in \mathbb{R}_{++} \tag{7.27}$$

leads to

$$\left.\begin{array}{l} yf'(xy) = g'(x) \\ xf'(xy) = h'(y) \end{array}\right\} \Rightarrow yh'(y) = xg'(x) = A \Rightarrow \left\{\begin{array}{l} f(x) = A\log(Dx), \\ g(x) = A\log(Bx), \\ h(x) = A\log(Cx). \end{array}\right. \tag{7.28}$$

Thus, the extra condition becomes $f(1) = g(1)h(1)$.
Finally, Pexider's equation IV

$$f(xy) = g(x)h(y); \quad x, y \in \mathbb{R}_{++}, \quad g(x), \quad h(x) \neq 0 \tag{7.29}$$

leads to

$$\left.\begin{array}{l} yf'(xy) = g'(x)h(y) \\ xf'(xy) = g(x)h'(y) \end{array}\right\} \Rightarrow \frac{yh'(y)}{h(y)} = \frac{xg'(x)}{g(x)} = C \Rightarrow \left\{\begin{array}{l} f(x) = Dx^C, \\ g(x) = Ax^C, \\ h(x) = Bx^C. \end{array}\right. \tag{7.30}$$

The extra condition is $f(1) = g(1)h(1)$.

Thus, we conclude the equivalence of the pairs (7.23)–(7.24), (7.25)–(7.26), (7.27)–(7.28) and (7.29)–(7.30), with the above indicated extra conditions. ∎

Example 7.6 (Other functional equations). In this section we apply this methodology to several equations.

The equation

$$f(Ax + By + C) = Af(x) + Bf(y) + D; \quad AB \neq 0 \tag{7.31}$$

can be solved in the same manner; that is,

$$\left.\begin{array}{l} Af'(Ax + By + C) = Af'(x) \\ Bf'(Ax + By + C) = Bf'(y) \end{array}\right\} \Rightarrow f'(x) = f'(y) = \alpha \Rightarrow f(x) = \alpha x + \gamma. \tag{7.32}$$

∎

Example 7.7 (Wilson's equation). Wilson's generalization of the D'Alembert functional equation is

$$f(x+y) + f(x-y) = 2f(x)g(y). \tag{7.33}$$

By differentiating twice, with respect to x and then with respect to y, we get

$$\left.\begin{array}{l} f''(x+y)+f''(x-y)=2f''(x)g(y) \\ f''(x+y)+f''(x-y)=2f(x)g''(y) \end{array}\right\} \Rightarrow \frac{f''(x)}{f(x)}=\frac{g''(y)}{g(y)}=c, \tag{7.34}$$

which leads to

$$\left\{\begin{array}{lll} f(x)=a\cosh(cx)+b\sinh(cx) & \Rightarrow & g(x)=\cosh(cx), \\ f(x)=a\cos(cx)+b\sin(cx) & \Rightarrow & g(x)=\cos(cx), \\ f(x)=a+bx & \Rightarrow & \left\{\begin{array}{lll} g(x)=1 & \text{if} & |a|+|b|\neq 0, \\ g \text{ arbitrary} & \text{if} & a=b=0. \end{array}\right. \end{array}\right. \tag{7.35}$$

■

Example 7.8 (A uniqueness lemma). Next we prove Lemma 6.1, i.e., the uniqueness of the representation of a function of the form $f[g(x)+h(y)]$, using this technique. Let us assume that we have two different representations for the same function; that is, the functional equation

$$f_1[g_1(x)+h_1(y)]=f_2[g_2(x)+h_2(y)].$$

Differentiating first with respect to x and then with respect to y yields

$$\begin{array}{l} f_1'[g_1(x)+h_1(y)]g_1'(x)=f_2'[g_2(x)+h_2(y)]g_2'(x), \\ f_1'[g_1(x)+h_1(y)]h_1'(y)=f_2'[g_2(x)+h_2(y)]h_2'(y), \end{array}$$

and assuming $f_1'(x)\neq 0$, $g_2'(x)\neq 0$ and $h_2'(x)\neq 0$, we obtain

$$\frac{g_1'(x)}{g_2'(x)}=\frac{h_1'(y)}{h_2'(y)}=\frac{1}{C} \Rightarrow \left\{\begin{array}{l} g_2(x)=Cg_1(x)+a, \\ h_2(x)=Ch_1(x)+b, \end{array}\right. \Rightarrow f_2(x)=f_1\left(\frac{x-a-b}{C}\right),$$

which is the result given by Lemma 6.1. ■

Example 7.9 (A functional equation of special importance). In this section we analyze the functional equation

$$h(x\Delta y)=\sum_{k=1}^{n} f_k(x)\, g_k(y)=\mathbf{g}^T(y)\mathbf{f}(x),$$

$$\mathbf{f}(x)=\begin{pmatrix} f_1(x) \\ f_2(x) \\ \dots \\ f_n(x) \end{pmatrix}, \quad \mathbf{g}(y)=\begin{pmatrix} g_1(y) \\ g_2(y) \\ \dots \\ g_n(y) \end{pmatrix}, \tag{7.36}$$

where Δ is any commutative internal law of composition defined on $\mathbb{R}$ and h, f_k and g_k $(k=1,2,...,n)$ are unknown real functions of a real variable such that the set of functions $\{f_1(x), f_2(x), \dots, f_n(x)\}$, on the one hand, and the set of functions $\{g_1(y), g_2(y), \dots, g_n(y)\}$, on the other, are linearly independent. We show, without loss of generality, that Equation (7.36) can be considerably simplified and we give several equivalent functional equations. Then we demonstrate that when Δ="+", it can be solved by its reduction to a homogeneous differential equation of order n with constant coefficients (see Aczél (1966), pp. 197-199). Conversely, every solution $h(x)$ of a homogeneous differential equation of order n with constant coefficients satisfies Equation (7.36). Finally, we show how functional equations can be used to identify the differential equation associated with a practical problem and to obtain exact discrete solutions. We also give an algorithm to obtain discrete exact solutions when the value of function $h(x)$ is known at $2n$ points and some method for identifying the coefficients of the associated differential equation. ■

The following theorem demonstrates that Equation (7.36) can be simplified if we take into account the commutative property of Δ.

Theorem 7.1 (Symmetry). *Functional Equation (7.36) can be written as*

$$h(x\Delta y) = \sum_{i,j=1}^{n} a_{ij}\, f_i(x)\, f_j(y) = \mathbf{f}^T(x)\mathbf{A}\mathbf{f}(y); \;\; a_{ij} = a_{ji}; \;\; \forall i,j, \tag{7.37}$$

where $\mathbf{A}$ *is the symmetric matrix of coefficients* a_{ij} $(i, j = 1, 2, \ldots, n)$. ■

Proof: Because of the commutativity of Δ , we have

$$\begin{aligned} h(x\Delta y) &= \sum_{i=1}^{n} f_i(x)g_i(y) = \sum_{i=1}^{n} f_i(y)g_i(x) \Rightarrow \\ &\Rightarrow \sum_{i=1}^{n} [\, f_i(x)g_i(y) - f_i(y)g_i(x)] = 0, \end{aligned} \tag{7.38}$$

which is an equation of the form

$$\sum_{i=1}^{m} p_i(x)\, q_i(y) = 0. \tag{7.39}$$

Thus, according to Theorem 4.5:

$$\begin{pmatrix} \mathbf{f}(x) \\ \mathbf{g}(x) \end{pmatrix} = \begin{pmatrix} \mathbf{I} \\ \mathbf{A} \end{pmatrix} \mathbf{f}(x), \;\; \begin{pmatrix} \mathbf{g}(y) \\ -\mathbf{f}(y) \end{pmatrix} = \begin{pmatrix} \mathbf{B} \\ -\mathbf{I} \end{pmatrix} \mathbf{f}(y), \tag{7.40}$$

where $\mathbf{A}$ and $\mathbf{B}$ are constant matrices such that

$$(\mathbf{I} \;\; \mathbf{A}^T) \begin{pmatrix} \mathbf{B} \\ -\mathbf{I} \end{pmatrix} = \mathbf{0} \quad \Rightarrow \quad \mathbf{B} = \mathbf{A}^T. \tag{7.41}$$

From (7.40) and (7.41), we get

$$\mathbf{g}(x) = \mathbf{A}\mathbf{f}(x) = \mathbf{B}\mathbf{f}(x) = \mathbf{A}^T\mathbf{f}(x) \;\; \Rightarrow \;\; \mathbf{A} = \mathbf{A}^T. \tag{7.42}$$

Finally, substitution of (7.42) into (7.36) leads to (7.37). ■

Next, we study the uniqueness of representation of (7.37), i.e.,we try to answer the following question: given $h(x)$, is there a unique set of functions $\{f_1(x), f_2(x), \ldots, f_n(x)\}$ and a unique matrix $\mathbf{A}$, such that (7.37) is true?, and if the answer to this question is negative, what is the relation between different solution sets of functions and matrices? The answers to the above questions are given by the following theorem.

Theorem 7.2 (Uniqueness). *If there are two sets of linearly independent functions $\{f_1(x), f_2(x), \ldots, f_n(x)\}$ and $\{f_1^*(x), f_2^*(x), \ldots, f_n^*(x)\}$ and two symmetric matrices $\mathbf{A}$ and $\mathbf{A}^*$ such that*

$$h(x\Delta y) = \sum_{i,j=1}^{n} a_{ij}\, f_i(x)\, f_j(y) = \sum_{i,j=1}^{n} a_{ij}^*\, f_i^*(x)\, f_j^*(y), \tag{7.43}$$

then there exists a regular constant matrix $\mathbf{B}$ of order n such that

$$\mathbf{f}^*(x) = \mathbf{B}\mathbf{f}(x), \quad \mathbf{A} = \mathbf{B}^T\mathbf{A}^*\mathbf{B}, \tag{7.44}$$

where $\mathbf{A}$ and $\mathbf{A}^*$ are the matrices of coefficients a_{ij} and a_{ij}^*, respectively. ■

Proof: Equation

$$\sum_{i,j=1}^{n} a_{ij}\, f_i(x)\, f_j(y) = \sum_{i,j=1}^{n} a_{ij}^*\, f_i^*(x)\, f_j^*(y) \tag{7.45}$$

can be written as

$$\sum_{i=1}^{n}\left[f_i(x)\sum_{j=1}^{n} a_{ij}\, f_j(y)\right] - \sum_{i=1}^{n}\left[f_i^*(x)\sum_{j=1}^{n} a_{ij}^*\, f_j^*(y)\right] = 0, \tag{7.46}$$

which is of the form (4.13). Thus, according to Theorem 4.5, we have

$$\begin{pmatrix}\mathbf{f}(x)\\ -\mathbf{f}^*(x)\end{pmatrix} = \begin{pmatrix}\mathbf{I}\\ -\mathbf{B}\end{pmatrix}\mathbf{f}(x), \quad \begin{pmatrix}\mathbf{A}\mathbf{f}(y)\\ \mathbf{A}^*\mathbf{f}^*(y)\end{pmatrix} = \begin{pmatrix}\mathbf{A}\\ \mathbf{C}\end{pmatrix}\mathbf{f}(y), \tag{7.47}$$

where $\mathbf{B}$ and $\mathbf{C}$ are non-singular ($|B| \neq 0 \neq |C|$) constant matrices satisfying the equation

$$(\mathbf{I} \quad -\mathbf{B}^T)\begin{pmatrix}\mathbf{A}\\ \mathbf{C}\end{pmatrix} = \mathbf{0} \quad \Rightarrow \quad \mathbf{A} - \mathbf{B}^T\mathbf{C} = \mathbf{0}. \tag{7.48}$$

Finally,from (7.47) and (7.48), we get

$$\mathbf{A}^*\mathbf{B}\mathbf{f}(x) = \mathbf{A}^*\mathbf{f}^*(x) = \mathbf{C}\mathbf{f}(x) = \mathbf{B}^{T^{-1}}\mathbf{A}\mathbf{f}(x) \Rightarrow \mathbf{A} = \mathbf{B}^T\mathbf{A}^*\mathbf{B}. \tag{7.49}$$

■

Corollary 7.1 (Alternative representation). *Functional Equation (7.36) can be written as*

$$h(x\Delta y) = \sum_{i=1}^{p} f_i(x)\, f_i(y) - \sum_{i=p+1}^{n} f_i(x)\, f_i(y) = \mathbf{f}^T(x) \begin{pmatrix} \mathbf{I}_p & \mathbf{0} \\ \mathbf{0} & -\mathbf{I}_q \end{pmatrix} \mathbf{f}(y), \tag{7.50}$$

where $p + q = n$. ■

Proof: Expression on the right of (7.49) shows that matrices $\mathbf{A}$ and $\mathbf{A}^*$ are congruent, but we know that any non-singular symmetric matrix $\mathbf{A}^*$ of rank n can be transformed, by congruence, to a matrix of the form

$$\mathbf{D} = \begin{pmatrix} \mathbf{I}_p & \mathbf{0} \\ \mathbf{0} & -\mathbf{I}_q \end{pmatrix}, \quad p + q = n, \tag{7.51}$$

and then (7.50) holds. ■

Theorem 7.3 (Equivalence). *The functional equation*

$$h(x + y) = \mathbf{f}^T(y)\, \mathbf{D}\mathbf{f}(x) \tag{7.52}$$

is equivalent to any homogeneous differential equation with constant coefficients. ■

Proof: Taking separate derivatives with respect to x and y in (7.52) and equaling we get

$$h'(x + y) = \mathbf{f}^T(y)\, \mathbf{D}\mathbf{f}'(x) = \mathbf{f}'^T(y)\, \mathbf{D}\mathbf{f}(x). \tag{7.53}$$

Due to the fact that the set of functions $\{f_1(y), f_2(y), \ldots, f_n(y)\}$ is linearly independent, there exist constants y_m $(m = 1, 2, \ldots, n)$ such that det $f_k(y_m) \neq 0$. Consequently, it can be written

$$h'(x + y_m) = \mathbf{f}^T(y_m)\mathbf{D}\mathbf{f}'(x) = \mathbf{f}'^T(y_m)\, \mathbf{D}\mathbf{f}(x), \quad m = 1, 2, \ldots, n, \tag{7.54}$$

which in matrix form becomes

$$\mathbf{f}'(x) = \mathbf{G}^{-1}\mathbf{G}'\mathbf{f}(x) = \mathbf{F}\mathbf{f}(x), \quad \mathbf{F} = \mathbf{G}^{-1}\mathbf{G}', \tag{7.55}$$

where $\mathbf{G}$ and $\mathbf{G}'$ are matrices with elements depending on $\mathbf{D}$ and $f_m(y_k)$ and $f'_m(y_k)$, respectively.

With $y = 0$ Expression (7.52) becomes

$$h(x) = \mathbf{f}^T(0)\mathbf{D}\mathbf{f}(x) = \mathbf{C}\mathbf{f}(x), \quad \mathbf{C} = \mathbf{f}^T(0)\mathbf{D}, \tag{7.56}$$

and taking derivatives and using (7.55) we get

$$\begin{array}{rcl} h'(x) & = & \mathbf{C}\mathbf{f}'(x) = \mathbf{C}\mathbf{F}\mathbf{f}(x), \\ h''(x) & = & \mathbf{C}\mathbf{F}\mathbf{f}'(x) = \mathbf{C}\mathbf{F}\mathbf{F}\mathbf{f}(x) = \mathbf{C}\mathbf{F}^2\mathbf{f}(x), \\ \ldots & & \ldots\ \ldots\ \ldots\ \ldots\ \ldots\ \ldots\ \ldots\ \ldots\ \ldots \\ h^{(n)}(x) & = & \mathbf{C}\mathbf{F}^n\mathbf{f}(x), \end{array} \tag{7.57}$$

which, in matrix form, can be written as

$$\mathbf{H}(x) = \begin{pmatrix} h'(x) \\ h''(x) \\ \dots \\ h^{(n)}(x) \end{pmatrix} = \begin{pmatrix} \mathbf{CF} \\ \mathbf{CF}^2 \\ \dots \\ \mathbf{CF}^n \end{pmatrix} \mathbf{f}(x) = \mathbf{Uf}(x). \tag{7.58}$$

Finally, from (7.56) and (7.58), taking into account that the above functions are linearly independent, we have

$$\mathbf{d}^T \begin{pmatrix} h(x) \\ h'(x) \\ h''(x) \\ \dots \\ h^{(n)}(x) \end{pmatrix} = 0, \tag{7.59}$$

where $\mathbf{d}$ is a constant nonzero column vector. Thus (7.59) is a homogeneous differential equation of order n with constant coefficients.

We now show that every solution $h(x)$ of a homogeneous differential equation of order n with constant coefficients satisfies Equation (7.52). In effect, every solution is of the form

$$h(x) = \sum_{k=1}^{m} P_k(x) \, \exp\{w_k x\}, \tag{7.60}$$

where w_k, $(k = 1, 2, ..., m)$ are complex constants (the roots of the characteristic equation) and $P_k(x)$ are polynomials of degree $(n_k - 1)$ where $\sum_{k=1}^{m} n_k = n$.

From (7.60), we get

$$\begin{aligned} h(x+y) &= \sum_{k=1}^{m} P_k(x+y) \exp\{w_k(x+y)\} \\ &= \sum_{k=1}^{n} \lambda_k x^{\alpha_k} y^{\beta_k} \exp\{w_k x\} \exp\{w_k y\} \\ &= \sum_{k=1}^{n} f_k(x) g_k(y). \end{aligned} \tag{7.61}$$

Thus, Equation (7.52) gives a representation of every solution $h(x)$ of a homogeneous differential equation of order n with constant coefficients. ∎

The interesting result is that the set of h solutions of (7.52) is the set of all solutions of homogeneous differential equations with constant coefficients.

7.3.2 Reduction to partial differential equations

In this section we solve some functional equations containing functions of several variables by their reduction to partial differential equations.

The general solution of a functional equation with functions of several variables depends on one or several arbitrary functions. If, by differentiation, we eliminate those arbitrary functions, we get a partial differential equation. Thus, the general reduction method consists of differentiating one or several times, depending on the number of arbitrary functions, and eliminating the arbitrary functions.

For example, if we want to solve the functional equation

$$f(x,y) + f(y,z) = f(x,z) \tag{7.62}$$

we differentiate with respect to x, y and z, independently, and we get

$$\begin{array}{rcl} f_1'(x,y) & = & f_1'(x,z) \\ f_2'(x,y) + f_1'(y,z) & = & 0 \\ f_2'(y,z) & = & f_2'(x,z) \end{array} \tag{7.63}$$

where the subindices refer to partial derivatives with respect to the indicated arguments.

From (7.63), we obtain

$$\begin{array}{lll} f_1'(x,y) = s'(x) & \Rightarrow & f(x,y) = s(x) + g(y), \\ f_2'(x,y) = -f_1'(y,z) & \Rightarrow & g'(y) = -s'(y) \quad \Rightarrow \quad g(y) = -s(y) + k, \end{array} \tag{7.64}$$

and we get $f(x,y) = s(x) - s(y) + k$, but substitution into (7.62) leads to $k = 0$. Thus, the general differentiable solution of (7.62) becomes

$$f(x,y) = s(x) - s(y), \tag{7.65}$$

that is, the same solution as the one obtained from Theorem 5.8. Note, however, that in that theorem no differentiability conditions were required.

In some cases, reduction of functional equations to partial differential equations is based on the following theorem.

Theorem 7.4 (Functional dependence). *The necessary and sufficient conditions for the real differentiable functions $f_i(x_1, x_2, \ldots, x_n)$, $(i = 1, 2, \ldots, n)$ to be functionally dependent, i.e., for the existence of a function*

$$f_n(x_1, x_2, ..., x_n) = \Phi[f_1(x_1, x_2, ..., x_n), ..., f_{n-1}(x_1, x_2, ..., x_n)]$$

in a neighborhood of $(a_1, a_2, \ldots, a_n)$ is the vanishing of the Jacobian

$$J = \frac{\partial[f_1, f_2, ..., f_n]}{\partial[x_1, x_2, ..., x_n]} = \begin{pmatrix} \dfrac{\partial f_1}{\partial x_1} & \dfrac{\partial f_1}{\partial x_2} & \cdots & \dfrac{\partial f_1}{\partial x_n} \\ \dfrac{\partial f_2}{\partial x_1} & \dfrac{\partial f_2}{\partial x_2} & \cdots & \dfrac{\partial f_2}{\partial x_n} \\ \vdots & \vdots & \cdots & \vdots \\ \dfrac{\partial f_n}{\partial x_1} & \dfrac{\partial f_n}{\partial x_2} & \cdots & \dfrac{\partial f_n}{\partial x_n} \end{pmatrix} \tag{7.66}$$

in $(a_1, a_2, \ldots, a_n)$. ■

Theorem 7.5 (Generalized associativity equation). *The general local solution of the functional equation*

$$G(x,y) = K[M(x,z), N(y,z)] \tag{7.67}$$

is

$$\begin{array}{rclrcl} G(x,y) &=& f^{-1}[p(x)+q(y)]; & K(x,y) &=& f^{-1}[l(x)+n(y)], \\ M(x,y) &=& l^{-1}[p(x)+r(y)]; & N(x,y) &=& n^{-1}[q(x)-r(y)], \end{array} \tag{7.68}$$

with continuously differentiable f, l, n, p, q and r and strictly monotonic f, p and q, if the domain of (7.68) is such that G, K, M and N possess continuous partial derivatives and if some constant b exists such that $G_2(x,y) \neq 0, K_2(x,y) \neq 0, N_1(y,b) \neq 0, M_2(x,b) \neq 0, M_1(x,b) \neq 0$ and $N_2(y,b) \neq 0$ and $M(x,b) = u$ and $N(y,b) = v$ can be solved for x and y, respectively. ■

Proof: If we differentiate, first with respect to x, then with respect to y and finally with respect to z, and we set $z = b$, we get

$$\begin{array}{rcl} G_1(x,y) &=& K_1[M(x,b), N(y,b)]M_1(x,b), \\ G_2(x,y) &=& K_2[M(x,b), N(y,b)]N_1(y,b), \\ 0 &=& K_1[M(x,b), N(y,b)]M_2(x,b) + K_2[M(x,b), N(y,b)]N_2(y,b), \end{array}$$

where the indices refer to partial derivatives with respect to the indicated arguments. If we assume $G_2(x,y) \neq 0, K_2(x,y) \neq 0, N_1(y,b) \neq 0, M_2(x,b) \neq 0$ and $N_2(y,b) \neq 0$, we divide the first equation by the second and we use the third equation, we obtain

$$\frac{G_1(x,y)}{G_2(x,y)} = -\frac{\dfrac{M_1(x,b)}{M_2(x,b)}}{\dfrac{N_1(y,b)}{N_2(y,b)}} \Rightarrow \frac{G_1(x,y)}{G_2(x,y)} = \frac{p'(x)}{q'(y)}; \quad \left\{ \begin{array}{rcl} p(x) &=& \displaystyle\int \frac{M_1(x,b)}{M_2(x,b)}dx, \\ q(y) &=& -\displaystyle\int \frac{N_1(y,b)}{N_2(y,b)}dy. \end{array} \right.$$

If, in addition to all the conditions above, we assume $M_1(x,b) \neq 0$, then $p'(x) \neq 0$ and $q'(y) \neq 0$. Therefore, $G_1(x,y) \neq 0$ and $p(x), q(y)$ and $G(x,y)$ must be strictly monotonic.

But, for the partial differential equation above, we have

$$\frac{G_1(x,y)}{G_2(x,y)} = \frac{p'(x)}{q'(y)} \Leftrightarrow \frac{\partial[G(x,y), p(x)+q(y)]}{\partial(x,y)} = 0,$$

which, according to Theorem 7.4, leads to

$$G(x,y) = g[p(x)+q(y)],$$

where g must be strictly monotonic.

If we replace $G(x,y)$ in the initial functional equation and set $z = b$ we get

$$g[p(x)+q(y)] = K[M(x,b), N(y,b)],$$

and if $M(x,b)=u$ and $N(y,b)=v$ can be solved for x and y, respectively we obtain

$$K(u,v)=g[l(u)+n(v)],$$

where the meaning of $l(u)$ and $n(v)$ becomes obvious.
Substituting once more into the initial equation and setting $y=b$, leads to

$$g[p(x)+q(b)]=g\{l[M(x,z)]+n[N(b,z)]\},$$

which gives

$$M(x,z)=l^{-1}\{p(x)+q(b)-n[N(b,z)]\}=l^{-1}[p(x)+r(z)].$$

Finally, a new substitution into the initial equation leads to

$$g[p(x)+q(y)]=g\{p(x)+r(z)+n[N(y,z)]\},$$

which implies

$$N(y,z)=n^{-1}[q(y)-r(z)].$$

Hence, with $g=f^{-1}$, (7.68) follows. ∎

The transformation equation

Let us consider the transformation equation

$$S[S(x,y),z]=S[x,N(y,z)]. \tag{7.69}$$

If we differentiate, first with respect to x, then with respect to y and we set $y=b$ we get

$$\begin{aligned} S_1[S(x,b),z]S_1(x,b) &= S_1[x,N(b,z)],\\ S_1[S(x,b),z]S_2(x,b) &= S_2[x,N(b,z)]N_1(b,z). \end{aligned}$$

If we assume $S_2(x,y)\neq 0, S_1(x,y)\neq 0$ and $N_1(b,z)\neq 0$, and we divide the first equation by the second we obtain

$$\frac{S_1[x,N(b,z)]}{S_2[x,N(b,z)]}=\frac{S_1(x,b)N_1(b,z)}{S_2(x,b)}.$$

If $N(b,z)=y$ can be solved for z, that is $z=\varphi(y)$, we can write

$$\begin{aligned} \frac{S_1(x,y)}{S_2(x,y)} &= \frac{S_1(x,b)N_1[b,\varphi(y)]}{S_2(x,b)}\\ &\Rightarrow \frac{S_1(x,y)}{S_2(x,y)}=\frac{k'(x)}{n'(y)};\quad \begin{cases} k(x) = \displaystyle\int \frac{S_1(x,b)}{S_2(x,b)}dx,\\ n(y) = \displaystyle\int \frac{1}{N_1[b,\varphi(y)]}dy. \end{cases} \end{aligned}$$

where, due to the above conditions, $S(x,y),k(x)$ and $n(y)$ must be strictly monotonic.

But, for the partial differential equation above we have

$$\frac{S_1(x,y)}{S_2(x,y)} = \frac{k'(x)}{n'(y)} \Leftrightarrow \frac{\partial[S(x,y), k(x)+n(y)]}{\partial(x,y)} = 0,$$

which, according to Theorem 7.4, leads to

$$S(x,y) = f[k(x)+n(y)],$$

where f must be strictly monotonic.

Replacing $S(x,y)$ into the initial functional equation and setting $x = b$ we get

$$f\{k[\theta(y)] + n(z)\} = \theta[N(y,z)]; \;\; \theta(y) = f[k(b)+n(y)],$$

from which we obtain

$$N(y,z) = \theta^{-1}\{f[k[\theta(y)] + n(z)]\} = \alpha[\beta(y) + n(z)],$$

where $\alpha = \theta^{-1} f$ and $\beta(y) = k[\theta(y)]$, and substitution into the initial functional equation leads to

$$f\left\{k\left[f[k(x)+n(y)]\right] + n(z)\right\} = f\left\{k(x) + n\left[\alpha[\beta(y)+n(z)]\right]\right\},$$

which can be written as

$$k\left\{f\left[k(x)+n(y)\right]\right\} - k(x) = n\left[\alpha[\beta(y)+n(z)]\right] - n(z) = d(y), \tag{7.70}$$

because the left term is a function of x and y and the right term is a function of y and z. Thus, both must be a function of y only (see Section 2.9). Expression (7.70) can be written as

$$\begin{array}{l} f[k(x)+n(y)] = k^{-1}[d(y)+k(x)], \\ \alpha[\beta(y)+n(z)] = n^{-1}[d(y)+n(z)], \end{array}$$

which by Lemma 6.1 implies

$$\begin{array}{l} k^{-1}(x) = f(x-b), \;\; d(y) = n(y)+b, \\ n^{-1}(x) = \alpha(x-a), \;\; d(y) = \beta(y)+a, \end{array}$$

and then

$$\begin{array}{l} S(x,y) = k^{-1}[k(x)+n(y)+b], \\ N(x,y) = n^{-1}[n(x)+n(y)+b], \end{array}$$

and calling $m(x) = n(x)+b$ we finally get

$$\begin{array}{rcl} S(x,y) & = & k^{-1}[k(x)+m(y)], \\ N(x,y) & = & m^{-1}[m(x)+m(y)]. \end{array}$$

Thus, we have proved the following theorem.

Theorem 7.6 (The transformation equation). *The general local solution of the functional equation*

$$S[S(x,y),z] = S[x, N(y,z)] \tag{7.71}$$

is

$$S(x,y) = k^{-1}[k(x) + m(y)]; \quad N(x,y) = m^{-1}[m(x) + m(y)], \tag{7.72}$$

with continuously differentiable and strictly monotonic k and m, if the domain of (7.71) is such that S and N possess continuous partial derivatives and if $S_2(x,y) \neq 0, S_1(x,y) \neq 0$ and a constant b exists such that $N_1(b,z) \neq 0$ and $N(b,z) = y$ can be solved for z, respectively. ■

The transitivity equation

Let us consider the transitivity equation

$$S(x,y) = S[S(x,z), S(y,z)].$$

Differentiating with respect to x, y and z, and setting $z = b$ we get

$$\begin{array}{rcl} S_1(x,y) & = & S_1[S(x,b),S(y,b)]S_1(x,b), \\ S_2(x,y) & = & S_2[S(x,b),S(y,b)]S_1(y,b), \\ 0 & = & S_1[S(x,b),S(y,b)]S_2(x,b) + S_2[S(x,b),S(y,b)]S_2(y,b). \end{array}$$

If $S_1(x,b) \neq 0$ and $S_2(x,y) \neq 0$, we obtain

$$\frac{S_1(x,y)}{S_2(x,y)} = -\frac{\dfrac{S_1(x,b)}{S_2(x,b)}}{\dfrac{S_1(y,b)}{S_2(y,b)}} = -\frac{p'(x)}{p'(y)}; \quad p(x) = \int \frac{S_1(x,b)}{S_2(x,b)} dx,$$

where $p(x)$ is strictly monotonic. This implies

$$\frac{\partial[S(x,y), p(x) - p(y)]}{\partial(x,y)} = 0 \quad \Rightarrow \quad S(x,y) = f[p(x) - p(y)],$$

and substituting this into the initial equation we get

$$p(x) - p\{f[p(x) - p(z)]\} = p(y) - p\{f[p(y) - p(z)]\} = r(z),$$

which can be written as

$$p^{-1}[p(x) - r(z)] = f[p(x) - p(z)] \Rightarrow \left\{ \begin{array}{rcl} p^{-1}(x) & = & f(x-a), \\ r(z) & = & p(z) + a, \end{array} \right.$$

and then

$$S(x,y) = p^{-1}[p(x) - p(y) + a] = m^{-1}[m(x) - m(y)],$$

where we have made $m(x) = p(x) - a$. Thus, we have the following theorem.

Theorem 7.7 (The transitivity equation). *The general local solution of the functional equation*

$$S(x,y) = S[S(x,z), S(y,z)] \tag{7.73}$$

is

$$S(x,y) = m^{-1}[m(x) - m(y)], \tag{7.74}$$

with continuously differentiable and strictly monotonic m, if the domain of (7.73) is such that S possesses continuous partial derivatives and if there exists b such that $S_1(x,b) \neq 0$ and $S_2(x,y) \neq 0$. ■

The bisymmetry equation

Next we solve the bisymmetry equation

$$S[S(x,y), S(u,z)] = S[S(x,u), S(y,z)].$$

Differentiating with respect to x, y, z and u and setting $u = z = a$ we get

$$\begin{aligned}
S_1[S(x,y), S(a,a)]S_1(x,y) &= S_1[S(x,a), S(y,a)]S_1(x,a), \\
S_1[S(x,y), S(a,a)]S_2(x,y) &= S_2[S(x,a), S(y,a)]S_1(y,a), \\
S_2[S(x,y), S(a,a)]S_2(a,a) &= S_2[S(x,a), S(y,a)]S_2(y,a), \\
S_2[S(x,y), S(a,a)]S_1(a,a) &= S_1[S(x,a), S(y,a)]S_2(x,a),
\end{aligned}$$

and assuming $S_1(x,y) \neq 0$ and $S_2(x,y) \neq 0$, we get

$$\frac{S_1(x,y)}{S_2(x,y)} = \frac{\dfrac{S_1(x,a)}{S_2(x,a)}\dfrac{S_1(a,a)}{S_2(a,a)}}{\dfrac{S_1(y,a)}{S_2(y,a)}} = \frac{cp'(x)}{p'(y)}, \quad \left\{\begin{array}{rcl} p(x) &=& \displaystyle\int \frac{S_1(x,a)}{S_2(x,a)}dx, \\ c &=& \dfrac{S_1(a,a)}{S_2(a,a)}, \end{array}\right.$$

where $p(x)$ is strictly monotonic. This implies

$$\frac{\partial[S(x,y), cp(x) + p(y)]}{\partial(x,y)} = 0 \;\Rightarrow\; S(x,y) = f[cp(x) + p(y)],$$

and substituting back into the initial equation and setting $u = a$, we get

$$\begin{aligned}
&f\{cp[f[cp(x) + p(y)]] + p[f[cp(a) + p(z)]]\} \\
&\qquad = f\{cp[f[cp(x) + p(a)]] + p[f[cp(y) + p(z)]]\},
\end{aligned}$$

which leads to

$$\begin{aligned}
&p^*[f[p^*(x) + p(y)]] - p^*[f[p^*(x) + p(a)]] \\
&\qquad = p[f[p^*(y) + p(z)]] - p[f[p^*(a) + p(z)]] = r(y),
\end{aligned}$$

where $p^*(x) = cp(x)$. Then we have

$$f[p^*(x) + p(y)] = p^{*-1}\left\{r(y) + p^*[f[p^*(x) + p(a)]]\right\},$$

which according to Lemma 6.1 leads to

$$\begin{aligned} f(x) &= p^{*-1}\left(\frac{x-a_1-b_1}{c_1}\right), \\ p(y) &= c_1 r(y) + a_1, \\ p^*(x) &= c_1 p^*\{f[p^*(x)+p(a)]\} + b_1. \end{aligned}$$

Thus, $S(x,y)$ becomes

$$S(x,y) = p^{-1}\left[\frac{cp(x)+p(y)-a_1-b_1}{cc_1}\right] = p^{-1}\left[\frac{p(x)}{c_1}+\frac{p(y)}{cc_1}+\frac{-a_1-b_1}{cc_1}\right],$$

and making

$$A = \frac{1}{c_1}; \ B = \frac{1}{cc_1}; \ C = -\frac{a_1+b_1}{cc_1},$$

we finally get

$$S(x,y) = p^{-1}[Ap(x)+Bp(y)+C].$$

Thus, we have the following theorem.

Theorem 7.8 (The bisymmetry equation). *The general local solution of the functional equation*

$$S[S(x,y),S(u,z)] = S[S(x,u),S(y,z)] \tag{7.75}$$

is

$$S(x,y) = p^{-1}[Ap(x)+Bp(y)+C], \tag{7.76}$$

with continuously differentiable and strictly monotonic p*, if the domain of (7.75) is such that* S *possesses continuous partial derivatives and if* $S_1(x,y) \neq 0$ *and* $S_2(x,y) \neq 0$. ■

The associativity equation

Given the associativity equation

$$F[F(x,y),z] = F[x,F(y,z)],$$

by differentiating with respect to x and y, and setting $y = b$ we can write

$$\begin{aligned} F_1[F(x,b),z]F_1(x,b) &= F_1[x,F(b,z)], \\ F_1[F(x,b),z]F_2(x,b) &= F_2[x,F(b,z)]F_1(b,z), \end{aligned}$$

where the subindices refer to the variable with respect to which we differentiate, and assuming $F_1(x,y) \neq 0$ and $F_2(x,y) \neq 0$ we get

$$\frac{F_1[x,F(b,z)]}{F_2[x,F(b,z)]} = \frac{F_1(x,b)}{F_2(x,b)}F_1(b,z).$$

If now $F(b,z) = u$ can be solved for u (i.e.,$z = \phi(u)$), we obtain

$$\frac{F_1(x,u)}{F_2(x,u)} = \frac{p'(x)}{q'(u)}\ ; \quad \begin{cases} p(x) &= \displaystyle\int \frac{F_1(x,b)}{F_2(x,b)}dx, \\ q(u) &= \displaystyle\int \frac{1}{F_1(b,\psi(u))}du, \end{cases}$$

where $p(x)$ and $q(u)$ are strictly monotonic. This implies

$$\frac{\partial[F(x,y), p(x)+q(y)]}{\partial(x,y)} = 0 \ \Rightarrow \ F(x,y) = g[p(x)+q(y)],$$

and substituting back into the initial equation and setting $y = b$, we get

$$g\{p[g[p(x)+q(b)]] + q(z)\} = g\{p(x) + q[g[p(b)+q(z)]]\},$$

which leads to

$$p[g[p(x)+q(b)]] - p(x) = q[g[p(b)+q(z)]] - q(z) = C,$$

that is

$$\begin{array}{l} g[p(x)+q(b)] = p^{-1}[p(x)+C] \ \Rightarrow \ g(u) = p^{-1}[u+C-q(b)], \\ g[p(b)+q(z)] = q^{-1}[q(z)+C] \ \Rightarrow \ g(u) = q^{-1}[u+C-p(b)], \end{array}$$

or

$$p(x) = g^{-1}(x) + C - q(b); \ \ q(y) = g^{-1}(y) + C - p(b).$$

Therefore, the associativity equation becomes

$$F(x,y) = g[g^{-1}(x) + g^{-1}(y) + A]; \ \ A = 2C - q(b) - p(b)$$

and calling

$$f(z) = g(z - A),$$

we finally obtain

$$F(x,y) = f[f^{-1}(x) + f^{-1}(y)].$$

Therefore the following theorem holds.

Theorem 7.9 (The associativity equation). *The general local solution of the functional equation*

$$F[F(x,y),z] = F[x,F(y,z)] \tag{7.77}$$

is

$$F(x,y) = f[f^{-1}(x) + f^{-1}(y)], \tag{7.78}$$

with continuously differentiable and strictly monotonic f, if the domain of (7.77) is such that S possesses continuous partial derivatives and if $F_1(x,y) \neq 0, F_2(x,y) \neq 0$ and $F(b,z) = u$ can be solved for u. ■

Theorem 7.10 (Generalized auto-distributivity equation). *If the domain of (7.79) is such that F, G, H, M and N have continuous partial derivatives for $z \neq 0$; if $H_1(x,y) \neq 0$, $H_2(x,y) \neq 0$, $F_1(x,c) \neq 0$, $M_1(x,c) \neq 0$ and $N_1(x,c) \neq 0$, if $M(x,a)$ and $N(x,a)$ are constant and if $M(x,c) = u$ and $N(y,c) = v$ have unique solutions ($c \neq 0$), then the general solution continuous, on a real rectangle, of the functional equation*

$$F[G(x,y),z] = H[M(x,z),N(y,z)], \tag{7.79}$$

where we assume $G \neq M, G \neq N, H \neq M$ and $H \neq N$, is

$$\begin{array}{rcl} F(x,y) & = & l[f(y)g^{-1}(x) + \alpha(y) + \beta(y)], \\ G(x,y) & = & g[h(x) + k(y)], \\ H(x,y) & = & l[m(x) + n(y)], \\ M(x,y) & = & m^{-1}[f(y)h(x) + \alpha(y)], \\ N(x,y) & = & n^{-1}[f(y)k(x) + \beta(y)], \end{array} \tag{7.80}$$

where g, h, k, l, m and n are arbitrary strictly monotonic and continuously differentiable functions, $f(a) = 0$ and f, α and β are arbitrary continuously differentiable functions.

The two sides of (7.79) can be written as

$$l\left\{f(z)[h(x) + k(y)] + \alpha(z) + \beta(z)\right\}.$$

■

7.4 From difference to differential equations

In this section we show that given a difference equation with constant coefficients we can obtain an equivalent differential equation, which will have the same solutions at the grid points. To this aim we need the following lemma.

Lemma 7.1 (Linear difference equation). *The general solution of the linear difference equation*

$$z(t + nu) = \sum_{s=0}^{n-1} a_s z(t + su) \tag{7.81}$$

is

$$z(t) = \sum_{i=1}^{m} Q_i(\frac{t}{u}) w_i^{\frac{t}{u}}, \tag{7.82}$$

where $Q_i(t)$ and w_i are the polynomials and their associated characteristic m different roots of the solution

$$g(t) = \sum_{i=1}^{m} Q_i(t) w_i^t. \tag{7.83}$$

of the difference equation

$$g(t+n) = \sum_{s=0}^{n-1} a_s g(t+s). \tag{7.84}$$

■

Proof: Assume that the solution of (7.84) is of the form (7.83). Letting

$$z(t) = g(\frac{t}{u}) \Leftrightarrow g(t) = z(ut) \tag{7.85}$$

and

$$t^* = \frac{t}{u}, \tag{7.86}$$

(7.81) transforms to

$$g(t^*+n) = \sum_{s=0}^{n-1} a_s g(t^*+s), \tag{7.87}$$

which is of the form (7.84) and has solution (7.83).

Thus, considering (7.85) and (7.86) we finally get (7.82).

■

The following Theorem and Algorithm solve our problem.

Theorem 7.11 (Equivalence of differential and difference equations). *For a differential equation with constant coefficients to have the same solution as a difference equation, the characteristic equation of the differential equation must have as roots the logarithms of the roots of the characteristic equation of the difference equation with the same multiplicities, i.e., their characteristic equations must be of the form*

$$\prod_{i=1}^{n} (x-\alpha_i) = 0 \text{ and } \prod_{i=1}^{n} [x - \exp(\alpha_i \Delta x)] = 0, \tag{7.88}$$

respectively. Thus, once one of both equations is known, the other can be immediately obtained from (7.88). ■

Proof: Assume that we have a linear differential equation in $z(t)$ with constant coefficients, i.e., with a general solution

$$z(t) = \sum_{i=1}^{m} P_i(t)\exp(w_i t) + h(t), \tag{7.89}$$

where

$$P_i(t) = \sum_{\ell=0}^{k_i-1} c_{\ell i} t^\ell \tag{7.90}$$

is a polynomial of degree $k_i - 1$ (where k_i is the order of multiplicity of the associated root of its characteristic equation), $w_i; i = 1, \dots, m$ are the roots (real or imaginary) of its characteristic equation, and $h(t)$ is a particular solution.

In order to have an equivalent difference equation both must have the same solution, so we take $Q_i(t) = P_i(ut)$ and $w_i = \exp(w_i u)$, $i = 1, \dots, m$, since

$$z(t) = \sum_{i=1}^{m} P_i(t) \exp(w_i t) + h(t) = \sum_{i=1}^{m} Q_i(\frac{t}{u}) w_i^{\frac{t}{u}} + h(t). \tag{7.91}$$

That is, the characteristic equation of the differential equation must have as roots the logarithms of the roots of the difference equation with the same multiplicities. ■

This suggests the following algorithm to obtain the equivalent differential equation associated with a given difference or functional equation.

Algorithm 7.1 (Obtaining a differential equation equivalent to a linear difference equation with constant coefficients).

- **Input:** *A linear difference equation with constant coefficients:*

$$z(t + nu) = \sum_{s=0}^{n-1} a_s z(t + su) + h(t). \tag{7.92}$$

- **Output:** *The equivalent differential equation.*

1. **Step 1**: *Find the roots $w_i : i = 1, \dots, m$ and their associated multiplicities k_i of the characteristic equation of (7.92):*

$$p^n - \sum_{s=0}^{n-1} a_s p^s = 0. \tag{7.93}$$

2. **Step 2**: *Obtain the characteristic equation of the equivalent linear differential equation, using as roots the logarithms of the above roots divided by u and the same multiplicities:*

$$\prod_{i=1}^{m} (q - \frac{1}{u} \log(w_i))^{k_i} = 0. \tag{7.94}$$

3. **Step 3**: *Expand the characteristic equation and calculate the corresponding coefficients $b_s; s = 1, \dots, n-1$:*

$$\prod_{i=1}^{m} (q - \frac{1}{u} \log(w_i))^{k_i} = q^n + \sum_{s=0}^{n-1} b_s q^s = 0. \tag{7.95}$$

4. **Step 4**: *Return the equivalent differential equation:*

$$z^{(n)} + \sum_{s=0}^{n-1} b_s z^{(s)} = h(t). \tag{7.96}$$

■

Example 7.10 (Differential equation equivalent to a difference equation). If we consider the difference equation

$$f(x+3) = 3f(x+2) - 3f(x+1) + f(x),$$

its characteristic equation

$$r^3 - 3r^2 + 3r - 1 = 0,$$

has the root $w_1 = 1$ with multiplicity $k_1 = 3$.

According to (7.94), the characteristic equation of the equivalent differential equation is

$$(q - \log 1)^3 = q^3 = 0.$$

Thus, the equivalent differential equation becomes

$$f'''(x) = 0.$$

■

Remark 7.1 *It is well known that there are polynomials in which the roots are quite sensitive to small changes in the coefficients (see Atkinson (1989)). This represents an important problem in the process of finding a differential equation using a set of observed data, because the coefficients of a difference equation are approximated, so the w_i roots in Algorithm 7.1 can contain small errors leading to significant errors in the differential equation. In order to avoid this problem, it is necessary to analyze the stability of the difference equation.*

Suppose that $C(p) = 0$ is the characteristic equation of (7.92) and w_i, $i = 1, \ldots, m$ are the associated roots with multiplicities k_i, respectively. We define a perturbation $C(p) + \epsilon D(p)$, where $D(p)$ is a polynomial with degree$(D) \leq$ degree(C), then to estimate the modified roots we use

$$w_i(\epsilon) \approx w_i + \gamma_i \epsilon^{1/k_i}, \tag{7.97}$$

where

$$\gamma_i^{k_i} = \frac{-k_i! C(w_i)}{D^{(k_i)}(w_i)}. \tag{7.98}$$

Example 7.11 (Stability). Consider again the difference equation in Example 7.10. A small perturbation of value 0.01 in the $f(x)$ coefficient leads to a stability coefficient $\gamma_1 = 0.107722 + 0.18658i$ and to the equation

$$0.00996273 f(x) + 0.0149362 f'(x) - 0.00995033 f''(x) + f'''(x) = 0.$$

The new characteristic equation, instead of a real multiple solution of multiplicity three, has two complex roots. In other words, the new functional form (with sines and cosines) of the solution has nothing to do with the old solution (with polynomial functions). ∎

The coincidence of the solutions of the differential and the difference equations at the common points is not casual. In fact, if we assume an infinitely differentiable solution of the functional equation

$$\sum_{i=0}^{n} a_i y^{(n-i)}(x) = 0, \tag{7.99}$$

using the Taylor expansion, we can write

$$y(x+\Delta_j) \approx \sum_{k=0}^{m} \frac{\Delta_j^k y^{(k)}(x)}{k}, \quad j=1,\ldots,n, \tag{7.100}$$

and by derivation of (7.99) $m-n$ times, we get

$$\sum_{i=0}^{n} a_i y^{(n-i+s)}(x) = 0, \quad s=0,1,\ldots,m-n, \tag{7.101}$$

The system (7.100)–(7.101), independent of the value of m, allows us to eliminate all m derivatives of y and to obtain a difference equation of order n. When m tends to infinity we get the coincidence of solutions. Note that the order of the difference equation remains constant when m increases.

7.5 From differential to functional equations

In this section, we start from a differential equation and we look for an equivalent functional equation (see Eichhorn (2002)).

We give two different solutions: the exact approach and the Taylor series approach

7.5.1 Exact approach

First we show that given a linear differential equation with constant coefficients we can obtain an equivalent functional equation, in the sense of having the same sets of solutions. To this end, we need a lemma.

Lemma 7.2 (Polynomial representation). *Every polynomial of degree n in t and u can be written as a sum with $n+1$ summands with two factors each, one a function of t and one a function of u:*

$$P_n(u,t) = \sum_{i=1}^{n+1} f_i(t) g_i(u). \tag{7.102}$$

∎

Proof: Given the polynomial of degree n, we group its monomials using the following rules:

Rule 1: If $\alpha \le \beta$, the monomial in $u^\alpha t^\beta$ is included in the $2\alpha+1$ summand.

Rule 2: If $\alpha > \beta$, the monomial in $u^\alpha t^\beta$ is included in the $2\beta+2$ summand.

Note that the summand number is determined by $\min(\alpha, \beta)$.

Since all monomials in the $2\alpha + 1$ summand have the common factor u^α, they can be written as $u^\alpha f_{2\alpha+1}(t)$. Similarly, since all monomials in the $2\beta + 2$ summand have the common factor t^β, they can be written as $f_{2\beta+2}(u)t^\beta$.

Since we are dealing with polynomials of degree n, for all monomials whose summand number is assigned by Rule 1 we have $\alpha \le n/2$, because if $\alpha > n/2$, then $\beta < \alpha$ and then we must use Rule 2 instead of Rule 1. Thus,

$$\alpha \le n/2 \Rightarrow 2\alpha + 1 \le n + 1,$$

which implies that the summand number obtained by Rule 1 cannot be greater than $n + 1$.

Similarly, any monomial whose summand number is assigned by Rule 2, must satisfy $\beta < n/2$, because if $\beta \ge n/2$, then $\alpha \le \beta$ and then we must use Rule 1 instead of Rule 2. Then, we have

$$\beta < n/2 \Rightarrow 2\beta + 2 < n + 2 \Rightarrow 2\beta + 2 \le n + 1,$$

which implies that the summand number obtained by Rule 2 cannot be greater than $n + 1$.

Thus, we have

$$P_n(x,t) = \sum_{\alpha=0}^{\lfloor n/2 \rfloor} u^\alpha f_{2\alpha+1}(t) + \sum_{\beta=0}^{\lfloor (n-1)/2 \rfloor} f_{2\beta+2}(u)t^\beta, \tag{7.103}$$

where $\lfloor x \rfloor$ is the maximum integer less than or equal to x. ■

The following theorem shows that if $z(t)$ satisfies a linear differential equation with constant coefficients it also satisfies a functional equation and, more importantly, provides a way to obtain a functional equation equivalent to a given differential equation.

Theorem 7.12 (From differential to functional equations). *If $z(t)$ satisfies a linear differential equation of order n with constant coefficients, then it also satisfies the functional equation*

$$z(t+u_n) = \sum_{s=0}^{n-1} \alpha_s(u_1, \dots, u_n) z(t+u_s) + \delta(t; u_1, \dots, u_n), \forall t, u_1, \dots, u_n, \tag{7.104}$$

where

$$\delta(t; u_1, \dots, u_n) = h(t + u_n) - \sum_{s=0}^{n-1} \alpha_s(u_1, \dots, u_n) h(t + u_s) \tag{7.105}$$

and $h(t)$ is a particular solution. ■

Proof: Since $z(t)$ satisfies a linear differential equation of order n with constant coefficients, it must be of the form:

$$z(t) = \sum_{i=1}^{m} P_i(t)\exp(w_i t) + h(t), \tag{7.106}$$

where

$$P_i(t) = \sum_{\ell=0}^{k_i-1} c_{\ell i} t^{\ell} \tag{7.107}$$

is a polynomial of degree $k_i - 1$ (where k_i is the order of multiplicity of the associated root of its characteristic equation), $w_i; i = 1, \ldots, m$ are the roots (real or imaginary) of its characteristic equation, and $h(t)$ is a particular solution.

Letting

$$z^*(t) = z(t) - h(t), \tag{7.108}$$

Expressions (7.106) and (7.107) lead to

$$z^*(t) = \sum_{i=1}^{m} \exp(w_i t) \sum_{\ell=0}^{k_i-1} c_{\ell i} t^{\ell}, \tag{7.109}$$

Using Lemma 7.2, we can write

$$P_i(t+u) = \sum_{s=1}^{k_i} f_{is}(t) g_{is}(u),$$

and then

$$\begin{aligned} z^*(t+u) &= \sum_{i=1}^{m} \exp(w_i(t+u)) \sum_{\ell=0}^{k_i-1} c_{\ell i}(t+u)^{\ell} \\ &= \sum_{i=1}^{m}\sum_{s=1}^{k_i} [\exp(w_i t) f_{is}(t)]\,[\exp(w_i u) g_{is}(u)] \\ &= \sum_{j=1}^{n} f_j^*(t) g_j^*(u), \end{aligned} \tag{7.110}$$

where $n = \sum_{i=1}^{m} k_i$, and $f_j^*(t)$ and $g_j^*(u)$ are functions of the form $\exp(w_i t) f_{is}(t)$ and $\exp(w_i u) g_{is}(u)$, respectively.

Using (7.110) for $u = 0, u_1, \ldots, u_n$ we get

$$\begin{aligned} z^*(t) &= \sum_{j=1}^{n} f_j^*(t) g_j^*(0) \\ z^*(t+u_1) &= \sum_{j=1}^{n} f_j^*(t) g_j^*(u_1) \\ \cdots &\quad \cdots \quad \cdots \\ z^*(t+u_n) &= \sum_{j=1}^{n} f_j^*(t) g_j^*(u_n), \end{aligned} \tag{7.111}$$

that is,

$$\begin{pmatrix} z^*(t) \\ z^*(t+u_1) \\ \dots \\ z^*(t+u_n) \end{pmatrix} = f_1^*(t) \begin{pmatrix} g_1^*(0) \\ g_1^*(u_1) \\ \dots \\ g_1^*(u_n) \end{pmatrix} + \dots + f_n^*(t) \begin{pmatrix} g_n^*(0) \\ g_n^*(u_1) \\ \dots \\ g_n^*(u_n) \end{pmatrix}, \tag{7.112}$$

which shows that the left hand side vector is a linear combination of the right hand side vectors, and then

$$D = \begin{vmatrix} z^*(t) & g_1^*(0) & \dots & g_n^*(0) \\ z^*(t+u_1) & g_1^*(u_1) & \dots & g_n^*(u_1) \\ \dots & \dots & \dots & \dots \\ z^*(t+u_n) & g_1^*(u_n) & \dots & g_n^*(u_n) \end{vmatrix} = 0. \tag{7.113}$$

Calculating the determinant in (7.113) by its first column we get

$$D = \sum_{s=0}^{n} \gamma_s(u_1, \dots, u_n) z^*(t+u_s) = 0, \tag{7.114}$$

where $u_0 = 0$. Without loss of generality, we can assume that $\gamma_n(u_1, \dots, u_n) \neq 0$, and then

$$z^*(t+u_n) = -\sum_{s=0}^{n-1} \frac{\gamma_s(u_1, \dots, u_n)}{\gamma_n(u_1, \dots, u_n)} z^*(t+u_s) = \sum_{s=0}^{n-1} \alpha_s(u_1, \dots, u_n) z^*(t+u_s), \tag{7.115}$$

where $\alpha_s(u_1, \dots, u_n) = \dfrac{-\gamma_s(u_1, \dots, u_n)}{\gamma_n(u_1, \dots, u_n)}$.

Finally, from (7.108) and (7.115) the value of $z(t+u_n)$ becomes

$$z(t+u_n) = \sum_{s=0}^{n-1} \alpha_s(u_1, \dots, u_n) z(t+u_s) + \delta(t; u_1, \dots, u_n). \tag{7.116}$$

where $\delta(t; u_1, \dots, u_n)$ is as given in (7.105). ∎

Example 7.12 (A simple example). Consider the differential equation

$$z''(x) + (a+b)z'(x) + abz(x) = 0, \tag{7.117}$$

with a general solution

$$z(x) = c_1 \exp(-ax) + c_2 \exp(-bx). \tag{7.118}$$

Writing (7.118) for x, $x+u_1$ and $x+u_2$ we get

$$\begin{array}{rcl} z(x) & = & c_1 \exp(-ax) + c_2 \exp(-bx) \\ z(x+u_1) & = & c_1 \exp(-a(x+u_1)) + c_2 \exp(-b(x+u_1)) \\ z(x+u_2) & = & c_1 \exp(-a(x+u_2)) + c_2 \exp(-b(x+u_2)), \end{array} \tag{7.119}$$

and eliminating c_1 and c_2 we obtain

$$\boxed{z(x+u_2) = \alpha_0(u_1,u_2)z(x)+\alpha_1(u_1,u_2)z(x+u_1)} \tag{7.120}$$

where

$$\begin{aligned} \alpha_0(u_1,u_2) &= \frac{\exp(au_1+bu_2)-\exp(bu_1+au_2)}{\exp((a+b)u_2)(\exp(au_1)-\exp(bu_1))}, \\ \alpha_1(u_1,u_2) &= \frac{\exp((a+b)u_1)(\exp(au_2)-\exp(bu_2))}{\exp((a+b)u_2)(\exp(au_1)-\exp(bu_1))}, \end{aligned} \tag{7.121}$$

which is the functional equation equivalent to differential Equation (7.117). ■

Example 7.13 (The vibrating mass example). Consider again the vibrating mass given at the introduction.

The general solution of Equation (7.1) is

$$z(t) = z_h(t) + z_p(t), \tag{7.122}$$

where $z_h(t)$ is the general solution of the homogeneous equation and $z_p(t)$ is a particular solution. Suppose (case of regular damping $c^2 < 4km$) that the associated polynomial has two complex roots $a \pm bi$, then

$$z_h(t) = c_1\exp(at)\cos(bt) + c_2\exp(at)\sin(bt). \tag{7.123}$$

Taking u_1 and u_2 arbitrary real numbers, we get

$$\begin{aligned} z_h(t) &= c_1\exp(at)\cos(bt)+c_2\exp(at)\sin(bt), \\ z_h(t+u_1) &= c_1\exp(a(t+u_1))\cos(b(t+u_1)) \\ &\quad +c_2\exp(a(t+u_1))\sin(b(t+u_1)), \\ z_h(t+u_2) &= c_1\exp(a(t+u_2))\cos(b(t+u_2)) \\ &\quad +c_2\exp(a(t+u_2))\sin(b(t+u_2)). \end{aligned} \tag{7.124}$$

Eliminating c_1 and c_2 from (7.124), we obtain Equation (7.2) with (7.3). This proves the statement made at the introduction of the chapter. ■

7.5.2 Taylor series approach

We only study the case of the following linear ordinary differential equations

$$\sum_{i=0}^{n} a_i(x)f^{(n-i)}(x) = h(x), \tag{7.125}$$

where f, h, a_i $(i = 0, \ldots, n)$ are infinitely differentiable functions in a certain domain D. Without loss of generality we can assume $a_0(x) = 1$.

We also assume that the value of f is known in n points of the domain

$$D : \{p_j = x + \Delta_j \ , j = 1, \dots, n\} .$$

Using the Taylor expansion we have

$$f(x + \Delta_j) = \sum_{k=0}^{m} \frac{\Delta_j^k f^{(k)}(x)}{k!} + O(\Delta_j^{(m+1)}(x)), \ \ \forall\, j = 1, \dots, n, \tag{7.126}$$

where we assume $m > n$.
By differentiating $(m - n - 1)$ times Equation (7.125), we get

$$\sum_{i=0}^{n+k-1} A_{ki}(x) f^{(n-i+k-1)}(x) = h^{(k-1)}(x); \ \ k = 1, \dots, m - n + 1, \tag{7.127}$$

where the upper index denotes the order of differentiation and the functions A_{ki} $(k = 1, 2, \dots, m - n + 1)$ are given by :

$$\begin{aligned} A_{1i}(x) = &\quad a_i(x); \ \ i = 0, \dots, n \\ A_{(k+1)i}(x) \quad = &\begin{cases} A_{ki}(x) & \text{if} \ \ i = 0, \\ A_{ki}(x) + A'_{k(i-1)}(x) & \text{if} \ \ i = 1, \dots, n + k - 1, \\ A'_{k(n+k-1)}(x) & \text{if} \ \ i = n + k. \end{cases} \end{aligned}$$

It is worth mentioning that because $a_0(x) = 1$, the first coefficient of all equations in (7.127) is equal to 1.

Equations (7.126), without the complementary term, and (7.127) can be written, in matrix form, as

$$\begin{pmatrix} \mathbf{N} & \mathbf{D} \\ \mathbf{C} & \mathbf{B} \end{pmatrix} \cdot \mathbf{F} = \begin{pmatrix} \mathbf{H} \\ \Delta\mathbf{F} \end{pmatrix}, \tag{7.128}$$

where $\mathbf{N}, \mathbf{D}, \mathbf{C}, \mathbf{B}, \mathbf{F}, \mathbf{H}$, and $\Delta\mathbf{F}$ are the following matrices:

$$\mathbf{F} = \begin{pmatrix} f^{(m)}(x) \\ \vdots \\ \vdots \\ \vdots \\ f^{(n)}(x) \\ f^{(n-1)}(x) \\ \vdots \\ \vdots \\ f(x) \end{pmatrix}; \ \ \mathbf{H} = \begin{pmatrix} h(x) \\ h'(x) \\ \vdots \\ \vdots \\ \vdots \\ h^{(m-n)}(x) \end{pmatrix}; \ \ \Delta\mathbf{F} = \begin{pmatrix} f(x + \Delta_n) \\ \vdots \\ \vdots \\ f(x + \Delta_1) \end{pmatrix},$$

$$\mathbf{N}=\begin{pmatrix} 0 & 0 & \cdots & \cdots & \cdots & \cdots & 1 \\ 0 & 0 & \cdots & \cdots & \cdots & 1 & A_{21} \\ \cdots & \cdots & \cdots & \cdots & \cdots & \cdots & \cdots \\ 0 & 0 & \cdots & 1 & A_{k1} & \cdots & A_{k(k-1)} \\ \cdots & \cdots & \cdots & \cdots & \cdots & \cdots & \cdots \\ 0 & 1 & \cdots & \cdots & \cdots & \cdots & A_{(m-n)(m-n-1)} \\ 1 & A_{(m-n+1)1} & \cdots & \cdots & \cdots & \cdots & A_{(m-n+1)(m-n)} \end{pmatrix},$$

$$\mathbf{D}=\begin{pmatrix} A_{11} & \cdots & A_{1\,n} \\ \cdots & \cdots & \cdots \\ A_{kk} & \cdots & A_{k\,(k+n-1)} \\ \cdots & \cdots & \cdots \\ A_{(m-n)(m-n)} & \cdots & A_{(m-n)(m-1)} \\ A_{(m-n+1)(m-n+1)} & \cdots & A_{(m-n+1)\,m} \end{pmatrix},$$

$$\mathbf{C}=\begin{pmatrix} \dfrac{\Delta_n^m}{m!} & \dfrac{\Delta_n^{(m-1)}}{(m-1)!} & \cdots & \dfrac{\Delta_n^n}{n!} \\ \cdots & \cdots & \cdots & \cdots \\ \dfrac{\Delta_1^m}{m!} & \dfrac{\Delta_1^{(m-1)}}{(m-1)!} & \cdots & \dfrac{\Delta_1^n}{n!} \end{pmatrix};\quad \mathbf{B}=\begin{pmatrix} \dfrac{\Delta_n^{(n-1)}}{(n-1)!} & \cdots & \Delta_n & 1 \\ \cdots & \cdots & \cdots & \cdots \\ \dfrac{\Delta_1^{(n-1)}}{(n-1)!} & \cdots & \Delta_1 & 1 \end{pmatrix},$$

where the explicit dependence of matrices $\mathbf{N}$ and $\mathbf{D}$ on x has been omitted for the sake of clarity. From (7.128) we can eliminate all derivatives of the function f and get a functional equation. This is what we do in the following paragraphs.

First, we row-manipulate the matrix $(\mathbf{N}\ \mathbf{D})$ in order to transform the matrix $\mathbf{N}$ into an inverse unit diagonal matrix $\mathbf{P}$. These transformations produce some modifications in matrices $\mathbf{D}$ and $\mathbf{H}$, which become $\mathbf{D}^*$ and $\mathbf{H}^*$:

$$\mathbf{M}=\begin{pmatrix} \mathbf{N} & \mathbf{D} \\ \mathbf{C} & \mathbf{B} \end{pmatrix} \approx \begin{pmatrix} \mathbf{P} & \mathbf{D}^* \\ \mathbf{C} & \mathbf{B} \end{pmatrix};\ \mathbf{G}=\begin{pmatrix} \mathbf{H} \\ \Delta\mathbf{F} \end{pmatrix} \approx \begin{pmatrix} \mathbf{H}^* \\ \Delta\mathbf{F} \end{pmatrix},$$

where

$$\mathbf{P}=\begin{pmatrix} 0 & 0 & \cdots & 0 & 1 \\ 0 & 0 & \cdots & 1 & 0 \\ \cdots & \cdots & \cdots & \cdots & \cdots \\ 0 & 1 & \cdots & 0 & 0 \\ 1 & 0 & \cdots & 0 & 0 \end{pmatrix}.$$

Next, we transform matrix $\mathbf{C}$ into the null matrix by row-manipulations of matrices $\mathbf{M}$ and $\mathbf{G}$. It is easy to check that this is equivalent to making the following transformation

$$\mathbf{B}^* = \mathbf{B} - \mathbf{CPD}^* \quad \text{and} \quad \Delta\mathbf{F}^* = \Delta\mathbf{F} - \mathbf{CPH}^*.$$

With this, the system (7.128) becomes equivalent to the system

$$\begin{pmatrix} \mathbf{P} & \mathbf{D}^* \\ \mathbf{0} & \mathbf{B}^* \end{pmatrix} \cdot \mathbf{F} = \begin{pmatrix} \mathbf{H}^* \\ \Delta\mathbf{F}^* \end{pmatrix}. \tag{7.129}$$

From (7.129), we can write

$$\mathbf{B}^* \cdot \begin{pmatrix} f^{(n-1)}(x) \\ \vdots \\ f(x) \end{pmatrix} = \begin{pmatrix} f(x+\Delta_n) \\ \vdots \\ f(x+\Delta_1) \end{pmatrix} + \mathbf{K}, \tag{7.130}$$

where $\mathbf{K} = -\mathbf{CPH}^*$.
Now, from (7.130), we get

$$f(x) - \sum_{j=1}^{n} r^{n-j+1} \cdot f(x+\Delta_j) = \sum_{j=1}^{n} r^j \cdot k_j, \tag{7.131}$$

where $\begin{pmatrix} r^1 \dots & r^n \end{pmatrix}$ is the last row of the matrix $\mathbf{B}^{*-1}$, that is,

$$r^j = \frac{(-1)^{n+j} \cdot b^{j\,n}}{\det\ \mathbf{B}^*}, \quad \text{with } b^{\,jn} = \text{Adjoint}_{(j,n)} \text{ of } \mathbf{B}^*.$$

In this way, we obtain a difference Equation (7.131) of order n, which approximates (7.125).

In addition, once the manipulations have been performed for a given value of m, if one wants to carry out the same process for $m+p$, one can start from the manipulated matrices $\mathbf{N}^*$ and $\mathbf{D}^*$ instead of starting from the initial $\mathbf{N}$ and $\mathbf{D}$ matrices, with the corresponding saving in computational time.

If we increase the value of m we shall get a better approximation. However, this can be done without increasing the value of n. In other words, by increasing m we get a sequence of difference equations of order n which approximate the initial differential equation. In the limit, we shall obtain a difference equation which is an exact replicate of the starting differential equation in the sense that it gives the same solutions at the grid points.

But we can go even further, because Equation (7.131) can be interpreted as one functional equation in the variables $(x, \Delta_1, \Delta_2, ..., \Delta_n)$ and then we get a functional equation which is equivalent to (7.125).

Below, we give some examples.

Example 7.14 (A homogeneous differential equation with variable coefficients). We apply the above method to the following differential equation

$$xf'(x) - kf(x) = 0,$$

where k is a given constant.

In this case, because $n = 1$, we take a single point $x + \Delta$.

With $m = 3$, the matrices in (7.128) are

$$\mathbf{N} = \begin{pmatrix} 0 & 0 & 1 \\ 0 & 1 & -\dfrac{k}{x} \\ 1 & -\dfrac{k}{x} & \dfrac{2k}{x^2} \end{pmatrix}; \quad \mathbf{D} = \begin{pmatrix} -\dfrac{k}{x} \\ \dfrac{k}{x^2} \\ -\dfrac{2k}{x^3} \end{pmatrix}; \quad \mathbf{H} = \begin{pmatrix} 0 \\ 0 \\ 0 \end{pmatrix}$$

$$\mathbf{C} = \left(\frac{\Delta^3}{3} \quad \frac{\Delta^2}{2} \quad \Delta \right); \quad \mathbf{B} = (1); \quad \Delta\mathbf{F} = (f(x + \Delta))$$

After manipulating matrix $\mathbf{N}$ for the first time we get

$$\mathbf{D}^* = \begin{pmatrix} -\dfrac{k}{x} \\ -\dfrac{k(k-1)}{x^2} \\ -\dfrac{k(k-1)(k-2)}{x^3} \end{pmatrix}$$

and, after making the matrix $\mathbf{C}$ null, we have

$$\mathbf{B}^* = \left(1 + k\frac{\Delta}{x} + \frac{k(k-1)}{2!} \left(\frac{\Delta}{x}\right)^2 + \frac{k(k-1)(k-2)}{3!} \left(\frac{\Delta}{x}\right)^3 \right).$$

Due to the fact that we have a homogeneous differential equation, the matrices on the right hand side do not suffer any transformation and we get the difference equation

$$f(x) = \mathbf{B}^{*-1} f(x + \Delta).$$

It is easy to check that when we increase the value of m the added terms in matrix $\mathbf{B}$ are of the form

$$\frac{k(k-1)\dots(k-m+1)}{m\,!} \left(\frac{\Delta}{x}\right)^m.$$

Thus, in the limit, we have

$$\mathbf{B}^* = \left(1 + \frac{\Delta}{x}\right)^k$$

and then the functional equation equivalent to the initial differential equation is

$$f(x) = \left(1 + \frac{y}{x}\right)^{-k} f(x + y),$$

and the difference equation

$$f(x) = \left(1 + \frac{\Delta}{x}\right)^{-k} f(x + \Delta).$$

■

Example 7.15 (A complete differential equation with variable coefficients). We now apply the above method to the equation

$$f'(x) - \frac{k}{x}\, f(x) = x^2,$$

which is a complete equation associated with the homogeneous equation in Example 7.14.

For $m = 3$, all matrices are of the same form as before with the exception of matrix $\mathbf{H}$, which now becomes

$$\mathbf{H} = \begin{pmatrix} x^2 \\ 2x \\ 2 \end{pmatrix}.$$

Thus, after the first manipulation we get

$$\mathbf{H}^* = \begin{pmatrix} x^2 \\ (k+2)x \\ k^2+2 \end{pmatrix},$$

and after making the matrix $\mathbf{C}$ null, we obtain

$$\Delta\mathbf{F}^* = \Delta\mathbf{F} + \mathbf{K}, \;\; \mathbf{K} = \left(-\Delta\, x^2 - \frac{\Delta^2}{2!}\,(\,k+2)\,x - \frac{\Delta^3}{3!}\,(k^2+2) \right),$$

where $\mathbf{D}^*$ and $\mathbf{B}^*$ are the matrices indicated in the previous example.

Thus, the approximate difference equation becomes

$$f(x) - \mathbf{B}^{*-1}\, f(x+\Delta) = \mathbf{B}^{*-1}\mathbf{K}.$$

Finally, after some calculations, for m going to infinity we get the functional equation

$$f(x) - \left(1 + \frac{y}{x}\right)^{-k} f(x+y) = \frac{x^3}{3-k}\, \left[1 - \left(1 + \frac{y}{x}\right)^{3-k}\right],$$

and the difference equation

$$f(x) - \left(1 + \frac{\Delta}{x}\right)^{-k} f(x+\Delta) = \frac{x^3}{3-k}\, \left[1 - \left(1 + \frac{\Delta}{x}\right)^{3-k}\right].$$

■

Example 7.16 (A homogeneous linear differential equation with constant coefficients). Let us now deal with the constant coefficients linear equation

$$f''(x) - f(x) = 0.$$

We consider $n = 2$, i.e.,$\{x + \Delta_1 , x + \Delta_2\}$, and m=6. Thus, we have

$$\mathbf{N} = \begin{pmatrix} 0 & 0 & 0 & 0 & 1 \\ 0 & 0 & 0 & 1 & 0 \\ 0 & 0 & 1 & 0 & -1 \\ 0 & 1 & 0 & -1 & 0 \\ 1 & 0 & -1 & 0 & 0 \end{pmatrix}; \quad \mathbf{D} = \begin{pmatrix} 0 & -1 \\ -1 & 0 \\ 0 & 0 \\ 0 & 0 \\ 0 & 0 \end{pmatrix}$$

$$\mathbf{C} = \begin{pmatrix} \frac{\Delta_2^6}{6!} & \frac{\Delta_2^5}{5!} & \cdots & \frac{\Delta_2^2}{2!} \\ \frac{\Delta_1^6}{6!} & \frac{\Delta_1^5}{5!} & \cdots & \frac{\Delta_1^2}{2!} \end{pmatrix}; \quad \mathbf{B} = \begin{pmatrix} \Delta_2 & 1 \\ \Delta_1 & 1 \end{pmatrix}$$

and $\mathbf{H}$ is the column null matrix of dimension 5.

After all the manipulations, we get

$$\mathbf{B}^* = \begin{pmatrix} \Delta_2 + \frac{\Delta_2^3}{3!} + \frac{\Delta_2^5}{5!} & 1 + \frac{\Delta_2^2}{2!} + \frac{\Delta_2^4}{4!} + \frac{\Delta_2^6}{6!} \\ \Delta_1 + \frac{\Delta_1^3}{3!} + \frac{\Delta_1^5}{5!} & 1 + \frac{\Delta_1^2}{2!} + \frac{\Delta_1^4}{4!} + \frac{\Delta_1^6}{6!} \end{pmatrix}.$$

And when m goes to infinity we obtain

$$\mathbf{B}^* = \begin{pmatrix} \frac{e^{\Delta_2} - e^{-\Delta_2}}{2} & \frac{e^{\Delta_2} + e^{-\Delta_2}}{2} \\ \frac{e^{\Delta_1} - e^{-\Delta_1}}{2} & \frac{e^{\Delta_1} + e^{-\Delta_1}}{2} \end{pmatrix}.$$

Thus, we get the functional equation

$$f(x) = \frac{-(e^y - e^{-y})\ f(x+z) + (e^z - e^{-z})\ f(x+y)}{e^{(z-y)} - e^{(y-z)}},$$

and the difference equation

$$f(x) = \frac{-(e^{\Delta_1} - e^{-\Delta_1})\ f(x+\Delta_2) + (e^{\Delta_2} - e^{-\Delta_2})\ f(x+\Delta_1)}{e^{(\Delta_2-\Delta_1)} - e^{(\Delta_1-\Delta_2)}}.$$

■

Example 7.17 (A complete equation). Finally we deal with the equation

$$f''(x) - f(x) = x^3,$$

whose homogeneous equation is that given in Example 7.16.

We start by taking again $m = 6$ and we get the same $\mathbf{N}, \mathbf{D}$ and $\mathbf{C}$ matrices and

$$\mathbf{H} = \begin{pmatrix} x^3 \\ 3x^2 \\ 6x \\ 6 \\ 0 \end{pmatrix}; \quad \mathbf{H}^* = \begin{pmatrix} x^3 \\ 3x^2 \\ 6x + x^3 \\ 6 + 3x^2 \\ 6x + x^3 \end{pmatrix}.$$

When m goes to infinity we observe the following: the manipulations on the matrix (**ND**) consist in adding to each row the row which is two places above it, that is,

$$h^*_{2k} = \sum_{i=0}^{n-1} h_{2(k-i)}, \qquad h^*_{2k+1} = \sum_{i=0}^{n} h_{2(k-i)+1},$$

and taking into account that $h_k = 0$ if $k > 4$, we have

$$h^*_j = \begin{cases} x^3 & \text{if} \quad j = 1, \\ 3x^2 & \text{if} \quad j = 2, \\ 6x + x^3 & \text{if} \quad j = 2k+1 \; k > 0, \\ 6 + 3x^2 & \text{if} \quad j = 2k \;, k > 1. \end{cases}$$

After making the matrix **C** null, we get

$$\Delta\mathbf{F}^* = \Delta\mathbf{F} - \mathbf{CPH}^* = \Delta\mathbf{F} + \begin{pmatrix} P(\Delta_2) \\ P(\Delta_1) \end{pmatrix},$$

where

$$P(\Delta) = -x^3\frac{\Delta^2}{2!} - 3x^2\frac{\Delta^3}{3!} - \sum_{n=2}^{\infty} x(6+x^2)\frac{\Delta^{2n}}{(2n)!} - \sum_{n=2}^{\infty}(6+3x^2)\frac{\Delta^{2n+1}}{(2n+1)!},$$

which can be written as

$$\begin{aligned} P(\Delta) &= -x^3\left(\frac{e^{\Delta}+e^{-\Delta}}{2} - 1\right) - 3x^2\left(\frac{e^{\Delta}-e^{-\Delta}}{2} - \Delta\right) \\ & \quad -6x\left(\frac{e^{\Delta}+e^{-\Delta}}{2} - 1 - \frac{\Delta^2}{2}\right) - 6\left(\frac{e^{\Delta}-e^{-\Delta}}{2} - \Delta - \frac{\Delta^3}{3}\right). \end{aligned}$$

Finally, we get the difference equation

$$f(x) - E_1\, f(x+\Delta_2) - E_2\, f(x+\Delta_1) = E_1\, P(\Delta_2) + E_2\, P(\Delta_1),$$

where

$$E_1 = \frac{-(e^{\Delta_1} - e^{-\Delta_1})}{e^{(\Delta_2-\Delta_1)} - e^{(\Delta_1-\Delta_2)}}; \quad E_2 = \frac{e^{\Delta_2} - e^{-\Delta_2}}{e^{(\Delta_2-\Delta_1)} - e^{(\Delta_1-\Delta_2)}}.$$

■

7.6 From functional to difference equations

In this section we show that given a functional equation we can obtain an equivalent difference equation, in the sense of having the same solutions at the grid points.

In this section we show how a difference equation can be obtained from functional Equation (7.50), where the operator Δ has been replaced by the addition operator.

In the following we call $h_n = h(n\lambda)$ and we consider that functions $h(x)$ and $f_i(x)$ $(i = 1, 2, \ldots, n)$ are defined on a discrete subset of $\mathbb{R}$ $\{n\lambda, n \in \mathbb{Z}\}$. In order to have a unique solution for (7.50) we also assume that $h_0, \ldots, h_{2n-1}$ are known. Note that we give $2n$ values because Equation (7.50) is equivalent not to a single differential equation but to the family of all differential equations and the constant coefficients should be determined by data.

From Equation (7.50), we have

$$h_{m+n} = \mathbf{f}^T(n\lambda) \begin{pmatrix} \mathbf{I}_p & \mathbf{0} \\ \mathbf{0} & -\mathbf{I}_q \end{pmatrix} \mathbf{f}(m\lambda), \tag{7.132}$$

where $y = m\lambda$ and $x = n\lambda$ and

$$\begin{pmatrix} h_m \\ h_{m+1} \\ \cdots \\ h_{m+n-1} \end{pmatrix} = \mathbf{F}(n\lambda) \begin{pmatrix} \mathbf{I}_p & \mathbf{0} \\ \mathbf{0} & -\mathbf{I}_q \end{pmatrix} \mathbf{f}(m\lambda), \tag{7.133}$$

where

$$\mathbf{F}(n\lambda) = \begin{pmatrix} f_1(0) & f_2(0) & \cdots & f_n(0) \\ f_1(\lambda) & f_2(\lambda) & \cdots & f_n(\lambda) \\ \cdots & \cdots & \cdots & \cdots \\ f_1[(n-1)\lambda] & f_2[(n-1)\lambda] & \cdots & f_n[(n-1)\lambda] \end{pmatrix}. \tag{7.134}$$

Due to the non-singularity of $\mathbf{F}$, from (7.132) and (7.133), we can write

$$h_{m+n} = \mathbf{K}(n\lambda) \begin{pmatrix} h_m \\ h_{m+1} \\ \cdots \\ h_{m+n-1} \end{pmatrix}, \quad \mathbf{K}(n\lambda) = \mathbf{f}^T(n\lambda)\mathbf{F}^{-1}(n\lambda), \tag{7.135}$$

and taking into account (7.133) for $m = n$, and (7.135) we have

$$\mathbf{K}(n\lambda) = \begin{pmatrix} h_n & h_{n+1} & \cdots & h_{2n-1} \end{pmatrix} \left(\mathbf{F}(n\lambda) \begin{pmatrix} \mathbf{I}_p & \mathbf{0} \\ \mathbf{0} & -\mathbf{I}_q \end{pmatrix} \mathbf{F}^T(n\lambda) \right)^{-1}, \tag{7.136}$$

but making $m = 0, 1, 2, \ldots, n-1$ in (7.133), we get

$$\mathbf{F}(n\lambda) \begin{pmatrix} \mathbf{I}_p & \mathbf{0} \\ \mathbf{0} & -\mathbf{I}_q \end{pmatrix} \mathbf{F}^T(n\lambda) = \begin{pmatrix} h_0 & h_1 & \cdots & h_{n-1} \\ h_1 & h_2 & \cdots & h_n \\ \cdots & \cdots & \cdots & \cdots \\ h_{n-1} & h_n & \cdots & h_{2n-2} \end{pmatrix}, \tag{7.137}$$

and then

$$\begin{aligned} h_{m+n} &= \\ &= (h_n h_{n+1} \ldots h_{2n-1}) \begin{pmatrix} h_0 & h_1 & \cdots & h_{n-1} \\ h_1 & h_2 & \cdots & h_n \\ \cdots & \cdots & \cdots & \cdots \\ h_{n-1} & h_n & \cdots & h_{2n-2} \end{pmatrix}^{-1} \begin{pmatrix} h_m \\ h_{m+1} \\ \cdots \\ h_{m+n-1} \end{pmatrix}, \end{aligned} \tag{7.138}$$

for $m \geq n$, which is a difference equation of order n.

Hence, exact discrete solutions of Equation (7.52) can be obtained by the difference Equation (7.138). Consequently, *given a homogeneous differential equation of order n, there exists a difference equation of the same order such that their solutions coincide at the common points.* Note that the difference Equation (7.138) depends on λ and the coefficients of the differential equation.

Equation (7.138) can be written as

$$h(y+n\lambda) = \{h(n\lambda)h[(n+1)\lambda]\dots h[(2n-1)\lambda]\}$$

$$\times \begin{pmatrix} h(0) & h(\lambda) & \dots & h[(n-1)\lambda] \\ h(\lambda) & h(2\lambda) & \dots & h(n\lambda) \\ \dots & \dots & \dots & \dots \\ h[(n-1)\lambda] & h(n\lambda) & \dots & h[(2n-2)\lambda] \end{pmatrix}^{-1} \begin{pmatrix} h(y) \\ h(y+\lambda) \\ \dots \\ h[y+(n-1)\lambda] \end{pmatrix}, \tag{7.139}$$

which is a functional equation equivalent to (7.36) with $\Delta \equiv$"+".

From Theorem 7.12 we immediately get the following corollary, which shows that if $z(t)$ satisfies a linear differential equation with constant coefficients it also satisfies a difference equation and, more importantly, it provides a way of obtaining one from the other and vice versa.

Corollary 7.2 (From differential to difference equations). *If $z(t)$ satisfies a linear differential equation of order n with constant coefficients, then it satisfies the difference equation*

$$z(t+nu) = \sum_{s=0}^{n-1} \alpha_s(u)z(t+su) + \delta(t,u), \tag{7.140}$$

where

$$\delta(t,u) = h(t+nu) - \sum_{s=0}^{n-1} \alpha_s(u)h(t+su). \tag{7.141}$$

and $\alpha_s(u)$ are given functions. ∎

Proof: Letting $u_j = ju; j = 1, \dots, n$ in (7.104) and (7.105), we get (7.140) and (7.141). ∎

Since we use the functional Equation (7.104) we show how to go from a functional equation to a difference equation.

Example 7.18 (From differential to functional equations). (see Castillo et al. (1989)) Let us consider the differential equation of a string on an elastic foundation with no load on it:

$$h''(x) - \frac{K}{T}h(x) = 0, \tag{7.142}$$

where $h(x)$ is the vertical displacement of the string at the point x, K is the Winkler constant and T is the horizontal tension in the string. To simplify we assume that $K/T = 1$.

The exact solution for the case $\{h(0) = 0, h(1) = 1\}$ is

$$h(x) = \frac{\exp(x) - \exp(-x)}{\exp(1) - \exp(-1)}. \tag{7.143}$$

An approximation to Equation (7.142), by means of the finite difference method, is

$$h_{n+1} - [\,2 + (\Delta x)^2]\, h_n + h_{n-1} = 0, \quad h_n = h(n\,\Delta x), \tag{7.144}$$

which has as its characteristic equation and roots

$$\begin{aligned} &r^2 - (2 + \Delta^2 x)\, r\ + 1 = 0, \\ &r_1 = \frac{2 + \Delta^2 x + \Delta x\sqrt{4 + \Delta^2 x}}{2}, \quad r_2 = \frac{2 + \Delta^2 x - \Delta x\sqrt{4 + \Delta^2 x}}{2}. \end{aligned} \tag{7.145}$$

Thus, its general solution is

$$h_n = C_1 \left(\frac{2 + \Delta^2 x + \Delta x\sqrt{4 + \Delta^2 x}}{2}\right)^n + C_2 \left(\frac{2 + \Delta^2 x - \Delta x\sqrt{4 + \Delta^2 x}}{2}\right)^n, \tag{7.146}$$

which for $\{h(0) = 0, h(1) = 1\}$ becomes

$$\begin{aligned} h_n = &\frac{1}{\Delta x\sqrt{4 + \Delta^2 x}} \left(\frac{2 + \Delta^2 x + \Delta x\sqrt{4 + \Delta^2 x}}{2}\right)^n - \\ &- \frac{1}{\Delta x\sqrt{4 + \Delta^2 x}} \left(\frac{2 + \Delta^2 x - \Delta x\sqrt{4 + \Delta^2 x}}{2}\right)^n. \end{aligned} \tag{7.147}$$

Assume now that we do not know what the differential equation governing the problem of a string on an elastic foundation with no load on it is, but we run an experiment and we measure the displacements at equally spaced points, say $h_0, h_1, h_2, \ldots, h_p$. The recurrence formula (7.138) with $n = 1, 2, 3, \ldots$ allows us to obtain the value of the first integer n compatible with the $h_0, h_1, h_2, \ldots, h_p$ values and the coefficients of the difference Equation (7.138). Then, from (7.88), the differential equation governing our problem can be easily obtained.

As an example, let us assume that we know the exact values

$$h_i = \frac{e^{i\Delta x} - e^{-i\Delta x}}{e - e^{-1}}; \quad i = 0, 1, ..., 5. \tag{7.148}$$

Then, Equation (7.138) for $n = 1$ is not satisfied for h_2, but the same equation for $n = 2$ becomes

$$h_{m+2} = (e^{\Delta x} + e^{-\Delta x})h_{m+1} - h_m, \tag{7.149}$$

Table 7.1: Three different solutions of the example equation.

x	Exact solution (7.143)	Exact solution of difference eq. (7.149)	Approximate solution of difference eq. (7.144)
0.0	0.00000	0.00000	0.00000
0.4	0.34952	0.34952	0.34952
0.8	0.75571	0.75571	0.75496
1.2	1.28443	1.28443	1.28119
1.6	2.02141	2.02141	2.01241
2.0	3.08616	3.08616	3.06562
2.4	4.65131	4.65131	4.60933
2.8	6.97065	6.97065	6.89052

which is satisfied for h_4 and h_5 and then we can conclude that $n = 2$ and that (7.149) is the difference equation leading to exact values for all the discrete points.

The characteristic roots of (7.149) are $e^{\Delta x}$ and $e^{-\Delta x}$ and, according to (7.88), the characteristic roots of the associated differential equation are $+1$ and -1. Thus, Equation (7.142) with $k/T = 1$ is implied.

Note that (7.149) shows that (7.143) is one solution of the functional equation

$$h(x + 2y) = (e^y + e^{-y})h(x + y) - h(x). \tag{7.150}$$

Table 7.18 shows different exact and approximate solutions obtained by (7.143), (7.144) and (7.149) for $\Delta x = 0.4$.

■

Example 7.19 (The vibrating mass example). Consider again the vibrating mass example.

In the case of equally spaced data, making $u_1 = u$ and $u_2 = 2u$, as indicated in the introduction to the paper, we obtain Equation (7.4):

$$z(t + 2u) = \alpha_0(u)z(t) + \alpha_1(u)z(t + u) + \delta(t), \tag{7.151}$$

where, from (7.129), we get

$$\begin{array}{rcl} \alpha_0(u) & = & -\exp(2au), \\ \alpha_1(u) & = & 2\cos(bu)\exp(au). \end{array} \tag{7.152}$$

■

The previous examples show that the physical or engineering problems can be described either in terms of differential equations or in terms of functional equations. In the next section we propose an alternative way of describing these physical or engineering problems. The proposed method will be illustrated by applying it to the case of static beams.

7.7 A new approach to physical and engineering problems

Many physical problems have been represented by differential equation models. These equations are the result of a balance or equilibrium stated on a differential (very small) piece in the neighborhood of a given point, thus reflecting the fact that the balance or equilibrium condition holds for such a point. Once this equation is assumed to hold for all the points of a finite or infinite continuous, we get the corresponding mathematical model in terms of differential equations.

Since, given a physical or engineering problem, obtaining the functional equations associated with the corresponding differential equations is a difficult problem, in this section we present an alternative to this way of stating physical problems. We consider a discrete piece (not necessarily infinitesimal) and establish the same equilibrium or balance for the whole piece, thus reflecting the fact that the balance or equilibrium condition holds for such a piece. Once this equation is assumed to hold for all the pieces or a continuous set of pieces of a finite or infinite continuous, we get the corresponding mathematical model in terms of functional equations. This new statement of the problem allows new numerical and exact methods to be used, and these can be more efficient than those associated with the differential equation approach.

Thus, we shall show that we can state the same physical or engineering property in two different ways: differential and functional form. In the following subsection we illustrate the method by its application to the case of static beams. We have chosen this very simple example to illustrate the new methodology based on functional equations.

7.7.1 An illustrative example: the case of static beams

In this section we illustrate the previous methods by their application to the case of beams.

Classical approach: Differential equations

In the classical approach, the equilibrium equations are stated for differential pieces. In Figure 7.3 we show one of such pieces.

The equilibrium of vertical forces leads to

$$q(x+dx) = q(x) + p(x)dx \Rightarrow q'(x) = p(x), \tag{7.153}$$

where $q(x)$ and $p(x)$ are the shear and the load at the point x, respectively, and the equilibrium of moments

$$m(x+dx) = m(x) + q(x)dx + \frac{p(x)dxdx}{2} \Rightarrow m'(x) = q(x), \tag{7.154}$$

where $m(x)$ is the bending moment at x.

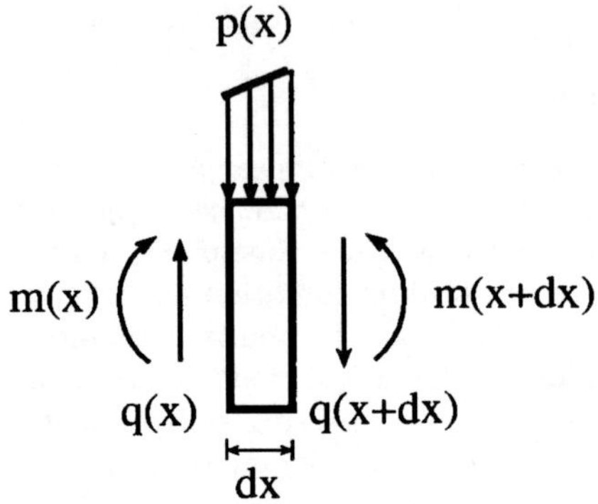

Figure 7.3: Illustration of the classical equilibrium of a differential piece.

Let us consider now the well known strength of materials relation

$$m(x) = EIz''(x), \tag{7.155}$$

where $z(x)$ is the deflection of the beam.

From (7.153), (7.154) and (7.155) we get the well known differential equation

$$EIz^{(IV)}(x) = p(x). \tag{7.156}$$

Calling $w(x) = z'(x)$ to the rotation of the beam at point x, from equations (7.153), (7.154) and (7.155) we get the system of differential equations

$$\begin{aligned} q'(x) &= p(x), \\ m'(x) &= q(x), \\ w'(x) &= \frac{m(x)}{EI}, \\ z'(x) &= w(x), \end{aligned} \tag{7.157}$$

which is the usual mathematical model in terms of differential equations, when we are interested in q, m, w and z. The system (7.157) of first order differential equations is equivalent to the fourth order differential equation (7.156).

New approach: Functional equations

In the new approach, the equilibrium equations are stated for discrete pieces. In Figure 7.4 we show one of such pieces.

The equilibrium of vertical forces leads to

$$q(x+u) = q(x) + A(x,u), \tag{7.158}$$

where

$$A(x,u) = \int_{x}^{x+u} p(s)ds. \tag{7.159}$$

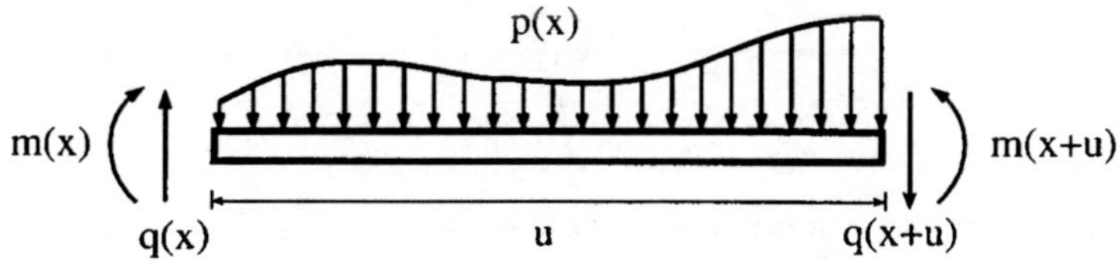

Figure 7.4: Illustration of the equilibrium of a discrete piece.

and the equilibrium of moments

$$m(x+u) = m(x) + uq(x) + B(x,u), \tag{7.160}$$

where

$$B(x,u) = \int_x^{x+u} (x+u-s)p(s)ds. \tag{7.161}$$

Now using Equation (7.155) we get

$$\begin{aligned} w(x+u) &= w(x) + \frac{1}{EI}\int_x^{x+u} m(s)ds \\ &= w(x) + \frac{1}{EI}\int_x^{x+u} m(x) + (s-x)q(x) + B(x,s-x)ds \\ &= w(x) + \frac{1}{EI}\left[m(x)u + q(x)\frac{u^2}{2} + C(x,u)\right], \end{aligned} \tag{7.162}$$

where

$$C(x,u) = \int_x^{x+u} B(x,s-x)ds. \tag{7.163}$$

In addition we have

$$\begin{aligned} z(x+u) &= z(x) + \int_x^{x+u} w(s)ds \\ &= z(x) + \int_x^{x+u} \left\{w(x) + \frac{1}{EI}\left[m(x)(s-x) \right.\right. \\ &\quad \left.\left. + q(x)\frac{(s-x)^2}{2} + C(x,s-x)\right]\right\} ds \\ &= z(x) + w(x)u + \frac{1}{EI}\left[m(x)\frac{u^2}{2} + q(x)\frac{u^3}{6} + D(x,u)\right], \end{aligned} \tag{7.164}$$

where

$$D(x,u) = \int_{x}^{x+u} C(x,s-x)ds. \quad (7.165)$$

Thus, we get the system of functional equations

$$\begin{aligned} q(x+u) &= q(x) + A(x,u), \\ m(x+u) &= m(x) + uq(x) + B(x,u), \\ w(x+u) &= w(x) + \frac{1}{EI}\left[m(x)u + q(x)\frac{u^2}{2} + C(x,u)\right], \\ z(x+u) &= z(x) + w(x)u + \frac{1}{EI}\left[m(x)\frac{u^2}{2} + q(x)\frac{u^3}{6} + D(x,u)\right], \end{aligned} \quad (7.166)$$

where

$$\begin{aligned} A(x,u) &= \int_{x}^{x+u} p(s)ds, \\ B(x,u) &= \int_{x}^{x+u} (x+u-s)p(s)ds, \\ C(x,u) &= \int_{x}^{x+u} B(x,s-x)ds, \\ D(x,u) &= \int_{x}^{x+u} C(x,s-x)ds, \end{aligned} \quad (7.167)$$

which is equivalent to the system of differential equations (7.157).

Note that the functions $A(x,u)$, $B(x,u)$, $C(x,u)$, $D(x,u)$, $E(x,u)$ and $F(x,u)$ become known as soon as the load function $p(x)$ is known and that in some cases we can solve the problem with $1,2,3$ or all equations in (7.166), depending on the boundary conditions. Note also that the system (7.166) can be considered as a system of difference equations (simply make $u = \Delta x$) which gives the exact solution at the interpolating points.

Writing the second equation in (7.166) for two values, u and u_1, we obtain

$$u_1 m(x+u) + (u-u_1)m(x) - um(x+u_1) + uB(x,u_1) - u_1 B(x,u) = 0, \quad (7.168)$$

which is a functional equation in $m(x)$.

Similarly, we can write the last equation in (7.166) for three different values of u and eliminate $w(x), m(x)$ and $q(x)$ to obtain a functional equation in $z(x)$. For example, if we write this equation for $u, 2u, 3u$ and $4u$, we get the (fourth order) functional equation

$$\begin{aligned} z(x+u) &= \frac{z(x)}{4} + \frac{3z(x+2u)}{2} - \frac{z(x+3u)+z(x+4u)}{4} \\ &\quad + \frac{4D(x,u) - 6D(x,2u) + 4D(x,3u) - D(x,4u)}{EI}, \end{aligned} \quad (7.169)$$

which is equivalent to the system of (first order) functional equations (7.166).

Equations (7.168) and (7.169) can also be interpreted as finite difference equations. In this case, they give the exact solution at the interpolating points.

Example 7.20 (Beams with uniform load). Assume that we have a beam with uniform load p. Then, we have

$$\begin{array}{ll} A(x,u) = -pu, & B(x,u) = -pu^2/2, \\ C(x,u) = -pu^3/6, & D(x,u) = -pu^4/24, \end{array} \tag{7.170}$$

and the system of functional equations (7.166) becomes

$$\left.\begin{aligned} q(x+u) &= q(x) - pu, \\ m(x+u) &= m(x) + uq(x) - \frac{pu^2}{2}, \\ w(x+u) &= w(x) + \frac{1}{EI}\left[-p\frac{u^3}{6} + um(x) + q(x)\frac{u^2}{2}\right], \\ z(x+u) &= z(x) + uw(x) + \frac{1}{EI}\left[-p\frac{u^4}{24} + m(x)\frac{u^2}{2} + q(x)\frac{u^3}{6}\right]. \end{aligned}\right\} \tag{7.171}$$

Now, making $x = 0$ and $u = s$, we analyze the following cases :

1. **Simply supported beam:**

$$m(0) = z(0) = 0, \quad m(s) = z(s) = 0. \tag{7.172}$$

2. **Cantilever beam:**

$$w(0) = z(0) = 0, \quad m(s) = q(s) = 0. \tag{7.173}$$

3. **Supported cantilever beam:**

$$w(0) = z(0) = 0, \quad m(s) = z(s) = 0. \tag{7.174}$$

4. **Clamped at both ends beam:**

$$w(0) = z(0) = 0, \quad w(s) = z(s) = 0. \tag{7.175}$$

Each of the above cases gives a system of 4 equations in 4 unknowns in the set $\{q(0), q(s), m(0), m(s), w(0), w(s), z(0), z(s)\}$. These systems are shown in Table 7.2. Solving these systems and substitution of their solutions into (7.171) with $x = 0$ leads to the shear, moment, rotation and deflection laws shown in Table 7.3. ■

Table 7.2: Resulting system of equations for different types of beams.

Type of beam	System of equations
Simply supported beam	$\begin{aligned} q(s) &= q(0) - ps \\ 0 &= sq(0) - \frac{ps^2}{2} \\ w(s) &= w(0) + \frac{1}{EI}\left[-p\frac{s^3}{6} + q(0)\frac{s^2}{2}\right] \\ 0 &= +sw(0) + \frac{1}{EI}\left[-p\frac{s^4}{24} + q(0)\frac{s^3}{6}\right] \end{aligned}$
Cantilever beam	$\begin{aligned} 0 &= q(0) - ps \\ 0 &= m(0) + sq(0) - \frac{ps^2}{2} \\ w(s) &= \frac{1}{EI}\left[-p\frac{s^3}{6} + sm(0) + q(0)\frac{s^2}{2}\right] \\ z(s) &= \frac{1}{EI}\left[-p\frac{s^4}{24} + m(0)\frac{s^2}{2} + q(0)\frac{s^3}{6}\right] \end{aligned}$
Supported cantilever beam	$\begin{aligned} q(s) &= q(0) - ps \\ 0 &= m(0) + sq(0) - \frac{ps^2}{2} \\ w(s) &= \frac{1}{EI}\left[-p\frac{s^3}{6} + sm(0) + q(0)\frac{s^2}{2}\right] \\ 0 &= \frac{1}{EI}\left[-p\frac{s^4}{24} + m(0)\frac{s^2}{2} + q(0)\frac{s^3}{6}\right] \end{aligned}$
Clamped at both ends beam	$\begin{aligned} q(s) &= q(0) - ps \\ m(s) &= m(0) + sq(0) - \frac{ps^2}{2} \\ 0 &= \frac{1}{EI}\left[-p\frac{s^3}{6} + sm(0) + q(0)\frac{s^2}{2}\right] \\ 0 &= \frac{1}{EI}\left[-p\frac{s^4}{24} + m(0)\frac{s^2}{2} + q(0)\frac{s^3}{6}\right] \end{aligned}$

Table 7.3: Shear, moment, rotation and deflection laws for different types of beams.

	Simply supported beam	Cantilever beam
Shear	$\frac{p(s-2u)}{2}$	$p(s-u)$
Moment	$\frac{pu(s-u)}{2}$	$\frac{-p(u-s)^2}{2}$
Rotation	$\frac{p(s-2u)(-s^2-2su+2u^2)}{24EI}$	$\frac{pu(-3s^2+3su-u^2)}{6EI}$
Deflection	$\frac{pu(u-s)(s^2+su-u^2)}{24EI}$	$\frac{pu^2(-6s^2+4su-u^2)}{24EI}$
	Supported cantilever beam	Clamped at both ends beam
Shear	$\frac{p(5s-8u)}{8}$	$\frac{p(s-2u)}{2}$
Moment	$\frac{p(s-4u)(u-s)}{8}$	$\frac{p(-s^2+6su-6u^2)}{12}$
Rotation	$\frac{pu(-6s^2+15su-8u^2)}{48EI}$	$\frac{p(s-2u)u(u-s)}{12EI}$
Deflection	$\frac{p(s-u)u^2(2u-3s)}{48EI}$	$\frac{-pu^2(u-s)^2}{24EI}$

Exercises

7.1 Solve the functional equation

$$f(xy) = f(x)g(y) + h(y); \quad x, y \in \mathbb{R}_{++}$$

in the class of differentiable functions by its reduction to a differential equation.

7.2 Find an exact difference equation equivalent to the differential equation

$$Af''(x) + f'(x) + f(x) = 0,$$

where A is a given non-zero constant, using:

(a) the system (7.100)–(7.101)

(b) Equation (7.139).

7.3 Find the differential equation equivalent to the difference equation

$$f(x+3) - 4f(x+2) + 5f(x+1) - 2f(x) = 0.$$

7.4 Using the technique in Theorem 7.12 and Example 7.12, obtain the functional equation equivalent to the differential equation

$$y^{(iv)}(x) = 0.$$

7.5 Solve Example 7.20 with a triangular load

$$p(u) = ku.$$

7.6 Solve the functional equation

$$f(x+y) = f(x) + f(y) + k$$

by its reduction to a differential equation.

7.7 Given the differential equation

$$f''(x) - 2f'(x) + f(x) = 0,$$

obtain an equivalent functional equation and an equivalent difference equation.

7.8 Given the differential equation

$$f''(x) - 2f'(x) + f(x) = -2\cos x,$$

obtain an equivalent functional equation and an equivalent difference equation.

7.9 Try to extend the methodology in Section 7.7 to the case of bending plates.

CHAPTER 8

Vector and matrix equations

8.1 Introduction

In previous chapters we have solved several real or complex functional equations of real or complex variables. In this chapter we extend some of these functional equations to the case of vector or matrix equations of vector or matrix variables. In some cases we have a direct extension because we can make use of the properties of groups, but in others this extension is much more complicated or becomes impossible. Sections 8.2, 8.3 and 8.4 are devoted to Cauchy's, Pexider's and Sincov's equations, respectively. In addition, some applications to the problem of aggregated allocation, to transition probabilities and to the characterization of all two-parameter families which are reproductive are included.

In this chapter vectors and matrices will be denoted by boldfaced letters.

8.2 Cauchy's equation

Cauchy's equation can be easily generalized to include vectors and matrices. The following theorem from Aczél (1966), pp. 348, gives its solution.

Theorem 8.1 (Cauchy's equation). *The general continuous solution of*

$$\mathbf{F}(\mathbf{x}+\mathbf{y}) = \mathbf{F}(\mathbf{x}) + \mathbf{F}(\mathbf{y}) \; ; \; \mathbf{x}, \mathbf{y} \in \mathbb{R}^{\mathbf{n}}, \tag{8.1}$$

where $\mathbf{x}$ *and* $\mathbf{y}$ *are now* n*-dimensional real vectors and* $\mathbf{F}(\mathbf{x})$ *is an* m*-dimensional real vector is*

$$\mathbf{F}(\mathbf{x}) = \mathbf{C}\mathbf{x}, \tag{8.2}$$

where $\mathbf{C}$ *is a constant* $m \times n$ *matrix.* ∎

Example 8.1 (Cauchy's equation). Theorem 5.1 is a particular case of Theorem 8.1. In effect, if we make $m = 1$, Expression (8.2) becomes $F(x) = \sum_{i=1}^{n} c_i x_i$, which coincides with (5.2). ∎

The extension of the remaining Cauchy equations is complicated because of the products involved. In this case, not all matrices possess an inverse and methods used in Chapters 2 and 3 are not valid any more. The following theorem gives the general continuous solution for a particular case.

Theorem 8.2 (Particular case of Cauchy's equation II). *The most general continuous-at-a-point and non identically zero solution of the functional equation*

$$F(\mathbf{x} + \mathbf{y}) = F(\mathbf{x})F(\mathbf{y}), \tag{8.3}$$

where $F : \mathbb{R}^n \to \mathbb{R}$ *is a real function of real variables, is given by*

$$F(\mathbf{x}) = \exp\left[\sum_{i=1}^{n} c_i x_i\right], \tag{8.4}$$

where c_i $(i = 1, 2, ..., n)$ *are arbitrary constants.*

Note that this theorem is in fact Theorem 5.5. ∎

Corollary 8.1 (Complex case). *The most general continuous-at-a-point and non identically zero solution of the functional equation (8.3), where now* $F : \mathbb{C} \to \mathbb{R}$, *is given by*

$$F(\mathbf{x}) = \exp[\mathbf{a}\mathbf{x} + \overline{\mathbf{a}\mathbf{x}}], \tag{8.5}$$

where $\mathbf{a}$ *is an arbitrary complex constant and* $\overline{\mathbf{a}\mathbf{x}}$ *is the conjugate of* $\mathbf{a}\mathbf{x}$. ∎

This corollary can also be derived from Theorem 5.6.

Theorem 8.3 (Particular case of Cauchy's equation III). *The most general solutions, continuous-in-a -neighborhood of a non-singular matrix, of the functional equation*

$$F(\mathbf{x}\mathbf{y}) = F(\mathbf{x})F(\mathbf{y}), \tag{8.6}$$

where $\mathbf{x}$ *and* $\mathbf{y}$ *are non-singular square matrices and* $F(\mathbf{x})$ *is a real number, are*

$$F(\mathbf{x}) = |\det \mathbf{x}|^{\alpha} \quad or \quad F(\mathbf{x}) = |\det \mathbf{x}|^{\alpha}\, sgn(\det \mathbf{x}), \tag{8.7}$$

where $\det \mathbf{x}$ *means the determinant of the matrix* $\mathbf{x}$, $sgn(x)$ *the sign of* x *and* α *is an arbitrary constant.* ∎

8.3 Pexider's equation

The general continuous solution of Pexider's equation for vectors is given by the following theorem.

Theorem 8.4 (Pexider's equation). *The general continuous solution of*

$$\mathbf{G}(\mathbf{x}+\mathbf{y}) = \mathbf{H}(\mathbf{x}) + \mathbf{K}(\mathbf{y}), \tag{8.8}$$

where $\mathbf{G}, \mathbf{H}$ *and* $\mathbf{K}$ *are* $n \times k$ *matrix functions and* $\mathbf{x}, \mathbf{y}$ *are* $m \times k$ *matrices, is*

$$\mathbf{G}(\mathbf{x}) = \mathbf{C}\mathbf{x} + \mathbf{a} + \mathbf{b}; \ \ \mathbf{H}(\mathbf{x}) = \mathbf{C}\mathbf{x} + \mathbf{a}; \ \ \mathbf{K}(\mathbf{x}) = \mathbf{C}\mathbf{x} + \mathbf{b}, \tag{8.9}$$

where $\mathbf{C}$, $\mathbf{a}$ *and* $\mathbf{b}$ *are constant* $n \times m, n \times k$ *and* $n \times k$ *matrices, respectively.* ■

Example 8.2 (Aggregated allocation). Aczél (1987b), page 2) A certain amount, s, of a quantifiable good is to be allocated to m $(m \geq 3)$ projects. For this purpose, a committee of n assessors is consulted and the $i-th$ assessor recommends delivering x_{ij} units of goods to project $j-th$, in such a manner that $\sum_{j=1}^{m} x_{ij} = s$ $(i = 1, 2, \ldots, n)$. The problem consists in synthesizing all recommendations into a single aggregate such that the following conditions be satisfied:

- (a) The aggregated allocation for the $j-th$ project depends only on the recommended allocations to that project; that is,

$$f_j(\mathbf{x}_j), \ \ \mathbf{x}_j = (\mathbf{x}_{1j}, \mathbf{x}_{2j}, \ldots, \mathbf{x}_{nj}).$$

- (b) The total amount allocated to all projects is constant and equal to s

$$\sum_{j=1}^{m} \mathbf{x}_j = \mathbf{s} \ \ \Rightarrow \ \ \sum_{j=1}^{m} f_j(\mathbf{x}_j) = s,$$

  where $\mathbf{s} = (s, s, ..., s)$.

- (c) There is a consensus on rejection

$$f_j(0, 0, \ldots, 0) = 0.$$

Substituting into (b) $\mathbf{x}_j = \mathbf{s}, \mathbf{x}_i = \mathbf{0}$ for $i \neq j$, in view of (c) leads to

$$f_j(\mathbf{s}) = s, \ \ j = 1, 2, \ldots, m.$$

If we substitute now $\mathbf{x}_j = \mathbf{z}, \mathbf{x}_r = \mathbf{s} - \mathbf{z}$ and $\mathbf{x}_i = \mathbf{0}$ ($i \neq j$ and $i \neq r$) we get

$$f_j(\mathbf{z}) = s - f_r(\mathbf{s} - \mathbf{z}); \ \ \mathbf{s} = (s, s, \ldots, s); \ \ \mathbf{z} = (z, z, \ldots, z).$$

Finally, substitution of $\mathbf{x}_j = \mathbf{x}$, $\mathbf{x}_k = \mathbf{y}$, $\mathbf{x}_r = \mathbf{s} - \mathbf{x} - \mathbf{y}$ and $\mathbf{x}_p = \mathbf{0}$ ($p \neq j$, $p \neq k$ and $p \neq r$), and taking into account the last expression, leads to

$$f_j(\mathbf{x}) + f_k(\mathbf{y}) = s - f_r(\mathbf{s} - \mathbf{x} - \mathbf{y}) = f_j(\mathbf{x} + \mathbf{y}),$$

which is Pexider's equation (8.8). Thus, its general continuous solution is

$$f_j(\mathbf{x}) = \mathbf{C}\mathbf{x} + \mathbf{a} + \mathbf{b} = \mathbf{C}\mathbf{x} + \mathbf{a}; \;\; f_k(\mathbf{x}) = \mathbf{C}\mathbf{x} + \mathbf{b} \;\; \Rightarrow \;\; \mathbf{b} = \mathbf{0}.$$

Now, consideration of (c) leads to $\mathbf{a} = \mathbf{0}$. Thus, we get

$$f_j(\mathbf{x}) = \sum_{i=1}^{n} c_i x_{ij}, \;\; j = 1, 2, \ldots, m.$$

Finally, condition (b) gives

$$s = f_j(\mathbf{s}) = \sum_{i=1}^{n} c_i s \Rightarrow \sum_{i=1}^{n} c_i = 1.$$

■

8.4 Sincov's equation and generalizations

The general solution of Sincov's equation for vectors is given by the following theorem.

Theorem 8.5 (Sincov's equation). *If there exist* $\mathbf{a}, \mathbf{b}$ *and* $\mathbf{c}$ *such that*

$$\det \mathbf{G}(\mathbf{a}, \mathbf{y}) \neq 0, \;\; \det \mathbf{H}(\mathbf{b}, \mathbf{c}) \neq 0, \tag{8.10}$$

then the general solution of

$$\mathbf{F}(\mathbf{x}, \mathbf{z}) = \mathbf{G}(\mathbf{x}, \mathbf{y})\mathbf{H}(\mathbf{y}, \mathbf{z}), \tag{8.11}$$

where $\mathbf{F}, \mathbf{G}$ *and* $\mathbf{H}$ *are square matrices, is*

$$\begin{aligned} \mathbf{F}(\mathbf{x}, \mathbf{z}) &= \mathbf{L}(\mathbf{x})\, \mathbf{N}(\mathbf{z}), \\ \mathbf{G}(\mathbf{x}, \mathbf{y}) &= \mathbf{L}(\mathbf{x})\, \mathbf{M}(\mathbf{y})^{-1}, \\ \mathbf{H}(\mathbf{y}, \mathbf{z}) &= \mathbf{M}(\mathbf{y})\, \mathbf{N}(\mathbf{z}), \end{aligned} \tag{8.12}$$

where $\mathbf{L}, \mathbf{M}$ *and* $\mathbf{N}$ *are arbitrary matrix functions and* $\mathbf{M}$ *is non-singular.* ■

Corollary 8.2 (Sincov's equation). *If there exists a vector* $\mathbf{a}$ *such that* $\det \mathbf{F}(\mathbf{a}, \mathbf{y}) \neq 0$, *then the general solution of the functional equation*

$$\mathbf{F}(\mathbf{x}, \mathbf{z}) = \mathbf{F}(\mathbf{x}, \mathbf{y})\mathbf{F}(\mathbf{y}, \mathbf{z})$$

is

$$\mathbf{F}(\mathbf{x}, \mathbf{y}) = \mathbf{M}(\mathbf{x})\, \mathbf{M}^{-1}(\mathbf{y}).$$

Proof: *This is a particular case of Theorem 8.5 with* $\mathbf{F} = \mathbf{G} = \mathbf{H}$. *Thus we have*

$$\mathbf{L}(\mathbf{x})\,\mathbf{N}(\mathbf{y}) = \mathbf{L}(\mathbf{x})\,\mathbf{M}(\mathbf{y})^{-1} = \mathbf{M}(\mathbf{x})\,\mathbf{N}(\mathbf{y}) \quad \Rightarrow \quad \mathbf{L} = \mathbf{M} \;\; \textit{and} \;\; \mathbf{N} = \mathbf{M}^{-1}.$$

■

■

Example 8.3 (Transition probabilities). (Aczél (1966), pp. 363) Let us consider a system with n possible states and let $f_{ij}(s,t)$ denote the probability of the system changing during the time interval (s,t) from state $i-th$ to state $j-th$. Then, under the Markov assumption of independence, the following Chapman-Kolmogorov equation must hold

$$f_{ij}(s,t) = \sum_{k=1}^{n} f_{ik}(s,u)\, f_{kj}(u,t) \;\text{ with }\; \sum_{j=1}^{n} f_{ij}(s,t) = 1, \;\; \forall i,$$

but this is equivalent to the equation

$$\mathbf{F}(s,t) = \mathbf{F}(s,u)\mathbf{F}(u,t),$$

and, according to Corollary 8.2,

$$\mathbf{F}(s,t) = \mathbf{M}(s)\,\mathbf{M}(t)^{-1},$$

where, for $\mathbf{F}$ to be a true transition matrix, $\mathbf{M}$ must be an arbitrary non-singular matrix, such that $\sum\limits_{j=1}^{n} M_{ij} = C$, $\forall i$ and C is a constant.

If, in addition, we want homogeneity, that is, $\mathbf{F}(s,t)$ to be dependent only on $(t-s)$, we must have

$$\mathbf{F}(s,t) = \mathbf{G}(t-s) = \mathbf{M}(s)\,\mathbf{M}(t)^{-1},$$

which is a multiplicative Pexider equation, because with $u = t-s, \mathbf{H}(t) = \mathbf{M}(t)^{-1}$ we get $\mathbf{H}(s+u) = \mathbf{H}(s)\mathbf{G}(u)$.

Making $v = -s$ and taking logarithms of the matrices involved, we get

$$\mathbf{G}(t+v) = \mathbf{M}(-v)\,\mathbf{M}(t)^{-1},$$

which implies

$$\log[\mathbf{G}(t+v)] = \log[\mathbf{M}(-v)] + \log[\mathbf{M}(t)^{-1}],$$

which is Pexider's equation (8.8). Thus, based on (8.9), we can write

$$\left.\begin{array}{rcl} \mathbf{G}(t) & = & \exp(\mathbf{C}t + \mathbf{a} + \mathbf{b}) \\ \mathbf{M}(-t) & = & \exp(\mathbf{C}t + \mathbf{a}) \\ \mathbf{M}(t)^{-1} & = & \exp(\mathbf{C}t + \mathbf{b}) \end{array}\right\} \Rightarrow \left\{\begin{array}{rcl} \mathbf{a} & = & -\mathbf{b} \\ \mathbf{G}(t) & = & \exp(\mathbf{C}t) \end{array}\right.$$

and then

$$\mathbf{F}(s,t) = \mathbf{G}(t-s) = \exp[\mathbf{C}(t-s)] = \mathbf{B}\exp[\mathbf{J}(t-s)]\mathbf{B}^{-1},$$

where $\mathbf{C} = \mathbf{B}\mathbf{J}\mathbf{B}^{-1}$ is the Jordan canonical form of $\mathbf{C}$. ■

Theorem 8.6 (Generalization of Sincov's Equation). *The general solution of the equation*

$$\mathbf{F}(\mathbf{x},\mathbf{z}) = \mathbf{H}[\mathbf{F}(\mathbf{x},\mathbf{y}),\mathbf{F}(\mathbf{y},\mathbf{z})], \tag{8.13}$$

where the domain of $\mathbf{F}$ *is an arbitrary set* $\mathbf{A}$ *and the range of* $\mathbf{F}$ *lies in an arbitrary group* $\mathbf{B}$ *with respect to the operation* $\mathbf{H}$, *is of the form*

$$\mathbf{F}(\mathbf{x},\mathbf{y}) = \mathbf{H}[\mathbf{N}(\mathbf{x}),\mathbf{N}(\mathbf{y})^{-1}], \tag{8.14}$$

where $\mathbf{N}(\mathbf{x})$ *is an arbitrary non-singular matrix function defined on* $\mathbf{A}$ *with values in* $\mathbf{B}$. ∎

Theorem 8.7 (A particular case). *If* $\mathbf{H}(\mathbf{u},\mathbf{v}) = \mathbf{w}$ *can be solved uniquely for* $\mathbf{u}$ *and* $\mathbf{F}(\mathbf{x},\mathbf{y}) = \mathbf{u}$ *and* $\mathbf{F}(\mathbf{t},\mathbf{a}) = \mathbf{v}$ *can be solved for* $\mathbf{y}$ *and* $\mathbf{t}$, *respectively, then the general solution of the functional equation*

$$\mathbf{F}(\mathbf{x},\mathbf{y}) = \mathbf{H}[\mathbf{F}(\mathbf{x},\mathbf{z}),\mathbf{F}(\mathbf{y},\mathbf{z})], \tag{8.15}$$

where $\mathbf{x},\mathbf{y}$ *and* $\mathbf{z} \in \mathbf{A}$ *and* $\mathbf{F}(\mathbf{x},\mathbf{y}),\mathbf{H}(\mathbf{u},\mathbf{v}) \in \mathbf{B}$ *is*

$$\mathbf{H}(\mathbf{u},\mathbf{v}) = \mathbf{G}(\mathbf{u},\mathbf{v}^{-1}); \;\; \mathbf{F}(\mathbf{x},\mathbf{y}) = \mathbf{G}[f(\mathbf{x}),f(\mathbf{y})^{-1}], \tag{8.16}$$

where $\mathbf{G}$ *is a group operation and* $\mathbf{v}^{-1}$ *is the inverse of* $\mathbf{v}$ *in that group operation.* ∎

Theorem 8.8 (Case of groups). *If* $\mathbf{A}$ *forms a quasigroup with respect to the operation* $\mathbf{F}(\mathbf{x},\mathbf{y})$ *and the transitivity relation*

$$\mathbf{F}(\mathbf{x},\mathbf{y}) = \mathbf{F}[\mathbf{F}(\mathbf{x},\mathbf{z}),\mathbf{F}(\mathbf{y},\mathbf{z})] \tag{8.17}$$

holds, then $\mathbf{A}$ *forms a group relative to the inverse operation* $\mathbf{x} = \mathbf{G}(\mathbf{z},\mathbf{y})$ *of* $\mathbf{F}(\mathbf{x},\mathbf{y}) = \mathbf{z}$ *and the general solution of (8.17) is*

$$\mathbf{F}(\mathbf{x},\mathbf{y}) = \mathbf{G}(\mathbf{x},\mathbf{y}^{-1}). \tag{8.18}$$

∎

Theorem 8.9 (Two particular Pexider equations). *The most general solution of the functional equations*

$$\mathbf{F}(\mathbf{x}\mathbf{y}) = \mathbf{F}(\mathbf{x})\,\mathbf{y}; \qquad \mathbf{F}(\mathbf{x}\mathbf{y}) = \mathbf{x}\,\mathbf{F}(\mathbf{y}), \tag{8.19}$$

where $\mathbf{x},\mathbf{y}$ *and* $\mathbf{F}$ *are square matrices of order* n, *are*

$$\mathbf{F}(\mathbf{x}) = \mathbf{C}\,\mathbf{x} \text{ and } \mathbf{F}(\mathbf{x}) = \mathbf{x}\,\mathbf{C}\,, \tag{8.20}$$

respectively. ∎

Theorem 8.10 (A particular case of the vector bisymmetry equation). *If the following conditions hold:*

- *(a) There exists a vector* $\mathbf{v}$ *such that* $\mathbf{K}(\mathbf{x},\mathbf{v}) = \mathbf{a}$; $\mathbf{L}(\mathbf{x},\mathbf{v}) = \mathbf{b}$*; i.e., they are constants,*

- *(b)* $\mathbf{K}(\mathbf{x},\mathbf{c}) = \mathbf{u}$; $\mathbf{L}(\mathbf{y},\mathbf{c}) = \mathbf{v}$ *have unique inverses for a constant vector* $\mathbf{c}$,

- *(c) the functions in (8.21) are differentiable, and*

- *(d)* $\mathbf{F}_2(\mathbf{z},\mathbf{v}) = \mathbf{u}$ *has a unique inverse,*

then the most general solution of the functional equation

$$\mathbf{F}\,[\mathbf{G}(\mathbf{x},\mathbf{y}),\mathbf{u}\,] = \mathbf{H}\,[\,\mathbf{K}(\mathbf{x},\mathbf{u}),\mathbf{L}\,(\mathbf{y},\mathbf{u})\,] \tag{8.21}$$

is

$$\begin{array}{rcl} \mathbf{F}[\mathbf{z},\mathbf{u}\,] &=& \mathbf{w}\,[\mathbf{C}(\mathbf{u})\mathbf{r}(\mathbf{z}) + \mathbf{a}(\mathbf{u}) + \mathbf{b}(\mathbf{u})], \\ \mathbf{K}(\mathbf{x},\mathbf{u}) &=& \mathbf{s}^{-1}\,[\,\mathbf{C}(\mathbf{u})\mathbf{p}(\mathbf{x}) + \mathbf{a}(\mathbf{u})], \\ \mathbf{L}(\mathbf{y},\mathbf{u}) &=& \mathbf{t}^{-1}[\,\mathbf{C}(\mathbf{u})\mathbf{q}(\mathbf{y}) + \mathbf{b}(\mathbf{u})], \\ \mathbf{G}(\mathbf{x},\mathbf{y}) &=& \mathbf{r}^{-1}\,[\,\mathbf{p}(\mathbf{x}) + \mathbf{q}(\mathbf{y})], \\ \mathbf{H}(\mathbf{u},\mathbf{v}) &=& \mathbf{w}\,[\,\mathbf{s}(\mathbf{u}) + \mathbf{t}(\mathbf{v})\,], \quad \left[\mathbf{C}(0) = \mathbf{0}, \quad \left|\dfrac{\partial \mathbf{C}(0)}{\partial u_1}\right| \neq 0\right]. \end{array} \tag{8.22}$$

For the proof of this theorem we refer the reader to Aczél (1966), page 372. ■

Exercises

8.1 Let $\underline{x} \in \mathbb{R}^n_+$ be the output of a firm that produces n commodities. Let $f(\underline{x})$ be the cost function of output $\underline{x}$. We are interested in the function which is additive; i.e.,

$$f(\underline{x} + \underline{y}) = f(\underline{x}) + g(\underline{y}),$$

that is, the cost of the sum of output $\underline{x} \geq 0$ and the increment output $\underline{y} \geq 0$ equals the sum of the cost of the output $\underline{x}$ and a non-negative real number depending on $\underline{y}$. Solve this equation and give it a physical interpretation.

8.2 The state of deterioration of the wearing surface of a bridge can be classified in 6 different states: smooth, fair, normal, rough, bad and unacceptable. Due to annual variations in traffic and weather conditions the transitions from states of lower to higher deterioration are probabilistic. Since deterioration facilitates roughness growth and aging of the surface material (hydration, freezing, thawing, etc.), the transition probabilities depend on state and time.

Based on Example 8.3, design a transition probability model; i.e., choose a reasonable $\mathbf{F}(s,t)$ matrix for this problem and discuss its properties.

Discuss whether or not the following matrix

$$F_{ij}(s,t) = \begin{cases} \binom{t-s}{j-i} p^{j-i}(1-p)^{t-s-j+i}, & \begin{cases} 0 \le j-i \le t-s, \\ i \le j < 6, \quad t > s, \end{cases} \\ 1 - \sum_{k=i}^{5} F_{ik}(s,t) \quad ; & j = 6, \\ 0; & \text{otherwise} \end{cases}$$

is an adequate choice for modelling this problem. If the answer is positive, write the model as

$$\mathbf{F}(s,t) = \mathbf{M}(s)\mathbf{M}(t)^{-1}.$$

8.3 Solve the functional equation

$$\mathbf{F}\,[\mathbf{G}(\mathbf{x},\mathbf{y}),\mathbf{u}\,] = \mathbf{H}\,[\,\mathbf{K}(\mathbf{x},\mathbf{u}),\mathbf{K}\,(\mathbf{y},\mathbf{u})\,]$$

based on Theorem 8.10.

8.4 Solve the functional equation

$$\mathbf{F}\,[\mathbf{G}(\mathbf{x},\mathbf{y}),\mathbf{u}\,] = \mathbf{H}\,[\,\mathbf{F}(\mathbf{x},\mathbf{u}),\mathbf{F}\,(\mathbf{y},\mathbf{u})\,]$$

based on Theorem 8.10.

8.5 Consider a system with three states 1, 2 and 3; let $f_{ij}(s,t)$ denote the probability of the system changing during the time interval (s,t) from state $i-th$ to state $j-th$, and assume the Markov assumption of independence. Give two possible probability transition matrices $\mathbf{F}_1$ and $\mathbf{F}_2$, one depending only on $(t-s)$ and one more general.

8.6 Derive the general solution of Cauchy's functional equation

$$\mathbf{G}(\mathbf{x}+\mathbf{y}) = \mathbf{G}(\mathbf{x}) + \mathbf{G}(\mathbf{y}),$$

from the general solution of the Pexider functional equation

$$\mathbf{G}(\mathbf{x}+\mathbf{y}) = \mathbf{H}(\mathbf{x}) + \mathbf{K}(\mathbf{y}),$$

8.7 Using Theorem 8.5 solve the functional equation

$$\mathbf{F}(\mathbf{x},\mathbf{z}) = \mathbf{G}(\mathbf{x},\mathbf{y})\mathbf{F}(\mathbf{y},\mathbf{z})$$

Part II

Applications of Functional Equations

CHAPTER 9

Functional Networks

9.1 Introduction

Neural networks (NN) have received a great deal of attention in the last few years (see Freeman and Skapura (1991), Hertz et al. (1991) and Anderson and Rosenberg (1988) for a survey). They consist of one or several layers of neurons connected by links. Each *computing unit* or *neuron* computes a scalar output from a weighted combination of inputs, coming from the previous layer, using a given scalar activation function f, which for standard NN is assumed the same for all the neurons. Thus, if we denote the inputs by $(x_1, \ldots, x_s)$ and the outputs by $(y_1, \ldots, y_k)$, the computation performed by unit j-th is $y_j = f\left(\sum_{i=1}^{s} w_{ji} x_i\right)$, where f is a specified monotone function (usually step or sigmoidal functions), and w_{ji} is a weight associated with each connection. That is, the input to a given unit is the weighted sum of its parent's outputs. Since the neural functions are given, the parameters of the neural network are the connection weights. Then, learning consists of obtaining the optimal weights to reproduce a given set of data.

In spite of their importance and diffusion in many fields, the neural networks paradigm has some limitations and it has been extended in several directions (see Castillo et al. (1999a), Chapter 2). In this chapter we introduce *functional networks*, a novel generalization of neural networks, where the activation functions are unknown multivariate functions from given families to be estimated during the learning process. This makes it possible to define arbitrary functional models and gives functional networks a power and flexibility that neural networks lack.

In functional networks not only are arbitrary neural functions allowed, but they are initially assumed to be multi-argument and vector-valued functions. An important characteristic of functional networks is the possibility of dealing with functional constraints determined by functional properties we may know about the model (e.g., associativity, distributivity, etc.). This can be done by considering coincident outputs (convergent links) of some selected neurons. This allows the value of these output units to be written in several different forms (one per different link) and leads to a system of functional equations, which can be directly written from the topology of the neural network. Solving this system of functional equations leads to a great simplification of the initial network topology and neuron functions, in the sense that some of the neurons can be removed or simplified (their associated number of arguments is reduced). In fact, the initial multidimensional functions can be written in terms of functions of fewer arguments, which in many cases become single argument functions.

In functional networks there are two types of learning to deal with domain and data knowledge, respectively:

1. *Structural learning*, which includes:
   (a) learning the initial topology of the network, based on some properties which are available to the designer, and
   (b) learning the posterior simplified topology using functional equations, leading to a simpler equivalent architecture.
2. *Parametric learning*, concerned with the estimation of the neuron functions. This can be done by considering linear combinations of given functional families and estimating the associated parameters from the available data. Note that this type of learning generalizes the idea of estimating the weights of the connections in a neural network.

For some examples of functional networks see Castillo (1998); Castillo and Gutiérrez (1998); Castillo et al. (1999a,b, 2000a,b,c,d, 2001).

This chapter gives an overview of functional networks. In particular, two important questions are discussed here: on one hand, we show that every problem that can be solved by a neural network can also be formulated by means of a functional network. On the other hand, we give some examples of problems that cannot be solved using neural networks, but which can be naturally formulated in terms of functional networks, implying that their functional representation is more general.

In Section 9.2 functional networks are motivated by means of two simple examples, which can be expressed using two well-known functional equations. In Section 9.3 the structure of functional networks is analyzed, giving a detailed description of its main components. Section 9.4 shows the main differences between functional and neural networks, while in Section 9.5 the main steps to follow when working with functional networks are described. Section 9.6 describes how the selection between different models can be made. In Section 9.7 some models of functional networks are given. Finally, with the aim of showing the power of functional networks, this approach is applied in Section 9.8 to several interesting problems, such as a time series analysis, a real economic example, chaotic map modelling, noise reduction and the retrieval of masked information.

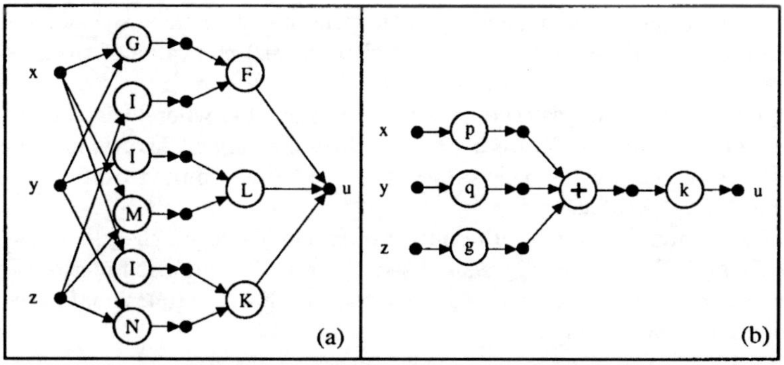

Figure 9.1: (a) Functional network for the concrete evaluation problem, and (b) equivalent simplified functional network.

Finally, the functional network methodology described in this chapter is applied in Sections 9.8.5 to model and predict the behavior of systems originally stated in terms of differential or difference equations. Thus, we consider the case of beams using two different approaches.

9.2 Motivating functional networks

To motivate functional networks we use two simple examples associated with concrete evaluation and Bayesian conjugate distribution problems, respectively.

Example 9.1 (Concrete evaluation). Suppose that we wish to predict the quality U of a given concrete based on three indicators X, Y and Z, taking numerical values x, y and z, respectively. Suppose also that such a prediction is $u = Q(x, y, z)$; that is, a function of the observed values for X, Y and Z, and that we are interested in determining the functional form of $Q(x, y, z)$.

With the aim of selecting the functional structure of $Q(x, y, z)$, we make some reasonable simplifying assumptions as follows.

If the values x and y, associated with indicators X and Y were known, we could calculate the influence of these two indicators on the quality $U = Q(x, y, z)$ by means of $G(x, y)$, a summary of the influences of indicators X and Y, and later incorporate the influence of the value z, of the indicator Z, by means of another function F to obtain $Q(x, y, z) = F(G(x, y), z)$. Suppose that the same argument is valid for any permutation of the X, Y and Z indicators. More precisely, we assume that functions F, G, H, K, L and M exist such that we have

$$u = Q(x, y, z) = F(G(x, y), z) = K(x, N(y, z)) = L(y, M(x, z)) \quad (9.1)$$

If the indicator variables are directly related to the quality being assessed, it is reasonable to assume that functions F, K, L, G, N and M are invertible (strictly monotonic) with respect to both variables. This means that the higher (lower) the level of one indicator, the higher (lower) the value of the predicted quality.

Equations (9.1) suggest the network in Figure 9.1, where I is used to refer to the identity function $I(x) = x$ and the three convergent arrows in the unit u are used to indicate coincident values, i.e., the values coming from each of the links must be the same.

The system of equations (9.1) puts strong conditions on the functions F, G, K, L, M and N. In fact, as we shall see, these two argument functions can be written in terms of single argument functions so that the functional network in Figure 9.1 can be simplified.

However, this network is not a neural network because:

1. The neuron functions are arbitrary.

2. Some neuron functions are multi-argument (e.g., F, G, K, L, M and N).

3. The outputs of neurons F, L and K are coincident, i.e., leading to the same output u.

Thus, neural networks are clearly inappropriate to reproduce this model.

The coincident connections in unit u or, equivalently, equations (9.1) put strong conditions on the neuron functions F, G, K, L, M and N. The methods of functional equations allow us to work with these equations and obtain the corresponding functional conditions. For instance, the system of functional equations (9.1), which has been previously analyzed in Section 6.4, has a general continuous solution (see Corollary 6.1):

$$\begin{aligned} F(x,y) &= k[f(x)+g(y)], \\ G(x,y) &= f^{-1}[p(x)+q(y)], \\ K(x,y) &= k[p(x)+n(y)], \\ N(x,y) &= n^{-1}[q(x)+g(y)], \\ L(x,y) &= k[q(x)+m(y)], \\ M(x,y) &= m^{-1}[p(x)+g(y)], \end{aligned} \tag{9.2}$$

where $k.g, p$ and q are arbitrary functions and f, m and n are arbitrary invertible functions. The first four equations in (9.2) are the solutions of the first functional equation in (9.1) (solved in Corollary 6.1), and the last four equations in (9.2) are the solutions of the last functional equation in (9.1).

Replacing this into (9.1) we get

$$Q(x,y,z) = k[p(x)+q(y)+g(z)]. \tag{9.3}$$

Equation (9.3) shows that the functional networks in Figures 9.1(a) and (b) are equivalent, in the sense of giving the same outputs for any given input. As we shall show, this fact leads to a simplification process that will be discussed in Section 9.5.

■

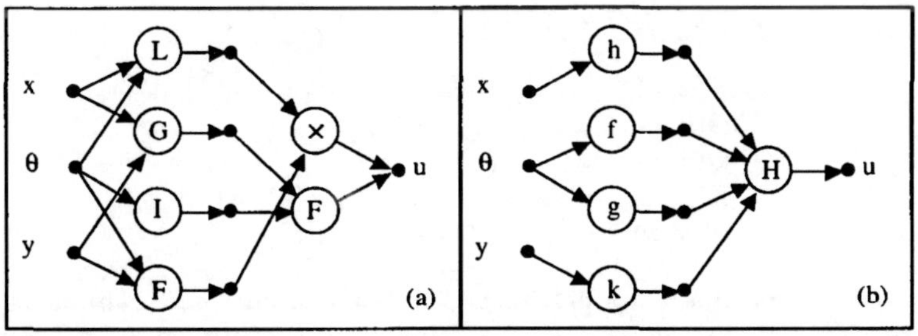

Figure 9.2: (a) Functional network associated with the Bayesian conjugate distributions problem, and (b) equivalent simplified functional network.

Example 9.2 (Bayesian Conjugate Distributions). Suppose that a random variable X belongs to a parametric family of distributions with likelihood function $L(x, \theta)$, where $\theta \in \Theta$ is a possibly vector-valued parameter. In Bayesian statistics a classical problem is to find a parametric family of probability density functions $F(\theta; \eta)$, with hyper-parameter η, so that both the prior probability density function $F(\theta; \eta)$ and the posterior probability density function $F(\theta; G(x; \eta))$ belong to the family. Bayes' theorem guarantees that the posterior density is proportional to $L(x; \theta)F(\theta; \eta)$, which leads to the functional equation

$$F(\theta; G(x; \eta)) = H(x; \theta)F(\theta; \eta), \tag{9.4}$$

where $H(x; \theta) = h(x)L(x; \theta)$, $h(x)$ is a function of x, and G gives the value of the new parameter, as a function of the sample value x and the old hyper-parameter value η. Here we have three functions, H, F, and G, each of which takes inputs and produces outputs, but the outputs are subject to the constraint given by (9.4); that is, the function F on the left-hand-side of (9.4) must be equal to the product of the functions H and F on the right-hand-side of (9.4).

Equation (9.4) suggests the network in Figure 9.2(a), which is not a neural network, because:

1. The neuron functions are different.
2. Some neuron functions are multi-argument (as F, G and L).
3. The outputs of neurons $\times$ and F in the second layer are coincident.

Thus, once again, neural networks are not appropriate to reproduce this model. ∎

Theorem 9.1 (Bayesian conjugate distributions theorem). *Under some regularity conditions (see Theorem 6.4), the general continuous solution of the functional equation*

$$F(G(x, y), z) = Q(M(x, z), N(y, z)), \tag{9.5}$$

where we assume $G \neq M, G \neq N, Q \neq M$ and $Q \neq N$, is

$$\begin{array}{rcl} F(x,y) & = & l[f(y)g^{-1}(x)+p(y)+q(y)], \\ G(x,y) & = & g[h(x)+k(y)], \\ Q(x,y) & = & l[m(x)+n(y)], \\ M(x,y) & = & m^{-1}[f(y)h(x)+p(y)], \\ N(x,y) & = & n^{-1}[f(y)k(x)+q(y)], \end{array} \tag{9.6}$$

where g, h, k, l, m and n are arbitrary strictly monotonic and continuously differentiable functions, and f, p and q are arbitrary continuously differentiable functions.

The two sides of (9.5) can be written as

$$l\{f(z)[h(x)+k(y)]+p(z)+q(z)\}.$$

■

Functional Equation (9.5) suggests a functional network such as that in Figure 9.2(a) with the only difference being that the neuron $\times$ must be replaced by Q.

The functional Equation (9.4) which has been obtained for the Bayesian conjugate case is a particular case of (9.5) with $Q(x,y) = xy$. A particular solution of the functional Equation (9.4) is (see Section 6.4):

$$\begin{array}{rcl} F(\theta;y) & = & f(\theta)^{k(y)}g(\theta). \\ L(x;\theta) & = & f(\theta)^{h(x)}. \\ G(x,y) & = & k^{-1}[h(x)+k(y)], \end{array} \tag{9.7}$$

showing that the two sides of (9.4) can be written as

$$f(\theta)^{h(x)+k(y)}g(\theta). \tag{9.8}$$

Thus, functional network in Figure 9.2(b) is equivalent to the functional network in Figure 9.2(a), where the meaning of H is obvious from (9.8).

The functional network associated with Theorem 9.1 arises in many different problems. The reader is referred to Exercises 9.3 and 9.4 at the end of this chapter for two more examples.

9.3 Elements of a functional network

A functional network consists of the following elements (see Figure 9.1(a)):

1. **Several layers of storing units.**

   (a) *One layer of input storing units.* This layer contains the input data. Input units are represented by small black circles with their corresponding names (x, y and z in Figure 9.1(a)).

(b) *No, one or several layers of intermediate storing units.* They are not neurons but units storing intermediate information. These layers are optional and contain units that store intermediate information produced by neuron units. Intermediate units are represented by small black circles (there is one layer with 6 intermediate units in the functional network in Figure 9.1(a)). These layers allow us to force the outputs of processing units to be coincident.

(c) *One layer of output storing units.* This layer contains the output data. Output units are also represented by small black circles with their corresponding names (u in Figure 9.1(a)).

2. **One or several layers of computing units.** A neuron is a computing unit which evaluates a set of input values, coming from the previous layer (of intermediate or input units) and gives a set of output values to the next layer (of intermediate or output units). To this end, each neuron has a neuron function which can be multivariate associated to it, and can have as many arguments as inputs. Each component (univariate) of a neural function is called a *functional cell.* Neurons are represented by circles with the name of the corresponding function inside. For example, the functional network in Figure 9.1(a) has 9 neurons G, I, I, M, I, N, F, L and K.

3. **A set of directed links.** They connect units in the input or intermediate layers to neuron units, and neuron units to intermediate or output units. Connections are represented by arrows, indicating the information flow direction. We remark here that information flows in only one direction, from the input layer to the output layer. Neurons receive information only from previous layers of the network, and output information to the next layer of neurons, or to the output units.

All these elements together form the *network architecture*, which defines the network topology and their associated functional capabilities. The network architecture refers to the organization of the neurons and the connections involved. Note that, as opposed to neural networks, in functional networks units are separated in two groups: storing and processing units.

9.4 Differences between neural and functional networks

It is natural to wonder what the differences between functional and neural networks are. We have already seen some of them when describing Examples 9.1 and 9.2. In this section, we discuss these differences and the advantages of using functional networks instead of standard neural networks.

1. In neural networks each neuron returns an output $y = f\left(\sum w_{ik}x_k\right)$ that depends solely on the value $\sum w_{ik}x_k$, where $x_1, x_2, \ldots, x_n$ are the received

inputs. Therefore, their neural functions have only one argument. On the contrary, as we have shown in the previous Examples 9.1 and 9.2, neural functions in functional networks can have several arguments. However, as we shall see, in many cases they can be equivalently replaced by functions of single variables.

2. In neural networks the neural functions are *univariate*: neurons can show different outputs, but all of them represent the same values. In functional networks, the neural functions can be *multivariate*.

3. In a given functional network the neural functions can be *different*, while in neural networks they are *identical*. In fact, arbitrary functions can be assumed for each neuron (see, for example, neurons F, G, I, K, L, M and N in Figure 9.1(a)).

4. In neural networks there are weights, which must be learned. These weights do not appear in functional networks (where neural functions are learned instead), since they can be incorporated into the neural functions.

5. Unlike standard neural networks, where the neuron functions are assumed to be fixed and known and only the weights are learned, in functional networks the functions are learned during the structural learning (which obtains the simplified network structure) and estimated during the parametric learning (which consists of obtaining the optimal neuron function from a given family).

6. In neural networks *the neuron outputs are different*, while in functional networks *neuron outputs can be coincident*. As we shall see, this fact leads to a set of functional equations, which have to be solved. For example, in Figure 9.1(a), the outputs of $m = 3$ neurons F, L and K are connected. Thus, they lead to a system of two functional equations (the last two equations in (9.1)).

   In general, these functional equations impose strong constraints, leading to a considerable reduction in the degrees of freedom of the neural functions. In most cases this implies that neural functions can be reduced in dimension or expressed as functions of smaller dimensions.

7. Intermediate layers of units are introduced in functional network architectures to allow several neuron outputs to be connected to the same units (this is not possible in neural networks).

8. Functional networks are extensions of neural networks.

   It is important to point out that neural networks are special cases of functional networks. For example, in Figure 9.3, a neural network and its equivalent functional network are shown. Note that weights are subsumed by the neural functions.

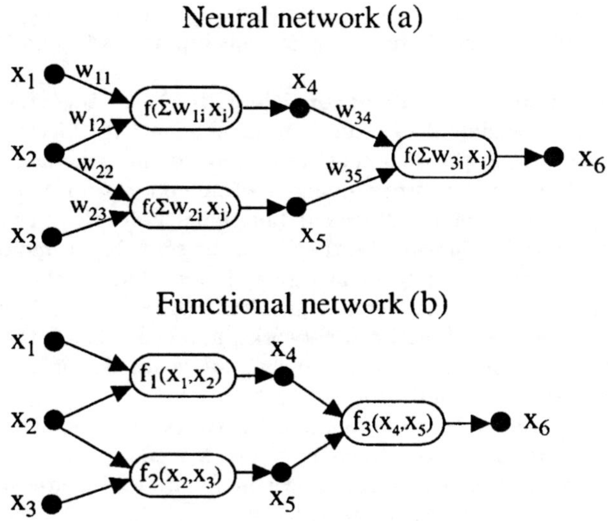

Figure 9.3: (a) neural network and (b) the corresponding functional network.

All these features show that the functional networks exhibit more interesting possibilities than standard neural networks. This implies that some problems (such as the ones introduced in Examples 9.1 and 9.2) require functional networks instead of neural networks in order to be solved.

9.5 Working with functional networks

In this section we describe the steps to be followed when working with functional networks. This new methodology requires the following steps:

Step 1 (Statement of the problem). An understanding of the problem to be solved. This is a crucial step.

Step 2 (Initial topology). Based on the knowledge of the problem, the topology of the initial functional network is selected. In neural networks such a selection is performed by trial and error using several topologies so as to make the error as small as possible with respect to the degrees of freedom. On the contrary, the selection of a functional network topology is based on the properties of the model, leading to a simple and unique network structure.

For example, the system of functional equations (9.1) leads to the functional network in Figure 9.1(a). Note that the above equations can be obtained from the network by considering the equality between the three values associated with the links connected to the output unit. We also remark that each of these values can be obtained in terms of the outputs of the preceding units writing the outputs of the neurons as functions of their inputs, and so on.

Step 3 (Simplification or structural learning). In this step, the initial functional network is simplified using functional equations. Given a functional network, an interesting problem consists of determining whether or not there exists another functional network giving the same output for any given input. This leads to the concept of equivalent functional networks. Two functional networks are said to be *equivalent* if they have the same input and output units and they give the same output for any given input. The practical importance of this concept is that we can define equivalent classes of functional networks; that is, sets of equivalent functional networks, and then choose the simplest in each class to be used in applications. *Functional equations constitute the main tool for simplifying functional networks.*

Equation (9.3) shows that the functional networks in Figures 9.1(a) and (b) are equivalent. However, note that the network topology and the functional structure of the functional network shown in Figure 9.1(b) are much simpler than those associated with Figure 9.1(a).

Step 4 (Uniqueness of representation). Before learning a functional network we need to be sure that there is a unique representation of it. In other words, for a given topology (structure), there are in some cases, several neuron functions leading to exactly the same output for any input. To avoid estimation problems we need to know what conditions must hold for uniqueness.

Step 5 (Data collection). For the learning to be possible, some information is required. In this step the data is collected.

Step 6 (Parametric learning). The neural functions of the network are estimated (learned) based on the given data. This is done by considering linear combinations of appropriate functional families and using some minimization method to obtain the optimal coefficients.

In functional networks, this learning process consists of obtaining the neural functions based on a set of data $D = \{(I_i, O_i)|i = 1, \dots, n\}$ given in the previous step, where I_i and O_i are the i-th inputs and outputs, respectively, and n is the sample size.

Usually, the learning process is based on minimizing the sum of squared errors of the actual and the observed outputs for the given inputs

$$Q = \sum_{i=1}^{n} (O_i - F(I_i))^2, \tag{9.9}$$

where F is the compound function given the outputs, as a function of the inputs, for the given network topology.

One learning alternative consists of approximating each neural function f_i by a linear combination of functions in a given family $\{\phi_{i1}, \dots, \phi_{im_i}\}$. Thus, the approximated neural function $f_i(\mathbf{x})$ becomes

$$\hat{f}_i(\mathbf{x}) = \sum_{j=1}^{m_i} a_{ij}\phi_{ij}(\mathbf{x}), \tag{9.10}$$

where $\mathbf{x}$ are the inputs associated with the i-th neuron. Note that the above function F includes all the neural functions in the network, and therefore it depends only on the coefficients a_{ij}, which are estimated in the learning process. Moreover, some of the neural functions f_i can be known, and then, the learning can be partial.

For certain network topologies and approximations, this learning leads to a linear system of equations, and then a single minimum is obtained. These methods are referred to as *linear methods.* In other cases, a non-linear system is obtained, and multiple minima may coexist. In these methods (usually referred to as *nonlinear methods*) the optimization process can be carried out by considering some standard gradient descendent method.

Step 7 (Model validation). The test for quality and/or the cross validation of the model is performed. Checking the obtained error is important to see whether or not the selected family of approximating functions are adequate. A cross validation of the model, with an alternative set of data, is also important. This makes it possible to decide on the existence of the over-fitting problem.

Step 8 (Use of the model). If the validation process is satisfactory, the model is ready to be used.

9.6 Model selection in functional networks

As we have seen in previous examples, in order to learn the resulting functional network, we can choose different sets of linearly independent functions to approximate their neuron functions.Thus, we need a model selection method to choose the best model according to a particular criterion of optimality.

The problem of model selection has been extensively analyzed from different points of view (see, for example, Akaike (1973), Atkinson (1978), Lindley (1968) and Stone (1974)). Here we use the Minimum Description Length (MDL) measure, which allows us to compare not only the quality of the different approximations, but also functional networks with different topologies.

The idea behind the MDL measure is to look for the minimum information required to store the given data set using the model. To this end, let us define the code length $L(x)$ of x as the amount of memory needed to store the information x. For instance, to store the data in the associative operation example we have two options:

Option 1: Store Raw Data. Store the triplets $\{(x_t, y_t, x_t \oplus y_t)|t \in T\}$. In this case, the initial description length (DL) of the data set is given by

$$DL = \sum_{t \in \mathcal{T}} \left[L(x_t) + L(y_t) + L(x_t \oplus y_t) \right]. \tag{9.11}$$

Option 2: Use a Model. By selecting a model, we try to reduce this length as much as possible. In this case, we can store the inputs $\{(x_t, y_t) | t \in \mathcal{T}\}$, the parameters of the model $\{c_i | i \in I\}$ and the residuals

$$e_t = \hat{f}((x_t \oplus y_t)) - \hat{f}(x_t) - \hat{f}(y_t), \; t \in \mathcal{T}, \tag{9.12}$$

where $\hat{f}$ is the approximate neuron function for the model. In this case, the description length becomes

$$DL_{\text{model}} = \sum_{t \in \mathcal{T}} \left[L(x_t) + L(y_t) \right] + \sum_{i \in I} L(\hat{c}_i) + \sum_{t \in \mathcal{T}} L(e_t | \text{model}). \tag{9.13}$$

Note that, since the ranges of the residuals e_t are smaller than the range of the data $x_t \oplus y_t$, the extra effort to store the parameters c_i in addition to the residuals e_t is compensated by the savings in storing $x_t \oplus y_t$. In addition, since the description length can be calculated for any model, the description length measure does not care about which model, or which dimension is used. This makes the minimum description length a convenient method for solving the model selection problem. Accordingly, the best functional network model for a given problem corresponds to the one with the minimum description length value.

To calculate the description length associated with a given model we need the code length of a given number, integer or real:

1. *Code length of an integer:* Since we measure the code length in bits, the code length of an integer n can be approximated by $\log_2(n)$.
2. *Code length of a real:* We can use two options.

   OPTION 1: Let $x = x_1 + x_2$ be a real number where x_1 is its integer part, and x_2 its fractional part. The integers $n_1 = x_1$ and an approximation,n_2, of x_2 are stored separately. Then, the code length $\mathcal{L}_1(x)$ is the sum of n_1 and n_2.

   OPTION 2: We compute the code length of a real as follows:

   - If $|x| < 10^{-q}; q \in \mathbb{N}$, we store a zero, and then we have $\mathcal{L}_2(x) = 1$;
   - otherwise, we store the integer part of $|x| \cdot 10^{-q}$ and then we have $\mathcal{L}_2(x) = \log_2(\lfloor |x| \cdot 10^{-q} \rfloor)$, where $q \in \mathbb{N}$ is a precision related integer, and $\lfloor x \rfloor$ is the integer part of x.

Suppose we know that the data comes from a model in the set of m families $\{f_m(\mathbf{x}|\boldsymbol{\theta}) | \boldsymbol{\theta} \in \boldsymbol{\Theta}; m \in \mathcal{M}\}$, where $\boldsymbol{\theta} = (\theta_1, \ldots, \theta_k)$, $\boldsymbol{\Theta}$ is the parameter space, and $\mathcal{M}$ is the model space. Suppose also that the model $f_m(\mathbf{x}|\boldsymbol{\theta})$ has associated probability $\pi_m(\boldsymbol{\theta})$. According to Rissanen (1989), p. 55, if $\hat{\theta}_j$ is estimated from all data points, the sample size is large, and the errors are normal, then the description length, using option 1 for storing real numbers, becomes

$$\mathcal{L}_1 = -\log \pi_m(\boldsymbol{\theta}) + \frac{k \log n}{2} + \frac{n}{2} \log \left(\frac{1}{n} \sum_{j=1}^{n} e_j(\boldsymbol{\theta})^2 \right), \tag{9.14}$$

where k is the number of parameters of the model. This MDL measure depends on both the data and the model, and consists of the sum of three terms. The first term represents the quality of the prior distribution given by the human expert, the second is a penalty for using complex models (those with a large number of parameters), and the third represents the quality of the fitted model (the smaller the errors, the better the model).

An alternative to (9.14), which consists of using option 2, leads to:

$$\mathcal{L}_2 = \sum_{i \in I} \mathcal{L}_2(\hat{c}_i) + \sum_{t \in T} \mathcal{L}_2(e_t|\text{model}). \tag{9.15}$$

where the first term penalizes the number of parameters, and the second the errors. Therefore, the measures in (9.14) and (9.15) allow:

1. A comparison of different sets $\{g_1(x), \ldots, g_k(x)\}$ of linearly independent approximating functions.

2. A selection of which of the functions in $\{g_1(x), \ldots, g_k(x)\}$ contribute more to the quality of the model, and which can be removed.

Computer implementations of model selection methods include:

1. **The exhaustive method.** This method calculates the values of $L(\mathbf{x})$ for all possible functional networks and all possible subsets of the approximating functions and chooses the one leading to the smallest value of $L(\mathbf{x})$. A clear disadvantage of this method is that it requires a lot of computational power.

2. **The forward-backward method.** This method starts with all models of a single parameter and selects the one leading to the smallest value of $L(\mathbf{x})$. Next, it incorporates one more parameter with the same criterion and the process continues until no improvement in $L(\mathbf{x})$ can be obtained by adding an extra parameter to the previous model. Then the inverse process is applied; that is, sequentially the parameter leading to the smallest value of $L(\mathbf{x})$ is removed until no improvement of $L(\mathbf{x})$ is possible. This double process is repeated until no further improvement in $L(\mathbf{x})$ is obtained either by adding or removing a single variable.

3. **The backward-forward method.** This method starts with the model with all parameters and sequentially removes the one leading to the smallest value of $L(\mathbf{x})$, repeating the process until there is no further improvement in $L(\mathbf{x})$. Next, the forward process is applied, but starting from this

model. The double process is repeated until no further improvement in $L(\mathbf{x})$ is obtained either by removing or adding a single variable.

9.7 Some examples of the functional network methodology

Now the functional network methodology has been carefully described in the previous section, its performance may be illustrated through some representative examples. For each example, a detailed analysis of the simplification, uniqueness of representation and learning process problems are presented.

9.7.1 The associative example

Suppose that we have the set of data points (x_1, x_2, y) shown in Table 9.1, obtained from a function $y = F(x_1, x_2)$[1]. Suppose also that we do not have any information about the form of the function, but we know that it is associative, i.e., the F function satisfies

$$F(F(x_1, x_2), x_3) = F(x_1, F(x_2, x_3)). \tag{9.16}$$

Let us solve the problem of finding the function F by using functional networks, following the steps indicated in the previous section.

Step 1 (Statement of the problem). We wish to reproduce the above associative operation F between two real numbers, i.e., the F function must satisfy (9.16). Note that $F(x_1, x_2)$ summarizes the contribution of x_1 and x_2 to $F(F(x_1, x_2), x_3)$. In fact, associativity means that we can operate consecutive pairs of operands in any order.

Step 2 (Initial topology). Equation (9.16) suggests the initial network topology shown in Figure 9.4(a).

Step 3 (Simplification or structural learning). Initially, it seems that a two-argument function F has to be learned. However, the functional Equation (9.16) puts strong constraints on it. In fact, the general solution of the functional Equation (9.16), according to Theorem 6.6, is:

$$F(x_1, x_2) = f^{-1}[f(x_1) + f(x_2)], \tag{9.17}$$

where $f(x)$ is an arbitrary continuous and strictly monotonic function, which can be replaced only by $cf(x)$, where c is an arbitrary constant.

Replacing (9.17) in (9.16), we can see that the two sides of (9.16) can be written as

$$f^{-1}[f(x_1) + f(x_2) + f(x_3)], \tag{9.18}$$

[1]We have used the operation $u = F(x_1, x_2) = f^{-1}(f(x_1)+f(x_2))$ with $f(x) = 0.2+\frac{x}{10}+e^x$.

Table 9.1: Data obtained from an associative operation F.

x_1	x_2	y	x_1	x_2	y	x_1	x_2	y
0.376	0.608	1.240	0.869	0.981	1.660	0.934	0.897	1.650
0.230	0.811	1.300	0.174	0.291	0.984	0.439	0.848	1.400
0.569	0.860	1.460	0.244	0.173	0.960	0.714	0.106	1.200
0.240	0.682	1.230	0.093	0.742	1.210	0.170	0.075	0.876
0.762	0.778	1.510	0.305	0.997	1.450	0.686	0.089	1.170
0.377	0.138	1.010	0.337	0.565	1.201	0.938	0.101	1.340
0.598	0.152	1.150	0.355	0.867	1.380	0.903	0.041	1.301
0.995	0.468	1.501	0.140	0.189	0.916	0.190	0.669	1.202
0.907	0.521	1.470	0.960	0.687	1.570	0.252	0.199	0.976
0.726	0.714	1.461	0.739	0.274	1.280	0.710	0.527	1.360
0.073	0.649	1.151	0.402	0.607	1.252	0.619	0.964	1.541
0.471	0.121	1.061	0.518	0.882	1.462	0.902	0.983	1.681
0.123	0.292	0.962	0.145	0.057	0.854	0.426	0.199	1.071
0.144	0.206	0.926	0.437	0.888	1.432	0.032	0.050	0.794
0.371	0.172	1.031	0.696	0.754	1.471	0.192	0.241	0.967
0.558	0.791	1.420	0.543	0.852	1.450	0.793	0.762	1.521
0.928	0.192	1.372	0.208	0.819	1.301	0.182	0.858	1.319
0.883	0.427	1.421	0.472	0.984	1.499	0.641	0.755	1.440
0.001	0.126	0.818	0.468	0.397	1.181	0.703	0.884	1.539
0.425	0.718	1.330	0.694	0.623	1.401	0.905	0.348	1.398
0.454	0.719	1.341	0.261	0.490	1.131	0.181	0.250	0.966
0.628	0.314	1.230	0.352	0.681	1.269	0.823	0.388	1.370
0.570	0.698	1.381	0.894	0.434	1.429	0.832	0.831	1.570
0.555	0.755	1.402	0.283	0.612	1.209	0.431	0.427	1.180
0.742	0.178	1.240	0.248	0.191	0.971	0.536	0.584	1.299
0.185	0.892	1.340	0.826	0.324	1.350	0.357	0.442	1.151
0.331	0.516	1.172	0.480	0.956	1.482	0.723	0.187	1.230
0.664	0.146	1.180	0.397	0.285	1.089	0.797	0.382	1.350
0.767	0.012	1.200	0.383	0.070	0.988	0.299	0.331	1.060
0.835	0.373	1.368	0.031	0.512	1.050	0.296	0.724	1.280
0.517	0.541	1.269	0.818	0.512	1.421	0.951	0.412	1.460
0.184	0.684	1.210	0.116	0.293	0.959	0.816	0.606	1.460
0.952	0.163	1.365	0.911	0.858	1.620	0.566	0.641	1.350

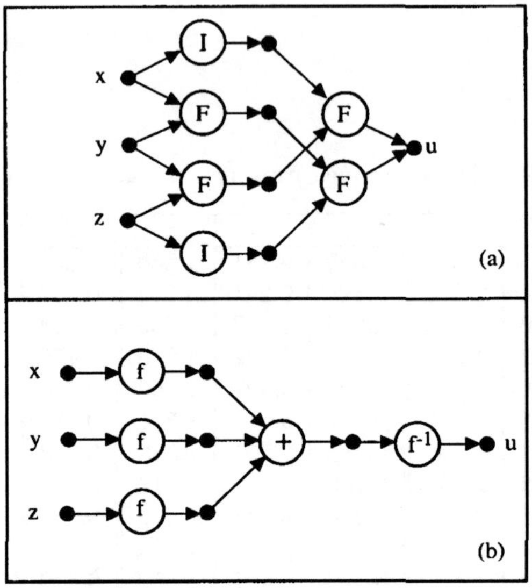

Figure 9.4: Illustration of the associativity functional network. (a) Initial network, and (b) equivalent simplified network.

and then, the functional network in Figure 9.4(a) is equivalent to the functional network in Figure 9.4(b), where only a one-argument function f need be learned.

Figure 9.5 shows a network able to perfectly reproduce any associative operation (see Equation (9.17)).

Two important conclusions can be derived from this theorem:

1. No other functional forms for F satisfy Equation (9.16). So, no other neurons can be replaced by neurons F.

2. The initial two-dimensional function F is completely determined by means of a unidimensional function f. Thus, the functional Equation (9.16) reduces the initial degrees of freedom of $F(\cdot,\cdot)$ from a two-argument function to a single-argument function f.

Step 4 (Uniqueness of representation). In this step, we analyze whether or not two different functional networks with the topology implied by (9.17) exist. In other words, whether there exist two different functions $f(\cdot)$ and $g(\cdot)$ such that

$$F(x_1, x_2) = f^{-1}[f(x_1) + f(x_2)] = g^{-1}[g(x_1) + g(x_2)]. \tag{9.19}$$

The solution to this functional equation is:

$$g(x) = cf(x),$$

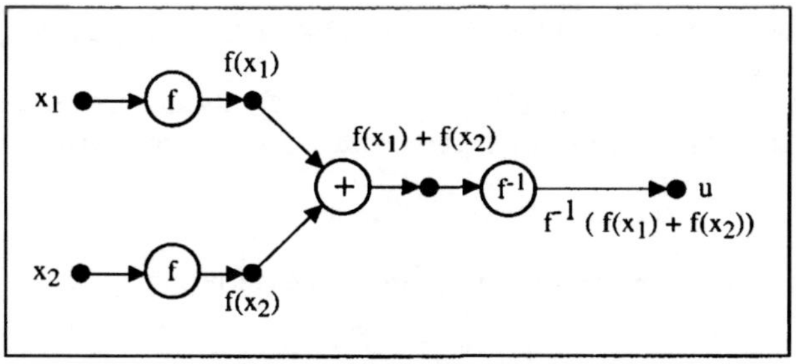

Figure 9.5: Graphical illustration of the network associated with $F(x_1, x_2) = u$, where F corresponds to an associative operation.

where c is an arbitrary constant. Thus, the functional structure of the functional network is determined up to a constant. In other words, the constant c is not identifiable from the functional equation alone; i.e., one extra condition is required to obtain c. However, no matter which value of c is used, expressions (9.17) and (9.18) give the same function; i.e., there are many different functional networks which are equivalent. Thus, a concrete value of c is not needed. However, to learn $f(\cdot)$ we need uniqueness and, thus, an extra constant.

Step 5 (Data collection). To learn this associative operation in the interval $(0, 5)$, we can take pairs of numbers in that interval and their operated values as triplets

$$\{(x_{1j}, x_{2j}, x_{3j}) | x_{3j} = F(x_{1j}, x_{2j}) = x_{1j} \oplus x_{2j}; j = 1, \ldots, n\}.$$

Note that, for convenience, we have used the notation x_1, x_2 and x_3 for x, y and z, respectively.

We can do this deterministically or we can simulate a set of triplets. Assume that we simulate 60 of these triplets and obtain the values in Table 9.1.

Step 6 (Learning). From (9.17) we get

$$u = F(x_1, x_2) \Leftrightarrow f(u) = f(x_1) + f(x_2), \tag{9.20}$$

an interesting relation to be exploited for learning $f(x)$.

Learning this network is equivalent to learning the function $f(x)$. To this end, we can approximate $f(x)$ by

$$\hat{f}(x) = \sum_{i=1}^{m} a_i \phi_i(x), \tag{9.21}$$

where the $\{\phi_i(x); i = 1, \ldots, m\}$ is a set of given linearly independent functions, capable of approximating $f(x)$ to the desired accuracy, and the coefficients a_i

are the parameters of the functional network; i.e., they play the role of the weights on a neural network.

To estimate the coefficients $\{a_i; i = 1, \ldots, m\}$, we use the collected data in the form of triplets (x_{1j}, x_{2j}, x_{3j}). According to (9.20) we must have

$$f(x_{3j}) = f(x_{1j}) + f(x_{2j}); \quad j = 1, \ldots, n, \tag{9.22}$$

thus, the error of the approximation can be measured by

$$e_j = \hat{f}(x_{1j}) + \hat{f}(x_{2j}) - \hat{f}(x_{3j}); \quad j = 1, \ldots, n. \tag{9.23}$$

To estimate the coefficients $\{a_i; i = 1, \ldots, m\}$, we minimize the sum of squared errors

$$Q = \sum_{j=1}^{n} e_j^2 = \sum_{j=1}^{n} \left(\sum_{i=1}^{m} a_i \left[\phi_i(x_{1j}) + \phi_i(x_{2j}) - \phi_i(x_{3j}) \right] \right)^2 \tag{9.24}$$

subject to

$$f(x_0) \equiv \sum_{i=1}^{m} a_i \phi_i(x_0) = \alpha, \tag{9.25}$$

where α is an arbitrary but given real constant, which is necessary to identify the otherwise unidentifiable constant c in Step 3.

Thus, using the Lagrange multipliers technique we build the auxiliary function

$$Q_\lambda = \sum_{j=1}^{n} \left(\sum_{i=1}^{m} a_i b_{ij} \right)^2 + \lambda \left(\sum_{i=1}^{m} a_i \phi_i(x_0) - \alpha \right), \tag{9.26}$$

where

$$b_{ij} = \phi_i(x_{1j}) + \phi_i(x_{2j}) - \phi_i(x_{3j}). \tag{9.27}$$

The minimum can be obtained by solving the following system of functional equations, where the unknowns are the coefficients a_i and the multipliers λ:

$$\begin{aligned} \frac{\partial Q_\lambda}{\partial a_r} &= 2 \sum_{j=1}^{n} \left(\sum_{i=1}^{m} a_i b_{ij} \right) b_{rj} + \lambda \phi_r(x_0) = 0; \quad r = 1, \ldots, m, \\ \frac{\partial Q_\lambda}{\partial \lambda} &= \sum_{i=1}^{m} a_i \phi_i(x_0) - \alpha = 0, \end{aligned} \tag{9.28}$$

which is a linear system of $m + 1$ equations and $m + 1$ unknowns having a unique solution, assuming the set of functions $\{\phi_i; i = 1, \ldots, m\}$ to be linearly independent. In matrix form, it can be written as

$$\begin{pmatrix} \mathbf{BB}^T & \phi_0^T \\ \phi_0 & 0 \end{pmatrix} \begin{pmatrix} \mathbf{a}^\mathrm{T} \\ \lambda \end{pmatrix} = \begin{pmatrix} 0 \\ \vdots \\ 0 \\ \alpha \end{pmatrix}, \tag{9.29}$$

where $\mathbf{B}$ is the matrix of coefficients b_{ij}, $\mathbf{B}^T$ is the transpose of $\mathbf{B}$, $\phi_0 = (\phi_1(x_0), \ldots, \phi_m(x_0))$ and $\mathbf{a} = (a_1, \ldots, a_m)$.

If we consider a polynomial family of functions $\phi = \{1, x, \ldots, x^m\}$ (i.e., a Taylor expansion) to approximate the neuron function f in (9.17) and apply the above learning algorithm, we get an approximate model for the neuron function. For example, if we consider the simplest case ($m = 1$), we get

$$\hat{f}(x) = 0.234 + 0.765x.$$

Then, the approximate model corresponding to the associative operation $F(x, y)$ is

$$\hat{F}(x,y) = \hat{f}^{-1}(\hat{f}(x) + \hat{f}(y)) = \frac{0.468 + 0.765(x+y)}{0.765} - 0.234.$$

Proceeding in the same way, we can obtain the approximate models associated with $m = 2$ and $m = 3$,

$$\hat{f}(x) = 0.423 + 0.196x + 0.381x^2,$$

and

$$\hat{f}(x) = 0.395 + 0.398x + 0.079x^2 + 0.177x^3,$$

respectively. Note that the above functions are invertible for positive values and the inverses can be easily obtained using, for example, the bisection method.

Step 7 (Model validation). The performance of the above models can be measured in terms of the RMSE (root mean squared error), as shown in Table 9.2. Note the remarkable improvement of the approximate model as the number of approximating functions m is increased.

With the aim of graphically illustrating the performance of these models, Figure 9.6 represents a scatter plot of the patterns shown in Table 9.1 versus the corresponding approximate values obtained with each of the above approximate models. It can be seen how the approximate model gets close to the data as more terms in the polynomial family are considered (Figures (b) and (c)).

To cross-validate the model, we have predicted the values of 1000 more couples and determined the prediction errors. The obtained RMS and maximum error values are shown in Table 9.2. A comparison between the errors for the training and testing data shows that they are comparable. Thus, we can conclude that no over-fitting occurs. Note that an error for the training data which is significantly smaller than the error for the testing data is a clear indication of over-fitting.

With the aim of comparing the performance of functional and neural networks we have used several neural networks with 2, 5 and 10 hidden units to fit the data shown in Table 9.1. For each of the above structures, ten networks were trained using 10000 iterations of the standard back-propagation algorithm with different values for the learning rate and momentum parameters. Table 9.3 shows the averaged RMSE and maximum errors obtained in each case.

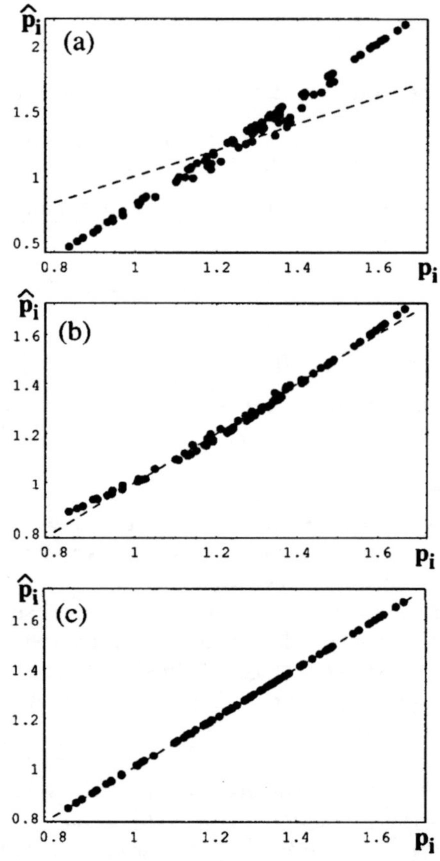

Figure 9.6: Actual (p_i) vs. predicted ($\hat{p}_i$) training patterns for: (a) $m = 1$ (the linear case), (b) $m = 2$, and (c) $m = 3$. The dashed line corresponds to the exact model $p_i = \hat{p}_i$.

Table 9.2: RMSE and maximum errors for five different approximate models. m indicates the number of approximating functions considered in each case and *par* the number of parameters of the resulting model. The 100 training triplets shown in Table 9.1 and 1000 additional test triplets have been considered for training and testing the models, respectively.

		Training		Test	
m	*par*	RMSE	Max	RMSE	Max
1	2	0.2128	0.4864	0.2135	0.5370
2	3	0.0167	0.0456	0.0163	0.0771
3	4	1.3310^{-3}	3.4910^{-3}	1.5410^{-3}	6.410^{-3}
5	6	8.3110^{-6}	2.8310^{-5}	8.5410^{-6}	3.8810^{-5}
10	11	5.6110^{-8}	1.1610^{-7}	5.6810^{-8}	1.1910^{-7}

Table 9.3: RMSE and maximum errors for three different network architectures. h indicates the number of hidden units and *par* the number of parameters of the resulting model. The 100 training triplets shown in Table 9.1 and 1000 additional test triplets have been considered for training and testing the models, respectively.

		Training		Test	
h	*par*	RMSE	Max	RMSE	Max
2	9	0.0625	0.0910	0.0739	0.1120
5	21	1.5810^{-3}	5.2210^{-3}	1.8710^{-3}	7.6310^{-3}
10	41	2.1410^{-4}	6.1610^{-4}	2.8910^{-4}	7.8710^{-4}

From Tables 9.2 and 9.3 we can see how the functional networks methodology outperforms neural networks. For example, comparable errors are obtained when using a functional network with 4 parameters and a neural network with 21 parameters. Moreover, when the number of parameters increases the performance of functional networks increases very quickly, whereas the performance of neural networks increases slowly.

The above results show that functional networks need a small number of parameters and provide us with a meaningful model to understand the problem. On the other hand, neural networks are black-boxes that give us no information about the structure underlying the data.

Step 8 (Use of the model). Once the model has been tested satisfactorily, it is ready to be used for the associativity operation.

9.7.2 The uniqueness model

In this example we consider the uniqueness model, an extension of the associative model presented in Section 9.7.1.

Step 1 (Statement of the problem). Consider the functional network shown in Figure 9.7, where the output z can be written as a function of the inputs x and y as:

$$z = f_3^{-1}(f_1(x) + f_2(y)). \tag{9.30}$$

In this example, we assume that we want to know whether or not there exist two different triplets of functions $\{f_1, g_1, h_1\}$ and $\{f_2, g_2, h_2\}$ such that their associated functional networks are equivalent. This is the same as considering the functional network in Figure 9.7, from which we can write:

$$f_3^{-1}[f_1(x) + f_2(y)] = g_3^{-1}[g_1(x) + g_2(y)], \tag{9.31}$$

This example tries to solve the uniqueness of Expression (9.31).

Step 2 (Initial topology). The initial topology of the functional network is shown in Figure 9.7.

Step 3 (Simplification of the Model). Note that no simplification is possible here, since no arrows convergent to storing units are included in the network.

Step 4 (Uniqueness of Representation). Suppose that there exist two different triplets of functions $\{f_1, f_2, f_3\}$ and $\{g_1, g_2, g_3\}$ such that their associated functional networks are equivalent. This is the same as considering the functional network in Figure 9.8 or writing:

$$F(x,y) = f_3^{-1}[f_1(x) + f_2(y)] = g_3^{-1}[g_1(x) + g_2(y)], \tag{9.32}$$

which is a functional equation with general solution given in Lemma 6.1:

$$\begin{array}{rcl} f_1(x) & = & a g_1(x) + b, \\ f_2(y) & = & a g_2(y) + c, \\ f_3^{-1}(u) & = & g_3^{-1}\left(\dfrac{u-b-c}{a}\right), \end{array} \tag{9.33}$$

where a, b and c are arbitrary constants. The meaning of equations (9.33) is that if we replace them in (9.32), we obtain the same $F(x, y)$, no matter the values of a, b and c we choose. Thus, constants a, b and c are not identifiable. Thus, to have uniqueness of solution we need to fix only functions f_1, f_2 and f_3 at a point. Note that these three conditions allow the values of a, b and c to be determined.

Step 5 (Data collection). For the learning to be possible, some information is required. In this case, the learning problem is really nothing more than estimating these functions from a data set consisting of triplets $\{(x_{1j}, x_{2j}, x_{3j}) | x_{3j} = F(x_{1j}, x_{2j}); j = 1, \ldots, n\}$.

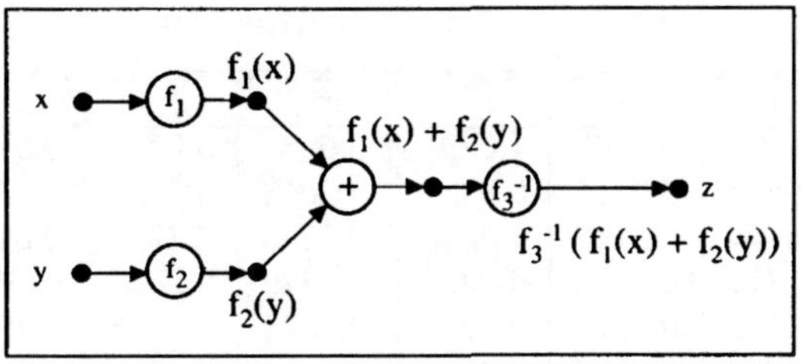

Figure 9.7: Functional network associated with the uniqueness Equation (9.30).

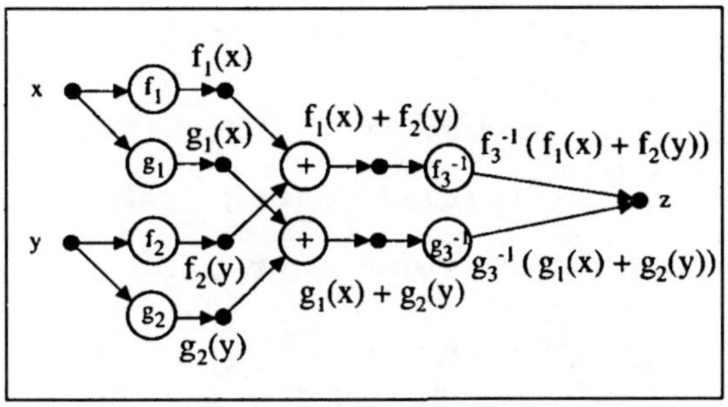

Figure 9.8: Functional network associated with the uniqueness Equation (9.31).

As an illustrative example consider the data in Table 9.4 obtained from an unknown operator given by the function $F(x, y)$, i.e., $x_{3j} = F(x_{1j}, x_{2j})$, $j = 1, \ldots, 14$.

Table 9.4: Set of simulated data.

x_{1j}	x_{2j}	x_{3j}	x_{1j}	x_{2j}	x_{3j}
0.9078	0.3165	1.0991	0.0281	0.6757	0.5170
0.0707	0.0285	0.0331	0.2018	0.5162	0.4569
0.6604	0.8147	1.0320	0.5256	0.3907	0.6061
0.8771	0.0558	0.8236	0.5379	0.2867	0.5414
0.4151	0.8210	0.7717	0.5973	0.4516	0.7294
0.4409	0.3413	0.4880	0.5810	0.9317	0.9960
0.5331	0.0248	0.3087	0.5529	0.2561	0.5337

Step 6 (Learning the Model). In this step, the neural functions must be estimated (learned). Note that in this case, learning the function $F(x, y) = f_3^{-1}[f_1(x) + f_2(y)]$ is equivalent to learning the functions $f_1(x), f_2(x)$ and $f_3(x)$. This can be done by considering the following equivalence:

$$x_{3j} = f_3^{-1}(f_1(x_{1j}) + f_2(x_{2j})) \Leftrightarrow f_3(x_{3j}) = f_1(x_{1j}) + f_2(x_{2j}); \;\; j = 1, \ldots, n.$$

This suggests approximating each of the functions f_s in (9.30) by considering a linear combination of known functions from a given family $\phi_s = \{\phi_{s1}, \ldots, \phi_{sm_s}\}$, $s = 1, 2, 3$. Then, we have

$$\hat{f}_s(x) = \sum_{i=1}^{m_s} a_{si}\phi_{si}(x); \quad s = 1, 2, 3.$$

Then, the error can be measured by

$$e_j = \hat{f}_1(x_{1j}) + \hat{f}_2(x_{2j}) - \hat{f}_3(x_{3j}); \;\; j = 1, \ldots, n. \tag{9.34}$$

Thus, we minimize the sum of square errors

$$Q = \sum_{j=1}^{n} e_j^2 = \sum_{j=1}^{n} \left(\sum_{s=1}^{3} \sum_{i=1}^{m_s} a_{si}\phi_{si}(x_{sj}) \right)^2. \tag{9.35}$$

Note that, for the sake of simplicity, the negative sign associated with the function f_3 in (9.34) has been included in the coefficients a_{3i}. As we have shown before, some auxiliary functional conditions have to be given in order to have uniqueness of solution for (9.30). In this case we consider the auxiliary conditions

$$\hat{f}_k(x_0) \equiv \sum_{i=1}^{m_k} a_{ki}\phi_{ki}(x_0) = \alpha_k \;\; ; \;\; k = 1, 2, 3, \tag{9.36}$$

where $\alpha_k,\ k = 1,2,3$ and x_0 are given constants.

Using the Lagrange multipliers we define the auxiliary function

$$Q_\lambda = \sum_{j=1}^{n}\left(\sum_{s=1}^{3}\sum_{i=1}^{m_s} a_{si}\phi_{si}(x_{sj})\right)^2 + \sum_{k=1}^{3}\lambda_k\left(\sum_{i=1}^{m_k} a_{ki}\phi_{ki}(x_0) - \alpha_k\right).$$

The minimum can be obtained by solving the following system of linear equations, where the unknowns are the coefficients a_{si} and the multipliers λ_t:

$$\left\{\begin{array}{l} \dfrac{\partial Q_\lambda}{\partial a_{tr}} = 2\displaystyle\sum_{j=1}^{n}\left(\sum_{s=1}^{3}\sum_{i=1}^{m_s} a_{si}\phi_{si}(x_{sj})\right)\phi_{tr}(x_{jt}) + \lambda_t\phi_{tr}(x_0) = 0; \\ \qquad\qquad\qquad\qquad\qquad\qquad t = 1,2,3\ ;\ r = 1,\ldots,m_t, \\ \dfrac{\partial Q_\lambda}{\partial \lambda_t} = \displaystyle\sum_{i=1}^{m_t} a_{ti}\phi_{ti}(x_0) - \alpha_t = 0;\ \ t = 1,2,3, \end{array}\right. \tag{9.37}$$

which is a system of $3m+3$ linear equations with $3m+3$ unknowns. Assuming the sets of approximating functions $\{\phi_{ij}; i = 1,\ldots,m\}$; $j = 1,2,3$ to be linearly independent, the system matrix is non-singular. Thus, it has a unique solution. This is a clear advantage of functional networks over standard neural networks, where the estimation leads to non–linear functions with many relative minima.

An interesting simplification consists of assuming f_3 to be a linear function $f_3(x) = a_{31}x + a_{32}$. In this case, the error can be measured directly in the variable scale in terms of

$$e_j = x_{3j} - \frac{\hat{f}_1(x_{1j}) - \hat{f}_2(x_{2j})}{a_{31}} - a_{32}\ ;\ j = 1,\ldots,n. \tag{9.38}$$

Then, the learning process reduces to solving the system of linear equations given in (9.37).

Example 9.3 (Estimation of a model). Let us consider again the data shown in Table 9.4 and apply the linear method to estimate a functional model. Suppose that we use the family of functions

$$(\phi_{s1}(x), \phi_{s2}(x), \phi_{s3}(x), \phi_{s4}(x)) = (1, x, x^2, \log(1+x));\ \ s = 1,2,3,$$

for all the neuron functions, and the conditions

$$(\alpha_1, \alpha_2, \alpha_3) = (1, \log 2, -1).$$

Then, solving the system of equations (9.37) using any standard algorithm we obtain

$$\begin{array}{llllllll} a_{11} & = 0; & a_{12} & = 0; & a_{13} & = 1; & a_{14} & = 0; \\ a_{21} & = 0; & a_{22} & = 0; & a_{23} & = 0; & a_{24} & = 1; \\ a_{31} & = 0; & a_{32} & = 1; & a_{33} & = 0; & a_{34} & = 0, \end{array}$$

which gives the exact function used to simulate the data in Table 9.4:

$$x_{3j} = F(x_{1j}, x_{2j}) = x_{1j}^2 + \log(1 + x_{2j});\ \ j = 1,\ldots,n \tag{9.39}$$

To illustrate the unidentifiable character of the constants a, b and c in (9.33), we consider the same case above but using

$$(\alpha_1, \alpha_2, \alpha_3) = (1, 1, 1).$$

Then, the a coefficients become

$$\begin{array}{llllllll} a_{11} = -3.328; & a_{12} = 0.000; & a_{13} = 4.328; & a_{14} = 0.000; \\ a_{21} = -2.000; & a_{22} = 0.000; & a_{23} = 0.000; & a_{24} = 4.328; \\ a_{31} = -5.328; & a_{32} = 4.328; & a_{33} = 0.000; & a_{34} = 0.000, \end{array}$$

which show that both solutions satisfy (9.33) with $a = -3.328, b = -2$ and $c = 4.328$. ■

Example 9.4 (Another choice from the family of functions). In the previous example we have considered the natural family of functions for estimating the neuron functions obtaining, therefore, the exact model. However, we can also consider any other family such as

$$(\phi_{s1}(x), \phi_{s2}(x), \phi_{s3}(x)) = (1, x, x^2); \quad s = 1, 2, 3,$$

for all the neuron functions. In this case, we shall obtain an approximate model. The a coefficients become

$$\begin{array}{lll} a_{11} = -3.369; & a_{12} = 0.015; & a_{13} = 4.355; \\ a_{21} = -1.963; & a_{22} = 3.954; & a_{23} = -0.991; \\ a_{31} = -5.352; & a_{32} = 4.308; & a_{33} = 0.043, \end{array}$$

With the aim of checking the validity of the obtained model, the RMS error was measured directly on the variable scale by inverting the function f_3. We obtained the RMS and maximum errors of 0.001 and 0.003, respectively, indicating a good approximation of the actual model. ■

One alternative to the preceding approach consists in assuming that $f_3(x) = x$ and then, we have

$$x_{3j} = f_1(x_{1j}) + f_2(x_{2j}); \quad j = 1, \ldots, n. \tag{9.40}$$

Thus, the error of the approximation can be measured by

$$e_j = \hat{f}_1(x_{1j}) + \hat{f}_2(x_{2j}) - x_{3j}; \quad j = 1, \ldots, n. \tag{9.41}$$

To estimate the coefficients $\{a_i; i = 1, \ldots, m\}$, we minimize the sum of square errors

$$Q = \sum_{j=1}^{n} e_j^2 = \sum_{j=1}^{n} \left(\sum_{i=1}^{m} \sum_{s=1}^{2} a_{si}\phi_{si}(x_{sj}) - x_{3j} \right)^2 \tag{9.42}$$

subject to

$$\hat{f}_1(x_1) \equiv \sum_{i=1}^{m} a_{1i}\phi_{1i}(x_0) = \alpha, \tag{9.43}$$

where x_0, x_1 and α are arbitrary but given real constants, which are necessary to identify the otherwise unidentifiable constant a in (9.33).
Thus, using the Lagrange multipliers technique we build the auxiliary function

$$Q_\lambda = \sum_{j=1}^{n} \left(\sum_{i=1}^{m} \sum_{s=1}^{2} a_{si}\phi_{si}(x_{sj}) - x_{3j} \right)^2 + \lambda \left(\sum_{i=1}^{m} a_{1i}\phi_{1i}(x_0) - \alpha \right). \tag{9.44}$$

The minimum corresponds to

$$\begin{aligned} \frac{\partial Q_\lambda}{\partial a_{1r}} &= 2\sum_{j=1}^{n} e_j\phi_{1r}(x_{1j}) + \lambda\phi_{1r}(x_0) = 0, \\ & \qquad r = 1, \ldots, m, \\ \frac{\partial Q_\lambda}{\partial a_{2r}} &= 2\sum_{j=1}^{n} e_j\phi_{2r}(x_{2j}) = 0, \\ & \qquad r = 1, \ldots, m, \\ \frac{\partial Q_\lambda}{\partial \lambda} &= \sum_{i=1}^{m} a_{1i}\phi_{1i}(x_0) - \alpha = 0, \end{aligned} \tag{9.45}$$

which is a system of $2m+1$ linear equations with $2m+1$ unknowns. Assuming the sets of approximating functions $\{\phi_{ij}; i = 1, \ldots, m\}$; $j = 1, 2, 3$ to be linearly independent, the matrix is non-singular. Thus, it has a unique solution.

Note finally that by using the linear method not only is the speed of learning much higher, but the uniqueness of a single solution (global optimum) is guaranteed.

9.7.3 The separable model

An interesting characteristic of functional networks is that the output of several neuron units can coincide in an intermediate or output unit, indicating that they must be equal. This occurs in the output unit z of the functional network in Figure 9.9, where the outputs of the two neurons in the previous layer coincide. These coincidences represent functional constraints in the model and lead to a functional equation or a system of functional equations. In this example, we have the functional equation

$$z = z(x, y) = \sum_{i=1}^{n} g_i(y) f_i(x) = \sum_{j=1}^{m} h_j(x) k_j(y), \tag{9.46}$$

which simply establishes the coincidence of both outputs.

An interesting family of functional network architectures with many applications is the so called *separable functional network*, which has associated the functional Expression (9.46) and combines the separate effects of input variables. Such a family of functional networks is analyzed in this section.

Step 1 (Statement of the problem). Separable functional networks, such as the one shown in Figure 9.9, arise in several practical problems where we require the output of our network $z = z(x, y)$ to satisfy the two following conditions:

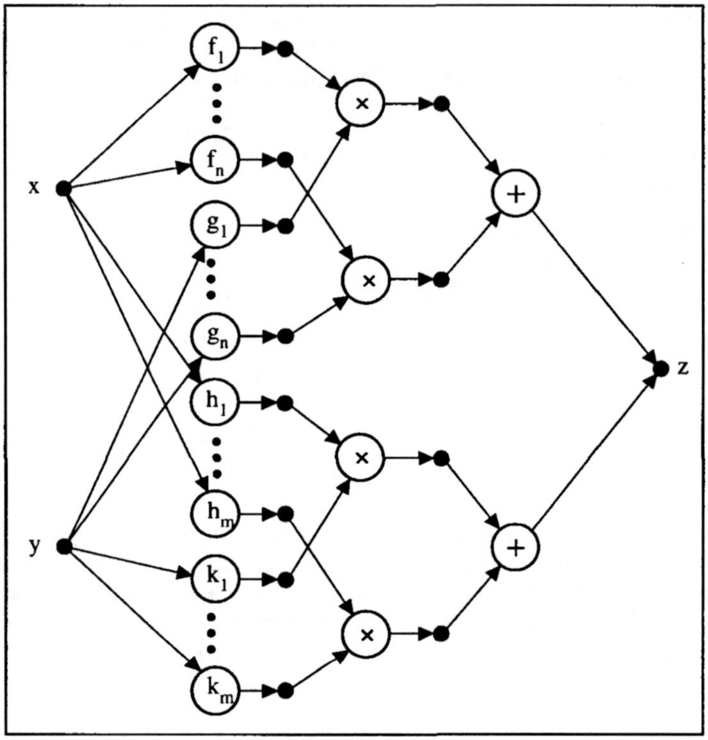

Figure 9.9: The general separable functional network architecture with two inputs and one output.

1. For any value of y, say y_0, the output must be a linear combination of a given set of functions $F = \{f_1(x), \ldots, f_n(x)\}$ with coefficients $G = \{g_1(y_0), \ldots, g_n(y_0)\}$.

2. For any value of x, say x_0, the output must be a linear combination of a given set of functions $K = \{k_1(y), \ldots, k_m(y)\}$ with coefficients $H = \{h_1(x_0), \ldots, h_m(x_0)\}$.

This property generalizes the class of models which combine the separate contribution of each of the independent variables $z = z(x, y) = f(x)g(y)$, but it holds in a large class of theoretical and practical dynamical systems.

Then Equation (9.46) itself represents the coincidence of both advantageous representations, and allows us to determine the form of the coefficients functions G and H when the sets F and K are given. However, many other possibilities exist because any subsets of the sets of functions F, G, H and K can be assumed known, and the relations among all of them must be determined for (9.46) to hold. This problem is solved in the Step 3.

Step 2 (Initial topology). The functional equation above immediately suggests the initial topology of the network (see Figure 9.9). As we already know this is called initial because it comes from the initial statement of the problem and because normally it can be simplified leading to a much simpler network (the final topology).

Step 3 (Simplification). Letting

$$\begin{array}{rcl} f_i(x) & = & h_{i-n}(x) \\ g_i(y) & = & -k_{i-n}(y) \end{array} \; ; \; i = n+1, \ldots, n+m. \tag{9.47}$$

Equation (9.46) can be written as

$$\sum_{i=1}^{k} f_i(x) g_i(y) = 0, \tag{9.48}$$

where $k = n + m$.

Without loss of generality, we can assume the two sets $\{f_1(x), \ldots, f_r(x)\}$ and $\{g_{r+1}(x), \ldots, g_k(x)\}$ to be sets of linearly independent functions. If they are not, we can write a new equation of the form (9.48) with a smaller value of k. Then, the general solution of (9.48) is (see Theorem 4.5):

$$\begin{array}{rclr} f_j(x) & = & \sum\limits_{s=1}^{r} a_{j-r,s} f_s(x), & j = r+1, \ldots, k, \\ g_s(y) & = & -\sum\limits_{j=r+1}^{k} a_{j-r,s} g_j(y), & s = 1, \ldots, r, \end{array} \tag{9.49}$$

where $a_{i,j}$, $i = 1, \cdots, k-r$; $j = 1, \cdots, r$ are arbitrary constants.

Replacing this in (9.47) we get

$$h_i(x) = \begin{cases} f_{i+n}(x) & \text{if} \quad 1 \le i \le r-n, \\ \sum\limits_{s=1}^{r} a_{i+n,s} f_s(x) & \text{if} \quad r-n < i \le m, \end{cases} \tag{9.50}$$

and

$$k_i(y) = \begin{cases} \sum\limits_{j=r+1}^{k} a_{j-r,i+n} g_j(y) & \text{if} \quad 1 \le i \le r-n, \\ -g_{i+n}(y) & \text{if} \quad r-n < i \le m. \end{cases} \tag{9.51}$$

This gives the desired relations between the sets F, G, H and K, and allows us to solve a wide range of problems playing with the values of r. For example, for $r = n$ we can assume that the sets of functions G and K are known, and determine the coefficient functions G and H (the problem stated above), but different values of r solve other interesting problems.

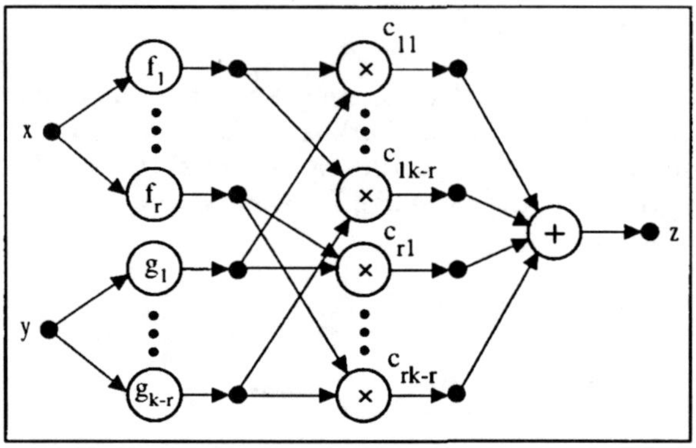

Figure 9.10: Functional network equivalent to the functional network in Figure 9.9.

Note also that, due to the symmetric role of the pairs of functions (F, G) and (K, H) in (9.46) we can replace their roles in (9.49), (9.50) and (9.51) and extend our conclusions.

Finally, replacing (9.49) in (9.46) we get

$$z = z(x, y) = \sum_{i=1}^{r} \sum_{j=1}^{k-r} c_{ij} f_i(x) g_j(y), \tag{9.52}$$

where c_{ij} are constants (parameters of the model).

The beauty of the functional network method is that we have not arbitrarily chosen (9.52), as a black box, but as the result of our desire for the solution $z(x, y)$ to be expressed in two compatible forms:

1. as a linear combination of the functions in F for each value of y, and

2. as a linear combination of the functions in K for each value of x.

Note that Equation (9.52) leads to a much simpler network (see Figure 9.10) than the initial network, because now the solution depends only on $k = n + m$ functions, while before there were $2k$ unknown functions. However, now we must learn the $r(k - r)$ constants c_{ij}; $i = 1, \ldots, r$; $j = 1, \ldots, k - r$.

Step 4 (Uniqueness of representation). In the uniqueness of representation problem, conditions for the neural functions of the simplified functional network must be obtained. In the case of the separable model, two cases can be considered:

1. *The $f_i(x)$ and $g_i(y)$ functions are given*: Assume that we have two con-

stant matrices C and C^* such that

$$z = F(x,y) = \sum_{i=1}^{r}\sum_{j=1}^{k-r} c_{ij} f_i(x) g_j(y) = \sum_{i=1}^{r}\sum_{j=1}^{k-r} c^*_{ij} f_i(x) g_j(y). \tag{9.53}$$

Then, we can write

$$\sum_{i=1}^{r}\sum_{j=1}^{k-r} (c_{ij} - c^*_{ij}) f_i(x) g_j(y) \equiv 0. \tag{9.54}$$

Since the sets of functions

$$\{f_i(x) | i = 1, \ldots, r\},$$

and

$$\{g_j(y) | j = 1, \ldots, k-r\},$$

are linearly independent, so is the set

$$\{f_i(x) g_j(y);\ \ i = 1, \ldots, r;\ \ j = 1, \ldots, k-r\},$$

and thus,

$$c_{ij} = c^*_{ij};\ \ i = 1, \ldots, r;\ \ j = 1, \ldots, k-r,$$

which proves that the representation in (9.52) is unique.

2. *The $f_i(x)$ and $g_i(y)$ functions are to be learned (unknown)*: in this case we approximate the functions $f_i(x), g_j(y)$ as follows:

$$\begin{aligned} f_i(x) &= \sum_{\ell_i=1}^{n_i} a_{i\ell_i} \phi_{\ell_i}(x), \\ g_j(y) &= \sum_{\ell_j=1}^{n_j} b_{j\ell_j} \psi_{\ell_j}(y). \end{aligned} \tag{9.55}$$

With this, (9.52) transforms to

$$z = F(x,y) = \sum_{i=1}^{n_i}\sum_{j=1}^{n_j} d_{ij} \phi_i(x) \psi_j(y), \tag{9.56}$$

which is of the same form as (9.52), but with the sets of functions $\phi_i(x)$ and $\psi_j(y)$ instead of the set of functions $f_i(x)$ and $g_j(y)$. Thus, this second case is reduced to the first.

Step 5 (Data collection). Suppose we are given the data set consisting of the triplets shown in Table 9.5 and we are interested in estimating a representative

Table 9.5: Triplets (x_0, x_1, x_2) obtained from a function $x_0 = f(x_1, x_2)$.

x_0	x_1	x_2	x_0	x_1	x_2
0.880	0.384	0.439	1.072	0.136	0.062
1.117	0.976	0.440	1.369	0.801	0.263
0.299	0.309	0.922	1.449	0.457	0.006
0.873	0.725	0.558	1.214	0.377	0.143
0.359	0.449	0.907	0.604	0.818	0.727
0.935	0.058	0.122	0.669	0.799	0.689
0.681	0.673	0.682	0.760	0.663	0.627
1.336	0.697	0.242	1.228	0.862	0.364
1.019	0.387	0.320	0.989	0.405	0.357
0.566	0.662	0.762	0.814	0.028	0.214

separable model of the form (9.46). For instance, if we choose the simplest case, $k = 2$ and $r = 1$ we get a functional network of the form:

$$x_0 = F(x_1, x_2) = f(x_1)g(x_2). \tag{9.57}$$

Step 6 (Learning the Model). In this case, the problem of learning the functional network associated with (9.46) does not involve auxiliary functional conditions and, therefore, a simple least squares method allows us to obtain the optimal coefficients c_{ij} from the available data consisting of triplets $\{(x_{0i}, x_{1i}, x_{2i});\ i = 1, \ldots, n\}$, where x_0, x_1 and x_2 refer to z, x and y, respectively. Then, the error can be measured by

$$e_i = x_{0i} - \sum_{i=1}^{r} \sum_{j=1}^{s-r} c_{ij} f_i(x_{1i}) g_j(x_{2i});\ \ i = 1, \ldots, n. \tag{9.58}$$

Thus, to find the optimum coefficients we minimize the sum of square errors $Q = \sum_{k=1}^{n} e_k^2$. In this case, the parameters are not constrained by extra conditions, so the minimum can be obtained by solving the following system of linear equations, where the unknowns are the coefficients c_{ij}:

$$\frac{\partial Q}{\partial c_{pq}} = 2 \sum_{k=1}^{n} e_k f_p(x_{1k}) g_q(x_{2k}) = 0;\ \ p = 1, \ldots, r;\ \ q = 1, \ldots, r - s. \tag{9.59}$$

Then, we can consider a polynomial family of functions, say $\{1, x, x^2\}$ for each of the functions, and obtain the optimal coefficients for approximating the given data.

Thus, to estimate the coefficients we solve the system (9.59), where the coefficients are the unknowns. Then, we get the model $x_0 = f(x_1, x_2) = 1 + x_1 - x_2 - x_1^2 x_2$, which was used to generate the data shown in Table 9.5.

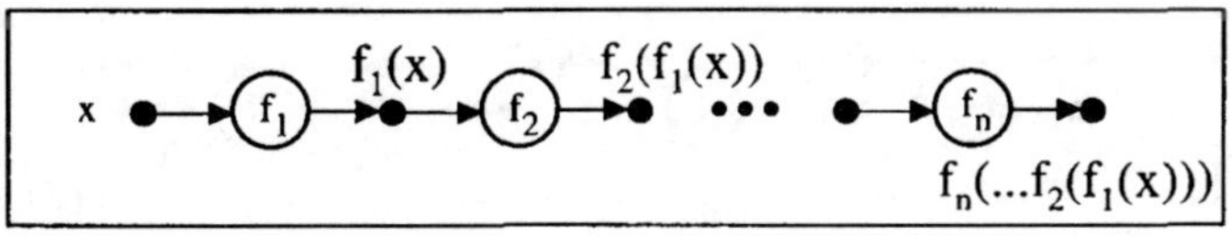

Figure 9.11: A serial functional network.

9.7.4 Serial functional model

Let us now center our attention on *serial functional networks*, which are sequences of one-layer functional units, as shown in Figure 9.11.

Step 1 (Statement of the problem) The output of a serial functional network can be written as:

$$y_{n+1} = f_n(f_{n-1}(\dots f_1(y_1)\dots)). \tag{9.60}$$

Of special interest is the simple case where $f_i = f, \forall i$; that is, when all neurons are identical. Then, Equation (9.60) becomes:

$$y_{n+1} = f(f(\dots f(y_1)\dots)) = f^{(n)}(y_1). \tag{9.61}$$

where $f^{(n)}(x)$ is the n-th f-iterate of f.

This case was analyzed in Aczél (1966) and applied to functional nets by Gómez-Nesterkín (1997), who reduced this equation to the translation functional equation.

Step 2 (Initial topology). The initial topology of the functional network is shown in Figure 9.11.

Step 3 (Simplification of the Model). Calling $F(x,n) = f^{(n)}(x)$ to the n-th f-iterate, it is clear that

$$F(x, m+n) = f^{(n+m)}(x) = f^{(n)}(f^{(m)}(x)) = F(F(x,m),n), \tag{9.62}$$

which proves that $F(x,n)$ satisfies the translation equation, the solution of which, as given by Theorem 5.10, is

$$f^{(n)}(x) = F(x,n) = g^{-1}[g(x)+n]. \tag{9.63}$$

This implies

$$f(x) = F(x,1) = g^{-1}[g(x)+1] \Leftrightarrow g(f(x)) = g(x)+1. \tag{9.64}$$

Thus, the problem of finding the n-th iterate is now a question of solving the equivalent functional equation

$$g(f(x)) = f(x+1), \tag{9.65}$$

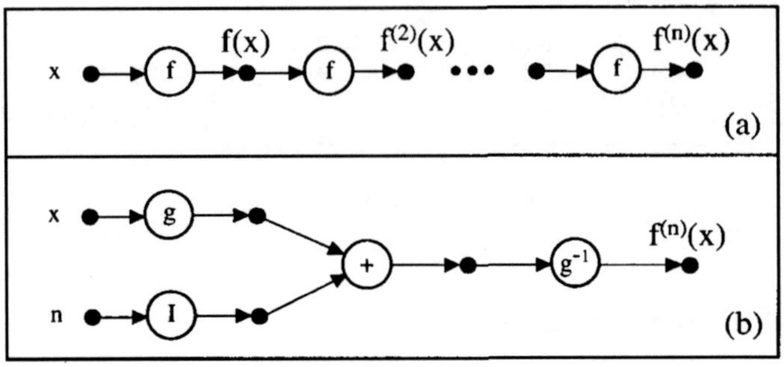

Figure 9.12: (a) Illustration of the n-th g-iterate and, (b) the associated iterator functional network.

or

$$f^{-1}(g(x)) = f^{-1}(x) + 1, \tag{9.66}$$

which is a particular case of the Abel Equation (see Section 5.4).

According to the above, there exists a correspondence between g and f. Given an invertible function $g(x)$, we can obtain $f(x)$ using (9.64) and its n-th iterate by (9.63). However, the problem of obtaining $g(x)$ when $f(x)$ is known is much more involved. An example is used below to illustrate how an approximate solution can be obtained.

Step 4 (Uniqueness of Representation). The uniqueness problem for the solution of the translation equation is given in Theorem 5.11. If

$$g^{-1}[g(x) + 1] = h^{-1}[h(x) + 1], \tag{9.67}$$

then

$$g(u) = h(u) + c, \tag{9.68}$$

where c is an arbitrary constant.

Step 5 (Data collection). Consider the function

$$f(x) = \log\left[1 + \exp(x)\right] \tag{9.69}$$

and assume that we are interested in calculating its n-th f-iterate $F(x, n) = f^{(n)}(x)$. To this end, according to (9.68), we can use the functional network in Figure 9.12(b).

To estimate the g function we have the data pairs in Table 9.6, where $y_t = f(x_t)$, and consider

$$g(x) = \sum_{i=1}^{8} c_i x^i, \tag{9.70}$$

Table 9.6: (x_t, y_t) data used to fit the approximate $g(x)$ function.

x_t	y_t	x_t	y_t	x_t	y_t
0.010	0.698	0.428	0.930	0.415	0.922
0.580	1.020	0.187	0.791	0.198	0.797
0.696	1.100	0.866	1.220	0.310	0.860
0.310	0.860	0.906	1.250	0.971	1.290
0.305	0.857	0.296	0.852	0.242	0.822
0.646	1.070	0.575	1.020	0.536	0.997
0.191	0.793	0.635	1.060	0.017	0.702
0.820	1.190	0.340	0.877	0.304	0.857
0.007	0.697	0.392	0.908	0.925	1.260
0.724	1.120	0.820	1.180	0.194	0.795

that is, we approximate $g(x)$ by a fourth degree polynomial. Note that the constant coefficient c_0, because of Theorem 5.11, can be assumed to be zero.

Step 6 (Learning the Model). From (9.63) we see that learning the function $f^{(n)}$ is now a question of estimating the function g from a given set of data pairs (x_i, y_i), $i = 1, \ldots, n$, where $y_i = f(x_i)$. From (9.67) we see that

$$g(y_i) = g(x_i) + 1; \quad \forall i = 1, \ldots, n. \tag{9.71}$$

Thus, we consider a linear combination of known functions

$$g(x) = \sum_{i=1}^{m} c_i \phi_i(x). \tag{9.72}$$

Then, the error for each data point in (9.71) is

$$e_i = g(y_i) - g(x_i) - 1 = \sum_{j=1}^{m} c_j(\phi_j(y_i) - \phi_j(x_i)) - 1. \tag{9.73}$$

Thus, to estimate the coefficients $c_1, \ldots, c_m$, we minimize

$$Q = \sum_{k=1}^{n} \left[\sum_{j=1}^{m} c_j(\phi_j(y_i) - \phi_j(x_i)) - 1 \right]^2, \tag{9.74}$$

which, in our example, means minimizing

$$Q = \sum_{t=1}^{20} \left[\sum_{i=0}^{8} c_i(y_t^i - x_t^i) - 1 \right]^2. \tag{9.75}$$

in order to estimate $c_1, \ldots, c_8$.

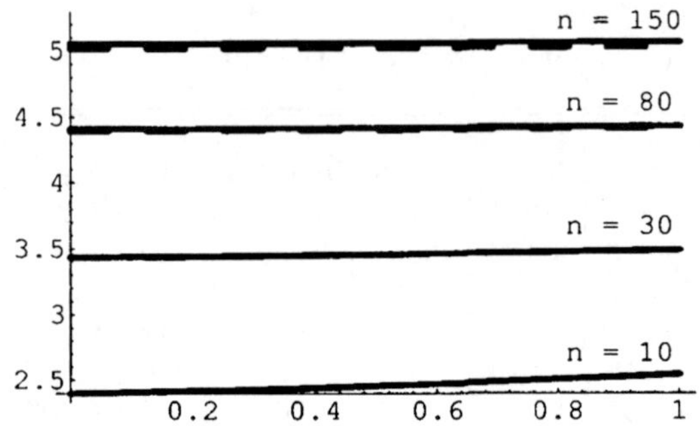

Figure 9.13: Exact (continuous lines) and approximated (dashed lines) iterates of orders 2, 4, 8, 16 and 32.

The resulting polynomial approximation becomes

$$\begin{aligned}\hat{g}(x) &= x + 0.5x^2 + 0.167x^3 + 0.0417x^4 + 0.0083x^5 \\ &\quad +0.00144x^6 + 0.000155x^7 + 0.0000444x^8.\end{aligned} \tag{9.76}$$

with $RMSE = 3.28 \times 10^{-6}$ and $\mathcal{L}_1 = -365.201$.
Figure 9.13 shows the exact (continuous lines) and approximated (dashed lines) f-iterates of orders 2, 4, 8, 16 and 32, where the approximated values have been calculated by (see Equation (9.63))

$$\hat{f}^{(n)}(x) = \hat{g}^{-1}[\hat{g}(x) + n]. \tag{9.77}$$

To illustrate the model selection methods described in Section 9.6, we apply them as follows:

- **The exhaustive method.** If we estimate all 256 possible models that can be obtained using as a basis eight-th degree polynomials, we reach the conclusion that the best model is

  $$\hat{g}(x) = x + 0.5x^2 + 0.165x^3 + 0.047x^4 + 0.0087x^6 - 0.00322x^7 + 0.000682x^8. \tag{9.78}$$

  with $RMSE = 4.3 \times 10^{-8}$ and $\mathcal{L}_1 = -496.792$.

- **The forward–backward method.** In Table 9.7 the forward–backward method is illustrated. In Step 1 we estimate all possible models with a single monomial in their bases. The best model corresponds to monomial x^2 ($RMSE = 0.1118$ and $\mathcal{L}_1 = -64.043$). Next we add one more monomial to the resulting basis. The monomial x is the best possible addition ($RMSE = 0.0248$ and $\mathcal{L}_1 = -107.499$). We continue making the best additions until no further improvement can be obtained. This occurs when the basis is $\{x^2, x, x^4, x^3, x^6, x^5, x^8\}$, with associated $RMSE = 5.5 \times 10^{-8}$ and $\mathcal{L}_1 = -489.394$. Now we run the backward procedure. No monomial

Table 9.7: Approximating functions, RSME and $\mathcal{L}_1$ obtained at different steps for the forward–backward method.

	Approximating functions	Quality measures	
Step	g(x)	RMSE	$\mathcal{L}_1$
1	$\{x^2\}$	1.1×10^{-1}	−64.043
2	$\{x^2, x\}$	2.5×10^{-2}	−107.499
3	$\{x^2, x, x^4\}$	9.7×10^{-4}	−202.926
4	$\{x^2, x, x^4, x^3\}$	8.5×10^{-5}	−274.418
5	$\{x^2, x, x^4, x^3, x^6\}$	1.9×10^{-6}	−386.595
6	$\{x^2, x, x^4, x^3, x^6, x^5\}$	1.0×10^{-7}	−472.299
7	$\{x^2, x, x^4, x^3, x^6, x^5, x^8\}$	5.5×10^{-8}	−489.394

Table 9.8: Approximating functions, RSME and $\mathcal{L}_1$ obtained at different steps for the backward–forward method.

	Approximating functions	Quality measures	
Step	g(x)	RMSE	$\mathcal{L}_1$
1	$\{x, x^2, x^3, x^4, x^5, x^6, x^7, x^8\}$	$3.28\ 10^{-6}$	−365.201
2	$\{x, x^2, x^3, x^4, x^6, x^7, x^8\}$	4.30×10^{-8}	−496.792

can be removed with improvement of the value of $\mathcal{L}_1$. Thus, the final model is

$$\hat{g}(x) = x+0.5x^2+0.167x^3+0.042x^4+0.0079x^5+0.00176x^6+0.00007377x^8. \tag{9.79}$$

with an $RMSE = 5.5 \times 10^{-8}$ and $\mathcal{L}_1 = -489.394$.

- **The backward–forward method.** In Table 9.8 the backward–forward method is illustrated. In Step 1 we estimate the complete model with associated $RMSE = 3.28 \times 10^{-6}$ and $\mathcal{L}_1 = -365.201$. Next, we estimate the eight models obtained by removing one monomial from the complete basis. The best option is to remove monomial x^5, with associated $RMSE = 4.3 \times 10^{-8}$ and $\mathcal{L}_1 = -496.792$. When we try to remove another monomial from the resulting basis, we get no improvement in the description length measure. Then, when we try to add a new monomial, we also get no improvement. Consequently, the best model is the one in (9.78); that is, the same as that obtained with the exhaustive method.

9.8 Some applications of functional networks

In this section some interesting examples of the application of the functional network methodology are introduced. First, we apply functional networks to the analysis of a real economic example and the problem of modelling chaotic maps are discussed. Finally, the application of functional networks to two interesting problems in the framework of chaotic dynamic systems, such as the noise reduction and the retrieval of masked information is also analyzed. As shown below, functional networks constitute a powerful tool for solving many different problems arising from time series, economics, dynamic systems, etc.

9.8.1 A real economical example

In this example we use real data from Alegre et al. (1995) corresponding to Spanish economic data. The data is shown in Table 9.9.

Table 9.9: Spanish economic data.

Year	Stock index (Madrid)	Consumer price index	Percentage of savings
1964	103.55	112.67	23.18
1965	108.64	123.22	22.35
1966	111.62	129.71	22.54
1967	111.86	138.22	21.89
1968	146.29	142.19	22.54
1969	215.14	147.07	24.37
1970	192.36	157.03	24.37
1971	220.88	172.18	24.36
1972	291.74	184.81	24.59
1973	328.94	211.06	24.99
1974	294.47	248.80	24.32
1975	306.28	283.85	23.29
1976	218.80	340.00	21.30
1977	147.23	429.79	20.75
1978	131.58	501.01	21.72
1979	109.98	578.82	20.40
1980	116.60	667.03	19.40

To reproduce these series, we use a uniqueness functional network of the type in Section 9.7.2 with $f_3(x) = x$:

$$s_t = f_1(m_t) + f_2(c_t), \tag{9.80}$$

where s_t is the percentage of savings, m_t is the general Madrid stock index, and c_t is the consumer price index.

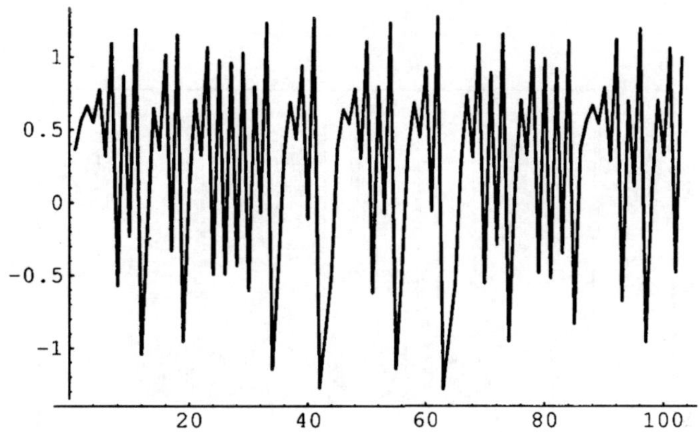

Figure 9.14: Percentage of savings using the functional network for the period 1964-1980 in Spain.

The estimates are obtained using the model (9.40) with the set of functions

$$\{\phi_{s1}(x), \phi_{s2}(x), \phi_{s3}(x), \phi_{s4}(x)\} = \{1, x, \log(x), x^2\}, \quad s = 1, 2. \tag{9.81}$$

The resulting approximating functions are:

$$f_1(x) \quad = \quad 1.09 \quad -0.089x \quad +13.33\log x \quad +0.000109x^2 \tag{9.82}$$

$$f_2(x) \quad = \quad 54.76 \quad +0.109x \quad -20.72\log x \quad -0.0000642x^2, \tag{9.83}$$

leading to a root mean squared error (RMSE) of 0.523661 and a largest prediction error of 1.019.

Figure 9.14 shows the estimated values of the percentage of savings.

The results are very good even though the selected functional network is simple.

9.8.2 Modelling chaotic maps

In this section we describe an application of the separable functional network architecture introduced in Section 9.7.3.

The cubic Holmes map

As a first example, consider the cubic Holmes map given by Holmes (1979):

$$x_n = (px_{n-1} - x_{n-1}^3 - 0.2x_{n-1}). \tag{9.84}$$

This system exhibits a great variety of behaviors when the parameter p is varied. In the following we consider the chaotic system associated with the value $p =$

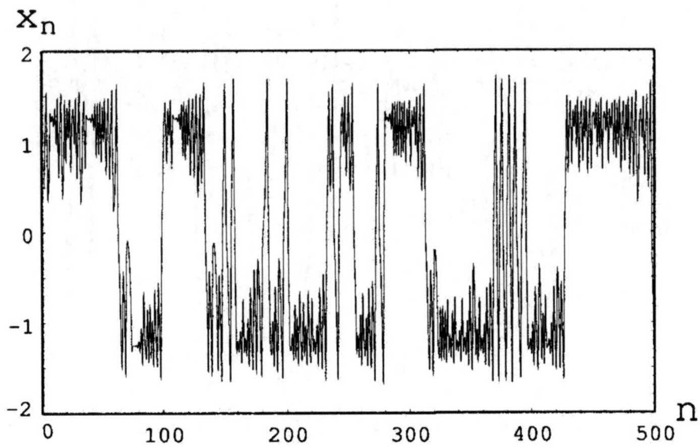

Figure 9.15: Time series for the x coordinate of a chaotic orbit of the Holmes map.

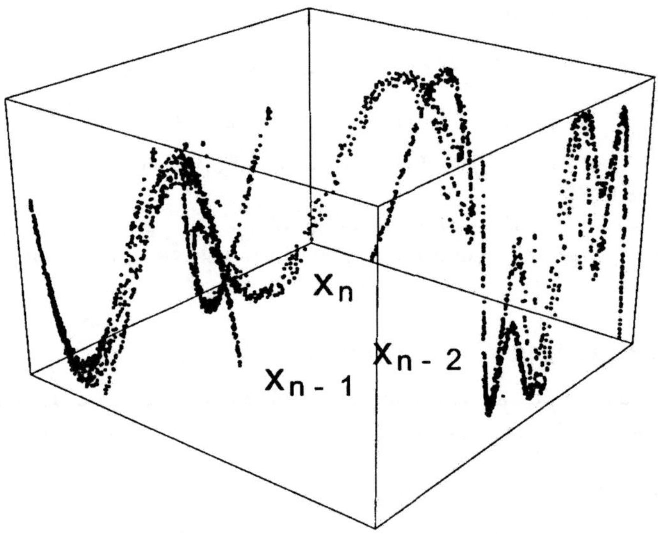

Figure 9.16: Embedding space (x_n, x_{n-1}, x_{n-2}) displaying the cubic relationship between the variables.

Table 9.10: Performance of several Fourier functional networks for the Holmes time series. The number of parameters and the RMSE and maximum errors obtained in each case are shown.

m	Parameters	Training Data		Test Data	
		RMSE	Max	RMSE	Max
1	$(2m)^2 = 4$	0.5300	2.5400	0.6100	2.5700
2	$(2m)^2 = 16$	0.0710	0.3350	0.0725	0.3630
3	$(2m)^2 = 36$	0.0037	0.0250	0.0041	0.0330
4	$(2m)^2 = 64$	$1.1\ 10^{-4}$	$5.5\ 10^{-4}$	$3.2\ 10^{-4}$	$5.3\ 10^{-3}$
5	$(2m)^2 = 100$	$9.3\ 10^{-5}$	$1.5\ 10^{-4}$	$1.2\ 10^{-4}$	$1.5\ 10^{-3}$

2.765. Suppose that we measure a 500 points time series of a single coordinate x corresponding to the initial conditions $x_0 = y_0 = 0.1$ (see Figure 9.15).

In spite of the seemingly stochastic behavior of the time series, which is characteristic of chaotic maps, the embedding space (x_n, x_{n-1}, x_{n-2}) reveals the actual deterministic nature of the system (see Figure 9.16). Suppose that we want to obtain a representative functional network model for the underlying dynamics. Then, we can use the functional network architecture shown in Figure 9.10 with $z = x_n$, and $(x, y) = (x_{n-1}, x_{n-2})$, $n = 3, \ldots, 500$, where each of the triplets is obtained from three consecutive terms of the time series.

Note that due to the cubic polynomial structure of the Holmes map, if we consider the polynomial base family $\{1, x, x^2, x^3\}$ for the functions f_i and g_j in Figure 9.10, then we obtain the exact Holmes map given in (9.84) after solving the corresponding system given in (9.59). However, if we use a different base family for the neuron functions such as a Fourier expansion given by $\{\sin(x), \ldots, \sin(mx), \cos(x), \ldots, \cos(mx)\}$ then we obtain an approximate model. Table 9.10 illustrates the quality of the approximation by giving the Root Mean Square Errors (RMSE) and maximum errors for different values of m for a 500 points time series.

We have also performed several experiments to check the quality of the approximation when varying the size of the data. For example, Table 9.11 shows the results obtained when approximating the Holmes map using the $m = 2$ model data sets of different sizes. In all the cases analyzed we found that small data sizes (of the order of the number of parameters) lead to over-fitting problems, as shown in Table 9.11 when comparing the training and test errors obtained for $n = 100$ (note that the number of parameters of the $m = 2$ model is 16). However, when the set of data is large enough to avoid the over-fitting problem, then the quality of the approximation reaches a stable value and no longer improves when the length of the time series is enlarged. For example, Table 9.11 shows that the models with $n = 500$ and $n = 2000$ have comparable training and test errors, indicating that both models capture the same deterministic structure underlying the two time series. Therefore, while other local methods require long time series to obtain an approximate model, the separable

Table 9.11: Performance of the $m = 2$ Fourier functional network for several time series of size n of the Holmes model. The number of parameters and the RMSE and maximum errors obtained in each case are shown.

	Training Data		Test Data	
n	RMSE	Max	RMSE	Max
100	0.0410	0.1270	0.1150	0.9650
200	0.0630	0.2520	0.0780	0.5890
500	0.0710	0.3350	0.0725	0.3630
2000	0.0700	0.3210	0.0720	0.3430

functional networks presented in this paper can be applied to small time series available in many practical situations.

With the aim of cross-validating the model, a test time series consisting of the next 1000 points of the orbit shown in Figure 9.15 have been considered. Table 9.11 shows the RMSE and maximum errors in this case (under the "Test Data" label). Note that the errors obtained for both the training and the test data are very similar, indicating that no over-fitting is produced during the learning process. Therefore, the resulting model can be used to accurately predict the future behavior of the system (forecasting).

Higher dimensional separable models: the Lorenz system

In Section 9.7.3 we have introduced the separable models considering the particular case of two inputs. However, proceeding in a similar way, it is easy to show that the general form for a separable functional network with inputs $\{x_1, \ldots, x_m\}$ is

$$z = z(x_1, \ldots, x_m) = \sum_{i=1}^{r_1} \cdots \sum_{j=1}^{r_m} c_{i\ldots j} f_i(x_1) \ldots g_j(x_m), \tag{9.85}$$

where $c_{i\ldots j}$ are the parameters of the model and

$$\{f_1, \ldots, f_{r_1}\}, \ldots, \{g_1, \ldots, g_{r_m}\}$$

sets of base functions. Note that the number of parameters of the model increases exponentially with the number of inputs (independent variables). Therefore, for high dimensional systems, the problem of model selection is very important, since it allow us to keep only the relevant terms in (9.85). Some preliminary results show that the minimum description length measure, which penalizes the number of parameters, gives excellent results for high dimensional systems.

In this section we shall use the well known Lorenz model to illustrate both the generalization of the separable model to higher dimensions and the application of this methodology to continuous systems. The Lorenz flow is given by the set of differential equations (see Lorenz (1963)):

$$(\dot{x}, \dot{y}, \dot{z}) = (\sigma(y - x), -xz + rx - y, xy - bz), \tag{9.86}$$

which we study for the parameter values $\sigma = 10$, $b = 8/3$, and $r = 28$. Considering the initial conditions $(x_0, y_0, z_0) = (-10, -5, 35)$ and using a fourth-order Runge-Kutta algorithm with a fixed time step $\Delta t = 10^{-3}$, we recorded a time series consisting of 20000 sample points for each of the variables after discarding the first transitory points of the orbit.

Instead of using the whole data set to train the separable functional network, we only use a random sample consisting of 2000 points out of the 20000. The rest of the points are used to cross-validate the model. As a first example, we use a quadratic polynomial basis family $\{1, x, x^2\}$ for each of the three neuron functions. In this case, after neglecting terms with coefficients smaller than a given threshold value we get a approximate model which accurately describes the dynamics of the Lorenz model:

$$\begin{aligned}
\hat{x} &= 0.990x + 0.0099y - 1.653 \times 10^{-9}x^2y - 4.961 \times 10^{-6}xz - \\
&\quad -1.643 \times 10^{-8}yz \\
\hat{y} &= -1.16 \times 10^{-7} + 0.027x + 1.516 \times 10^{-8}x^2 + 0.999y - \\
&\quad -3.768 \times 10^{-8}xy - 4.97694 \times 10^{-7}x^2y + 1.463 \times 10^{-8}y^2 - \\
&\quad -4.6186 \times 10^{-9}xy^2 + 1.628 \times 10^{-8}z - 9.9310^{-4}xz - \\
&\quad -4.93510^{-6}yz + 1.765 \times 10^{-9}xyz + 3.159 \times 10^{-9}xz^2 \\
\hat{z} &= 3.162 \times 10^{-8} + 1.404 \times 10^{-8}x + 1.382 \times 10^{-5}x^2 - \\
&\quad -5.599 \times 10^{-9}y + 9.933 \times 10^{-4}xy + 4.979 \times 10^{-6}y^2 + 0.997z - \\
&\quad -4.933 \times 10^{-7}x^2z - 5.247 \times 10^{-9}xyz,
\end{aligned} \tag{9.87}$$

which gives RMS and maximum errors $5.96 \times 10^{-8}, 2.25 \times 10^{-7}$ (for x), $5.03 \times 10^{-6}, 2.64 \times 10^{-5}$ (for y), and $6.70 \times 10^{-6}, 2.82 \times 10^{-5}$ (for z), respectively.

The errors obtained when cross-validating the model using the whole time series prove to be similar to the training errors.

Similar results can be obtained using different base functions. For example, if we use for the three neuron functions a trigonometric family of the form $\left\{1, \sin\left(\frac{x}{p}\right), \sin\left(\frac{2x}{p}\right)\right\}$, where p depends on the sampling time Δt of the time series (e.g., $p = 150$), then we obtain an approximate model equivalent to the above obtained quadratic model. In this case the RMS and maximum errors obtained are 1.67×10^{-7} and 9.12×10^{-7} (for x), 5.33×10^{-6} and 2.73×10^{-5} (for y), and 7.40×10^{-5} and 5.38×10^{-4} (for z), respectively.

With the aim of illustrating the prediction capabilities of the obtained models, we use the above approximate Fourier functional model to reconstruct the dynamics of the Lorenz model starting from an arbitrary orbit point. Figure 9.18 compares the actual and reconstructed dynamics and shows a large prediction time for the model.

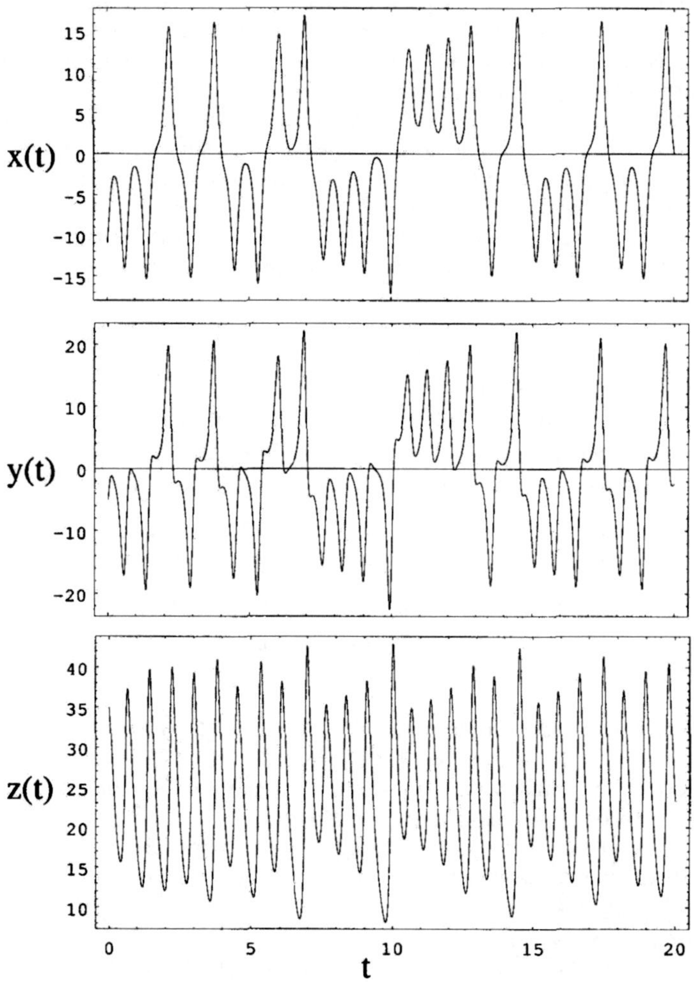

Figure 9.17: Time series of Lorenz system.

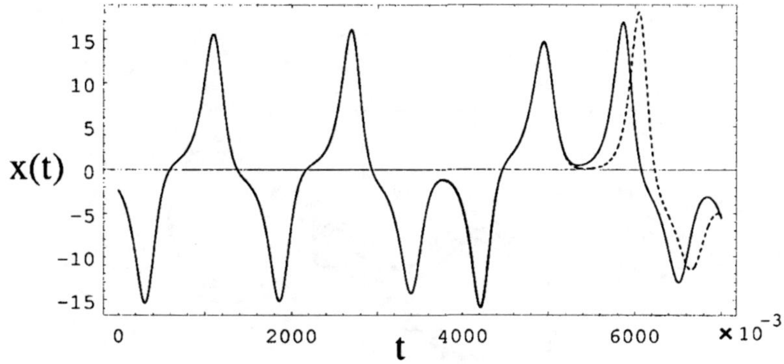

Figure 9.18: Actual (dashed line) and reconstructed (solid line) time series of Lorenz system.

Finally, we want to remark here that similar results are obtained when considering the time series of a single variable and using a delay reconstruction space calculated from the time series using some standard algorithm (Grassberger and Procaccia (1983)). We have checked the cases $d = 3$ and $d = 4$, obtaining similar results to those presented in this section.

9.8.3 Noise reduction

In this section we consider a simple application of the functional network methodology to reduce the observational noise contained in the data. Then, we consider the dependent variable z to be the combination of a deterministic component $z(x, y)$ plus a normally distributed noise with zero mean ξ_n, i.e., $z_n = z(x_n, y_n) + \xi_n$, although similar results have been obtained when considering other noise sources, such as multiplicative or dynamical noise (see Hamel (1990)). The learning algorithm described in Step 6 of Section 9.7.3 is a straightforward technique to reduce the noise contained in the data, since it is based on a global least squares method. So, when estimating the coefficients of the network functions, the noise contributions cancel out and the model automatically fits to the deterministic structure underlying the time series. Therefore, the value $\hat{z}_n$ predicted by the functional network for the input values x_n and y_n will give an estimation of the deterministic structure of the time series $z_n = z(x_n, y_n)$.

Figure 9.19 shows a noisy orbit computed by adding normally distributed noise with $\sigma = 0.1$ to the time series. The resulting functional network corresponding to a polynomial base function (i.e., the natural family of functions for the model) does not contain any contribution from the noise and, then, this is completely removed from the time series (see Figure 9.20).

If we use an approximate functional network considering, for example, a Fourier basis for the neuron functions, then it can be seen that when moderate

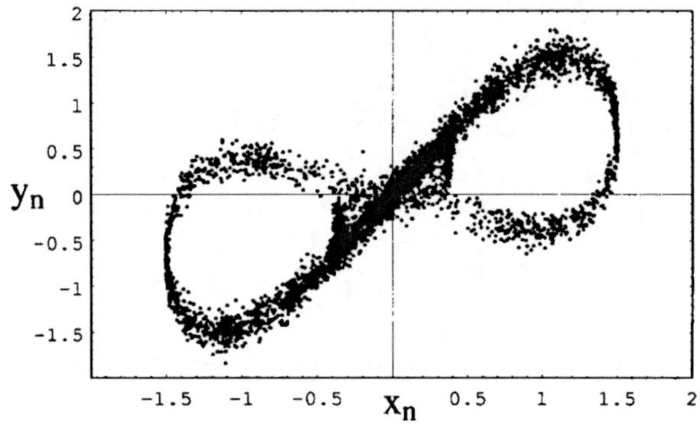

Figure 9.19: Phase space for a noisy time series of the Burger map with added normally distributed noise with $\sigma = 0.1$.

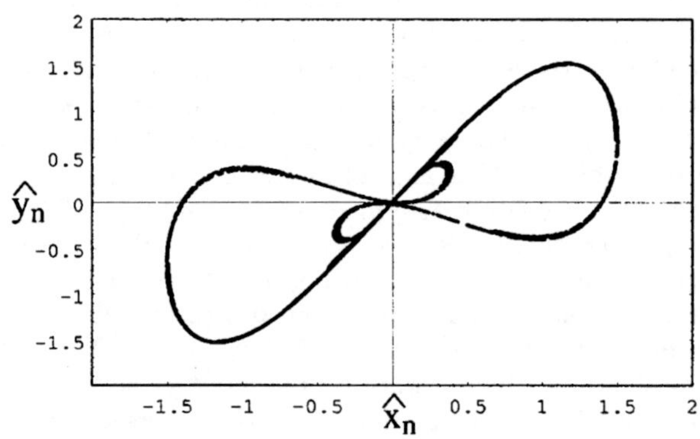

Figure 9.20: Phase space for the time series cleaned with a polynomial functional network.

m	Par.	Training Data		Test Data	
		RMSE	Max	RMSE	Max
1	4	0.178	0.515	0.167	0.517
2	16	$7.94\ 10^{-3}$	0.027	$7.61\ 10^{-3}$	0.029
3	36	$3.51\ 10^{-4}$	0.001	$3.93\ 10^{-4}$	0.002

Table 9.12: Performance of several Fourier functional networks for the Burger time series. The number of parameters and the RMSE and maximum errors obtained in each case are shown.

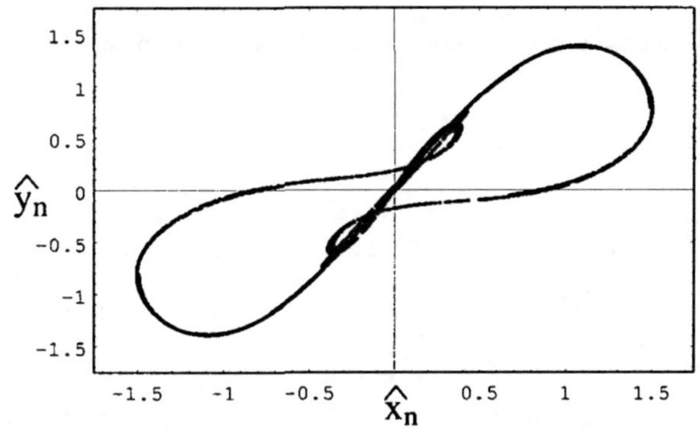

Figure 9.21: Phase space for the time series cleaned with a Fourier functional network with $m = 1$.

noise is added to the time series, the noise can also be cleaned off from the system and the actual deterministic dynamics be recovered. We have used the three approximate models shown in Table 9.12 to illustrate the quality of the resulting model for different values of m. For $m = 2$ and $m = 3$ the noise is also cleaned off from the orbit (see Figure 9.20). However, when we consider the case $m = 1$, which has associated errors larger than the noise intensity, then the resulting orbit still contains contributions from the noise and presents some differences with the original one (see Figure 9.21).

9.8.4 Retrieval of masked information

In this section we discuss the possibility of extracting information masked by chaos using the above functional models. The main idea of secure communications based on chaos consists of using a chaotic signal with broad-band power spectrum to mask a given message (see Pecora (1993) and references therein for a survey). The ability of two similar chaotic systems to synchronize is then used to recover the information masked in the chaotic carrier (see scheme in Figure 9.22).

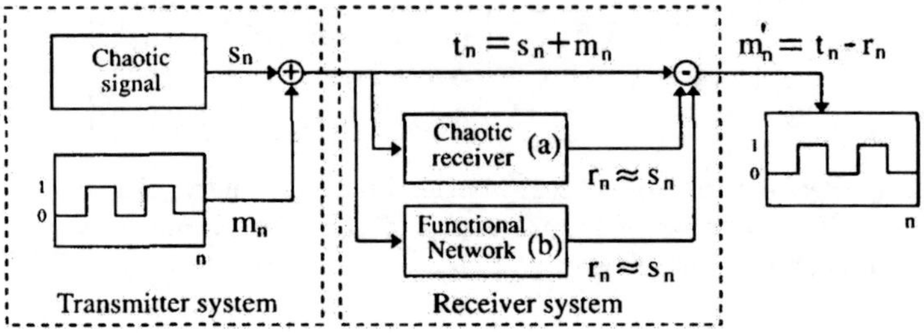

Figure 9.22: Scheme for secure communications based on chaos synchronization.

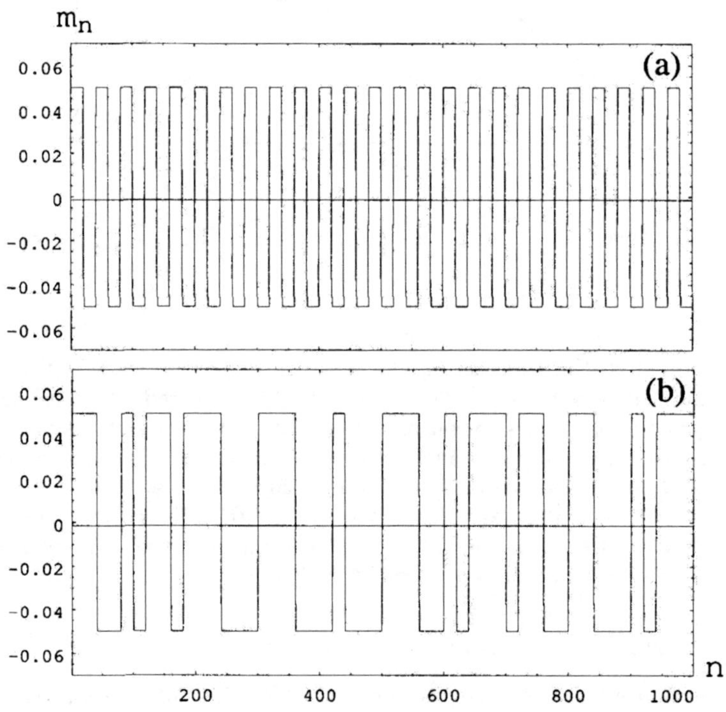

Figure 9.23: (a) Square bit stream message where each bit is represented by 20 consecutive sequence points and (b) random bit stream with 40 sequence terms for each bit.

Table 9.13: Performance of several polynomial and Fourier functional networks for the Burger time series with inserted message and different noise levels.

Noise	Polynomial		Fourier	
	RSME	MAX	RSME	MAX
0	0.050	0.062	0.049	0.078
0.01	0.051	0.091	0.050	0.096
0.02	0.053	0.113	0.052	0.119
0.05	0.072	0.231	0.070	0.224

Different approaches to implement this idea have been proposed in the literature. Some of these methods work by adding the message signal to the chaotic carrier (see, e.g., Cuomo and Oppenheim (1993)). However, it has been shown that this information can be unmasked by inferring the deterministic structure of the chaotic system by using some suitable return maps (Pérez and Cerdeira (1995)). As we shall see, functional network models provide an alternative automatic extracting method which is robust in the presence of noise. We use a simple masking model to illustrate this methodology, but more sophisticated techniques (such as the compound signals used to mask information in Murali and Lakshmanan (1998)) can also be considered.

Consider the bit streams shown in Figure 9.23, where each bit is transcribed as 20 and 40 consecutive values of the time series m_n in Figure 9.23(a) and (b), respectively. The value -1 is used to denote the bit 0 and the value 1 is used for the bit 1. Note that the series m_n is scaled by a factor 0.05 before it is added to the chaotic orbit obtained from the Burger map.

The message in Figure 9.23(a) is a squared signal with a characteristic power spectrum (see Figure 9.24(a)). In this example we shall use the Burger chaotic signal to mask the message. Figure 9.24(b) compares the power spectra of both the message and the chaotic signals. From this figure we can see how the broadban spectrum of the chaotic signal completely masks the message. The sequence transmitted is obtained by adding these two signals.

We can use a functional network model for estimating the deterministic structure of the transmitted signal (the chaotic component) and obtaining the message by subtracting the estimated from the transmitted signals. Figure 9.24(c) shows the power spectrum of the unmasked message obtained when using a Fourier functional network with $m = 7$ to infer the chaotic component of the signal. It can be shown that the noisy component of the power is very low. Figure 9.25 shows the message recovered by this procedure.

The above procedure is robust in the presence of noise (see Table 9.13). Figure 9.26 shows the messages unmasked from the received signal using polynomial (left column) and Fourier (right column) functional networks. Gaussian white noise with three different values for the standard deviation (0.01, 0.02, and 0.05) have been used to test the performance of the above decoding procedure in the presence of noise. From the figure it is clear that the decoded message degrades

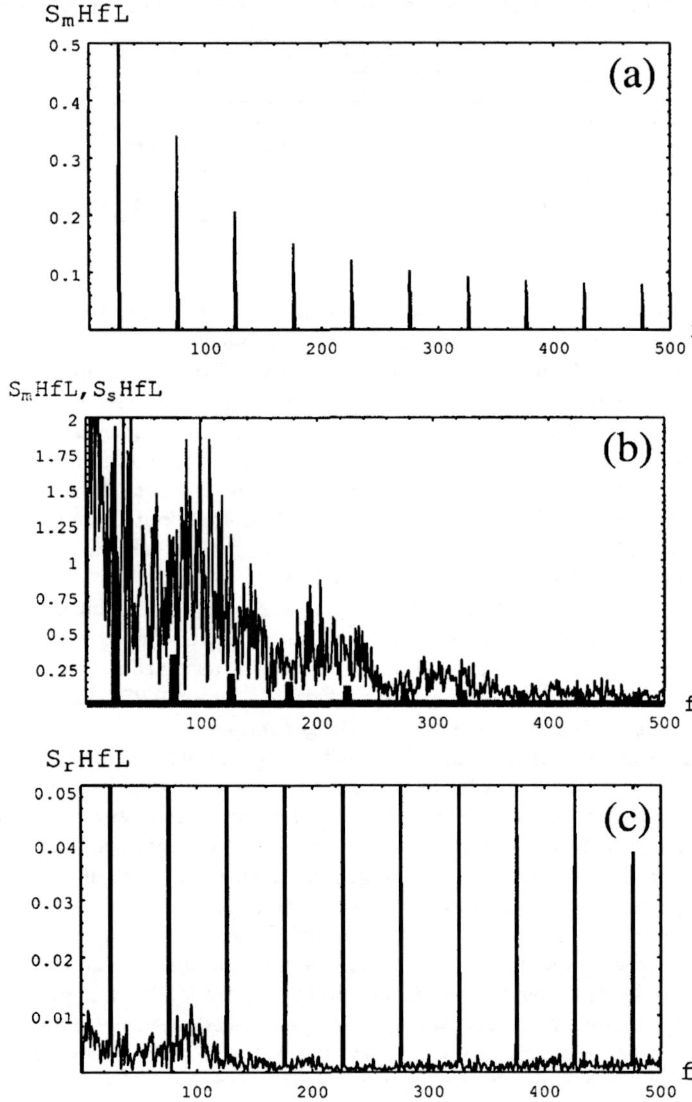

Figure 9.24: Power spectra of (a) the original message $S_n(f)$, (b) the message masked in the broad-band chaotic signal $S_s(f)$ and (c) the unmasked message $S_r(f)$.

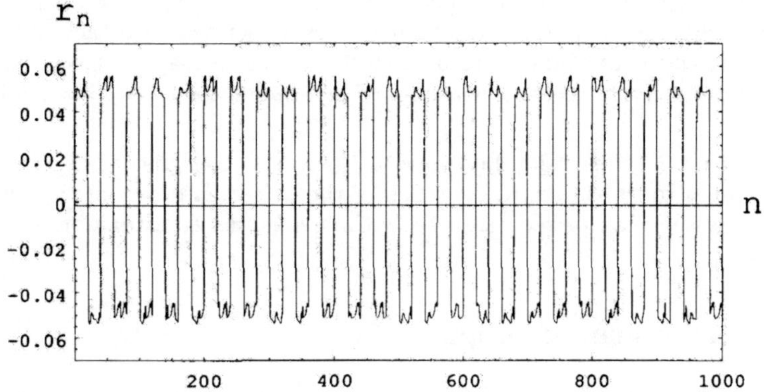

Figure 9.25: Recovered message after subtracting the values predicted by the functional network from the transmitted signal.

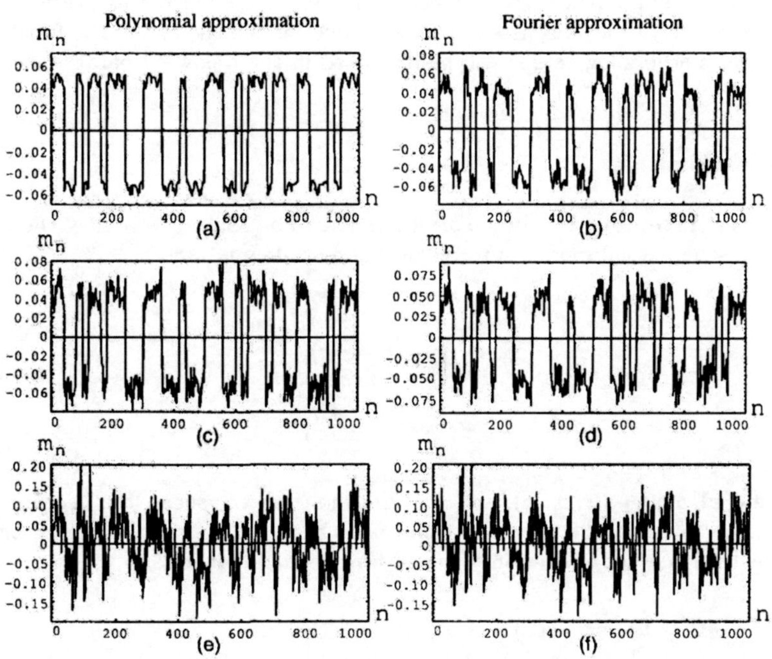

Figure 9.26: Messages unmasked from the received signal using polynomial (left column) and Fourier (right column) functional networks for three different noise levels: 0.01 ((a),(b)), 0.02 ((c),(d)), and 0.05 ((e),(f)).

with the noise level. To filter the decoded signal we propose using the median of the 20, or 40, consecutive values of time series associated with each of the bits as a robust estimate of the corresponding bit of the message. This has allowed us to recover the sent message without any error in all the cases shown in Figure 9.26.

Note that Figures 9.26(e),(f) have a high noise-to-signal ratio (the signal level is equal to the standard deviation of the noise) and even in this case the original message can be recovered using the median as an estimator.

9.8.5 The beam example

In this section we describe an example arising from civil engineering showing that some functional network architectures can be efficiently applied to model and predict the behavior of systems originally stated in terms of differential or difference equations (see Chapter 7). More precisely, we apply the functional networks model (already introduced in Chapter 9) to the case of beams using two different approaches. In both cases, the uniqueness of representation (step 4 of our model) is not discussed. Following Section 9.5 the rest of the functional networks methodology steps will be discussed in the following paragraphs.

First alternative

Step 1 (Statement of the problem). Castillo (1996) shows how the usual beam mathematical model in terms of differential equations:

$$\begin{aligned} q'(x) &= p(x), \\ m'(x) &= q(x), \\ w'(x) &= \frac{m(x)}{EI}, \\ z'(x) &= w(x), \end{aligned} \tag{9.88}$$

where $p(x), q(x), m(x), w(x)$, and $z(x)$ are the load, the shear, the bending moment, the rotation and the deflection of the beam at the point x, respectively, can be equivalently written in terms of functional equations:

$$\begin{aligned} q(x+u) &= q(x) + A(x,u), \\ m(x+u) &= m(x) + uq(x) + B(x,u), \\ w(x+u) &= w(x) + \frac{1}{EI}\left[m(x)u + q(x)\frac{u^2}{2} + C(x,u)\right], \\ z(x+u) &= z(x) + w(x)u, \\ &\quad + \frac{1}{EI}\left[m(x)\frac{u^2}{2} + q(x)\frac{u^3}{6} + D(x,u)\right], \end{aligned} \tag{9.89}$$

where

$$
\begin{aligned}
A(x,u) &= \int_{x}^{x+u} p(s)ds, \\
B(x,u) &= \int_{x}^{x+u} (x+u-s)p(s)ds, \\
C(x,u) &= \int_{x}^{x+u} B(x,s-x)ds, \\
D(x,u) &= \int_{x}^{x+u} C(x,s-x)ds.
\end{aligned} \tag{9.90}
$$

Step 2 (Initial topology). Assume that u is given; that is,one is interested in calculating the vector $(q(x+u), m(x+u), w(x+u), z(x+u), x+u))$ as a function of the vector $(q(x), m(x), w(x), z(x), x)$. Then, Expression (9.89) shows that the natural functional network associated with this problem is the one in Figure 9.27, where

$$F(q(x), x) = q(x) + A(x,u), \tag{9.91}$$

$$G(m(x), q(x), x) = m(x) + uq(x) + B(x,u), \tag{9.92}$$

$$H(w(x), m(x), q(x), x) = w(x) + \frac{\left[m(x)u + q(x)\frac{u^2}{2} + C(x,u)\right]}{EI}, \tag{9.93}$$

$$R(z(x), w(x), m(x), q(x), x) = z(x) + w(x)u + \frac{\left[m(x)\frac{u^2}{2} + q(x)\frac{u^3}{6} + D(x,u)\right]}{EI}, \tag{9.94}$$

$$S(x) = x + u. \tag{9.95}$$

Step 3 (Simplification). In this case simplification of the selected topology is not possible.

Step 5 (Data collection). Table 9.14 shows the vectors $(q(x), m(x), w(x), z(x), x)$ measured in a cantilever beam (see Figure 9.28) corresponding to a load which is unknown to the analyst (we have derived the data assuming a constant load).

Step 6 (Learning). Since we are interested in reproducing the set of data of several beams, for a given u, the neuron functions (9.91)-(9.95) can be approx-

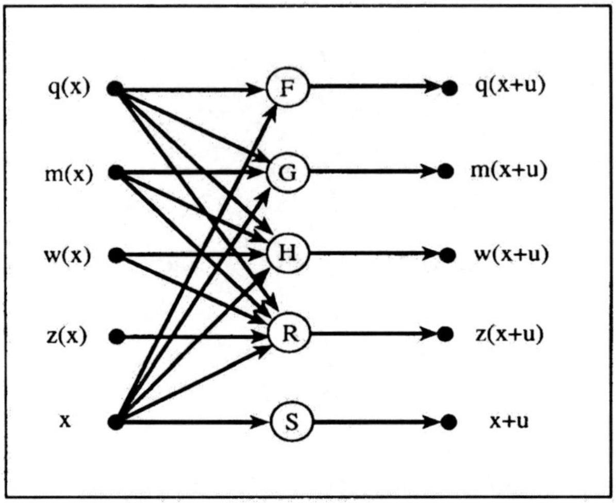

Figure 9.27: A functional network reproducing the beam equation problem (9.89).

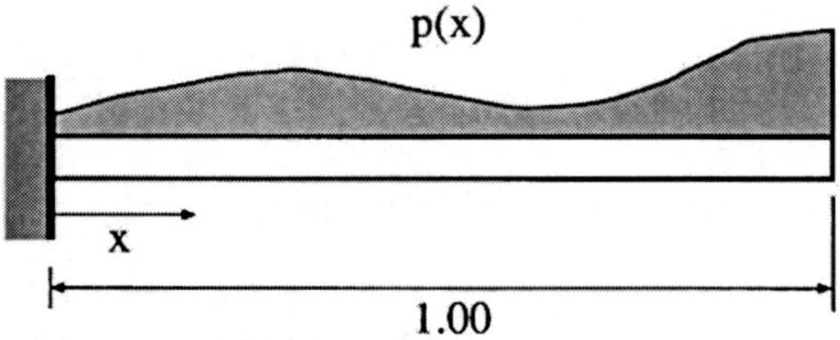

Figure 9.28: Cantilever beam and associated load.

imated, for example, by fourth degree polynomials; that is:

$$\begin{array}{rcl} A(x,u) & = & a_0 + a_1x + a_2x^2 + a_3x^3 + a_4x^4, \\ B(x,u) & = & b_0 + b_1x + b_2x^2 + b_3x^3 + b_4x^4, \\ C(x,u) & = & c_0 + c_1x + c_2x^2 + c_3x^3 + c_4x^4, \\ D(x,u) & = & d_0 + d_1x + d_2x^2 + d_3x^3 + d_4x^4. \end{array} \quad (9.96)$$

Table 9.14: (Shear, bending moment, rotation, deflection, location) vectors measured in the beam example.

$q(x)$	$m(x)$	$w(x)$	$z(x)$	x
1.632	-0.7642	0.000	0.0000	0.00
1.533	-0.6851	-362.1	-9.2180	0.05
1.437	-0.6109	-685.9	-35.570	0.10
1.343	-0.5414	-973.8	-77.210	0.15
1.251	-0.4765	-1228.0	-132.40	0.20
1.161	-0.4163	-1451.0	-199.50	0.25
1.073	-0.3604	-1645.0	-277.00	0.30
0.987	-0.3089	-1812.0	-363.60	0.35
0.902	-0.2617	-1955.0	-457.80	0.40
0.812	-0.2187	-2075.0	-558.70	0.45
0.739	-0.1797	-2174.0	-665.00	0.50
0.659	-0.1448	-2255.0	-775.80	0.55
0.581	-0.1138	-2320.0	-890.20	0.60
0.504	-0.0867	-2369.0	-1007.0	0.65
0.429	-0.0633	-2407.0	-1127.0	0.70
0.355	-0.0438	-2433.0	-1248.0	0.75
0.281	-0.0279	-2451.0	-1370.0	0.80
0.210	-0.0156	-2462.0	-1493.0	0.85
0.139	-0.0069	-2467.0	-1616.0	0.90
0.069	-0.0017	-2469.0	-1740.0	0.95
0.000	0.0000	-2470.0	-1863.0	1.00

To estimate the 20 parameters in (9.96) we have minimized the function

$$\begin{aligned} Q \;=\; & \sum_x \left[F(q(x),x) - q(x+u)\right]^2 + \\ & \sum_x \left[G(m(x),q(x),x) - m(x+u)\right]^2 + \\ & \sum_x \left[H(w(x),m(x),q(x),x) - w(x+u)\right]^2 + \\ & \sum_x \left[R(z(x),w(x),q(x),x) - z(x+u)\right]^2. \end{aligned} \tag{9.97}$$

with respect to $a_0,\ldots,a_4,b_0,\ldots,b_4,c_0,\ldots,c_4,d_0,\ldots,d_4$ to obtain:

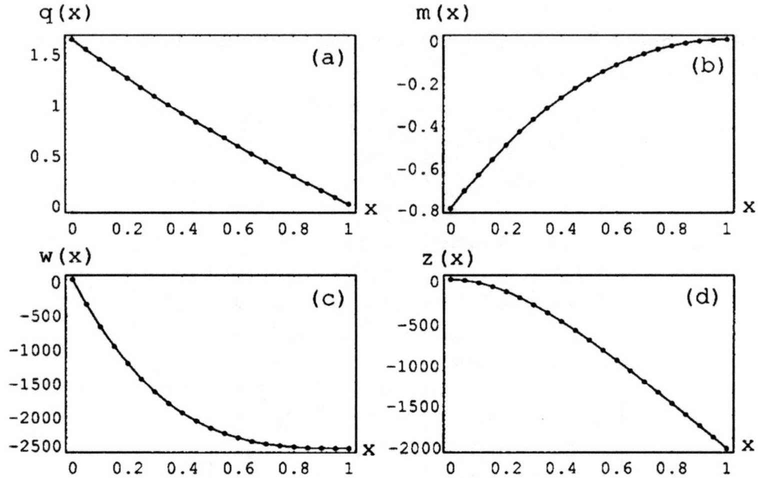

Figure 9.29: Exact (continuous lines) and predicted values (dots) of the shear, bending moment, rotation and deflection of the beam, calculated using the functional network model (9.91)–(9.95) with (9.96) and the exact previous values.

$$\begin{array}{lll}
a_0 = -0.0882, & a_1 = -0.0035, & a_2 = 0.0100, \\
a_3 = 0.0130, & a_4 = 0.0131, & b_0 = -0.0022, \\
b_1 = -0.000087, & b_2 = 0.00025, & b_3 = 0.00033, \\
b_4 = 0.00033, & c_0 = -0.000041, & c_1 = 0.000021, \\
c_2 = -0.000010, & c_3 = 3.163 \times 10^{-6}, & c_4 = -5.35 \times 10^{-7}, \\
d_0 = -5.18 \times 10^{-7}, & d_1 = 2.57 \times 10^{-7}, & d_2 = -1.27 \times 10^{-7}, \\
d_3 = 3.96 \times 10^{-8}, & d_4 = -6.72 \times 10^{-9}. &
\end{array} \tag{9.98}$$

Figure 9.29 shows the measured vectors of the beam and the predictions using the model above. In this case every vector

$$(q^*(x+u), m^*(x+u), w^*(x+u), z^*(x+u), x+u)$$

is predicted based on the previous observed vector

$$(q(x), m(x), w(x), z(x), x).$$

Note that we have used an asterisk to refer to predicted values based on measured values.

Step 7 (Model validation). Figure 9.30 shows the measured deflection of the beam and the prediction using the model above. In this case every vector

$$(q^{**}(x+u), m^{**}(x+u), w^{**}(x+u), z^{**}(x+u), x+u)$$

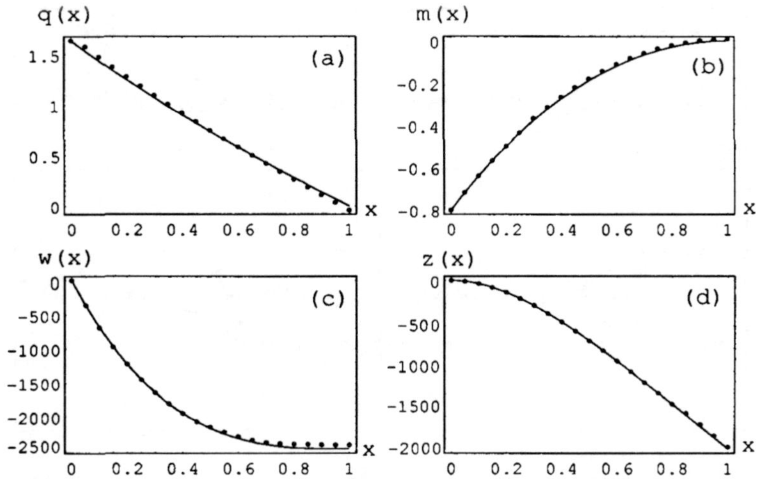

Figure 9.30: Exact (continuous lines) and predicted values (dots) of the shear, bending moment, rotation and deflection of the beam, calculated using the functional network model (9.91)–(9.95) with (9.96) and the predicted previous values.

is predicted based on the previous predicted vector

$$(q^{**}(x), m^{**}(x), w^{**}(x), z^{**}(x), x).$$

Note that we have used two asterisks to refer to predicted values based on predicted values.

Note that the error now accumulates and we can obtain important errors for the end predictions. Note also that the error increases from shear to deflections, that is, with the inverse of the order of the derivative. These errors can be reduced by increasing the degrees of the polynomials in (9.96).

Second alternative

Step 1 (Statement of the problem). Alternatively, from (9.88) we get the well known differential equation

$$EIz^{(iv)}(x) = p(x). \tag{9.99}$$

To obtain an equivalent functional equation in $z(x)$, we can write the last equation in (9.89) for three different values of u and eliminate $w(x), m(x)$ and $q(x)$. For example, if we write this equation for $u, 2u, 3u$ and $4u$, we get the functional equation

$$\begin{aligned} z(x+4u) = \; & 4z(x+u) - 6z(x+2u) + 4z(x+3u) - z(x) + \\ & + \frac{-4D(x,u) + 6D(x,2u) - 4D(x,3u) + D(x,4u)}{EI}, \end{aligned} \tag{9.100}$$

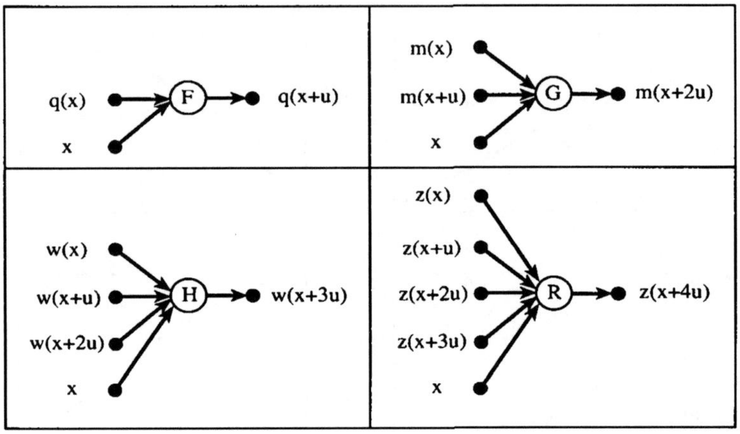

Figure 9.31: Functional networks used to predict $q(x), m(x), w(x)$ and $z(x)$.

which is equivalent to (9.99).

By a similar process, we can obtain the following functional equations for $w(x)$, $m(x)$ and $q(x)$:

$$\begin{aligned} w(x+3u) &= w(x) - 3w(x+u) + 3w(x+2u) \\ &\quad + \frac{[3C(x,u) - 3C(x,2u) + C(x,3u)]}{(EI)}, \end{aligned} \tag{9.101}$$

$$m(x+2u) = 2m(x+u) - m(x) - B(x,u), \tag{9.102}$$

$$q(x+u) = q(x) + A(x,u). \tag{9.103}$$

Equations (9.89) and (9.100)–(9.103) can also be interpreted as finite difference equations. In this case, they give the exact solution at the interpolating points.

The difference between systems (9.89) and (9.100)–(9.103) is that in (9.89) the unknown functions are coupled, while in (9.100) to (9.103) they are uncoupled.

Step 2 (initial topology). Figure 9.31 shows the initial topologies for these cases.

Note that the knowledge of the problem allows not only the topology of the network to be selected but also the required dimensions of the inputs. In many problems, the functional mapping between input and output of training instances is complicated and not clear enough to determine even the input patterns. Therefore, selection of an adequate set of basic functions is difficult for these cases. Hence, functional networks show a clear advantage over neural networks in these kinds of problems.

Table 9.15: Estimates of the fourth order polynomial coefficients corresponding to the functional networks in Figure 9.31 and Equations (9.100) to (9.103) when the $A(x,u), B(x,u), C(x,u)$ and $D(x,u)$ expressions are approximated by $c_0 + c_1x + c_2x^2 + c_3x^3 + c_4x^4$.

	$q(x)$	$m(x)$	$w(x)$	$z(x)$
c_0	−0.09877	−0.00488	−2.4100	−0.11908
c_1	0.04875	0.00238	1.1597	0.05656
c_2	−0.02420	−0.00118	−0.5767	−0.02815
c_3	0.00755	0.00037	0.1820	0.00893
c_4	−0.00128	0.00006	−0.0319	−0.00159

Step 3 (Simplification). In this case simplification of the selected topology is not possible.

Step 5 (Data collection). As in alternative one, we use the data in Table 9.14.

Step 6 (Learning). Now, we use equations (9.100)–(9.103). For example, in Equation (9.100), which corresponds to the functional network in the lower right corner of Figure 9.31, we can make

$$\begin{aligned} &\frac{-4D(x,u) + 6D(x,2u) - 4D(x,3u) + D(x,4u)}{EI} \\ &= c_0 + c_1x + c_2x^2 + c_3x^3 + c_4x^4. \end{aligned} \tag{9.104}$$

For estimating the parameters we minimize

$$\begin{aligned} Q \;=\; & \sum_x \left[z(x+4u) - 4z(x+u) + 6z(x+2u) - 4z(x+3u) + z(x)\right. \\ & \left. -c_0 - c_1x - c_2x^2 - c_3x^3 - c_4x^4\right]^2, \end{aligned} \tag{9.105}$$

with respect to $\{c_0, c_1, c_2, c_3, c_4\}$, to obtain

$$\begin{aligned} &c_0 = -0.119, \quad c_1 = 0.057, \quad c_2 = -0.028, \\ &c_3 = 0.0089, \quad c_4 = -0.0016. \end{aligned} \tag{9.106}$$

Similarly, we use the remaining functional networks in Figure 9.31 to reproduce the data in columns 1 to 3 of Table 9.14. The resulting coefficients of the fourth order polynomials used are given in Table 9.15.

Step 7 (Model validation). The measured shear, moments, rotations and deflections of the beam and the corresponding predictions using the network models in Figure 9.31 and the predicted values are practically equal to those in Figure 9.29. Note that now the predictions are much better than before (compare Figures 9.29 and 9.30).

Since the approximated functions in (9.96) and (9.104) depend only on the applied load, they are valid for any boundary conditions. Thus, once the coefficients in (9.96) and (9.104) have been obtained for a particular set of boundary conditions, the functional network can be used to predict deformations, rotations, moments and shear forces for any boundary conditions if the load remains unchanged. To this end, a trial and error method, based on changing the boundary conditions at one end, can be used until we get the desired boundary conditions at the other end. Note that this simple iterative and convergent method uses the same estimated functions $A(x,u), B(x,u), C(x,u)$ and $D(x,u)$ which were obtained based on different boundary conditions.

Exercises

9.1 Solve the functional equation

$$F[G(x,y),z] = K[x,N(y,z)],$$

and draw its associated and the resulting simplified functional networks.

Hint: Use the solution in (9.2).

9.2 Design a functional network to obtain the most general surface such that its intersections with planes parallel to the coordinate planes ZY and ZX are linear combinations of functions in the sets $\{\phi_1(x), \phi_2(x), \phi_3(x)\}$ and $\{\psi_1(y), \psi_2(y), \psi_3(y)\}$, respectively.

9.3 If $F_X(x)$ and $F_y(y)$ are the cumulative distribution functions of two independent random variables X and Y, the cumulative distribution function $F_Z(z)$ of $Z = \max(X,Y)$ is

$$F_Z(z) = F_X(z)F_Y(y).$$

(a) Write the functional equation for X and Z to belong to the same single parameter family, and draw the corresponding functional network.

(b) Write the functional equation for X, Y and Z to belong to the same single parameter family; i.e., for the single parameter family of distributions to be closed with respect to maximum operations, and draw the corresponding functional network.

9.4 We say that a family of random variables is reproductive under convolution if the sum of independent random variables of the family belongs to the family; that is, if the family is closed under sums. We know that the characteristic function $\phi_Z(t)$ of the sum $Z = X + Y$ of two independent random variables is the product of their characteristic functions $\phi_X(t)$ and $\phi_Y(t)$ of the summands X and Y:

$$\phi_Z(t) = \phi_X(t)\phi_Y(t),$$

Thus, reproductivity of parametric families can be written as a functional equation. Write the functional equation for the case of single parameter families. Draw the functional network associated with this equation. Simplify the network by solving the resulting functional equation as a particular case of (9.5).

Hint: Decompose the resulting functional equation in a complex function into two functional equations in two real functions.

9.5 Using the general solution (9.2) of the functional Equation (9.1), solve the system of functional equations

$$F[G(x,y),z) = F[x, N(y,z)] = F[y, M[x,z)].$$

9.6 Suppose that in the Model (9.30) we assume

$$f_1(x) = f_2(x) = \sum_{i=1}^{n} a_i \phi_i(x)$$

and

$$f_3(x) = \sum_{i=1}^{n} b_i \psi_i(x).$$

Derive the formulas required to learn the $a_i, b_i;\ i = 1, \ldots, n$ coefficients.

9.7 Simplify the separable model

$$f_1(x)\cos(y) = \sin(x) g_2(y)$$

using the techniques developed in Section 9.7.3.

9.8 Use a functional network to calculate the n-th iterate of

$$f(x) = \sqrt{1 + x^2}.$$

Based on the results in Section 9.7.4, obtain the exact solution.

9.9 Use the data in Table 9.16 to fit a model of the form

$$\begin{aligned} u &= f\left[g(x) + h(y)\right], \\ v &= g\left[h(x) + f(y)\right], \\ w &= h\left[f(x) + g(y)\right]. \end{aligned}$$

Compare the obtained results with the model corresponding to

$$\begin{aligned} u &= \cos\left[x + \exp(y)\right], \\ v &= \exp(x) + \cos(y), \\ w &= \exp\left[cos(x) + y\right]. \end{aligned}$$

9.10 Show that the functional networks shown in Figure 9.32 are equivalent.

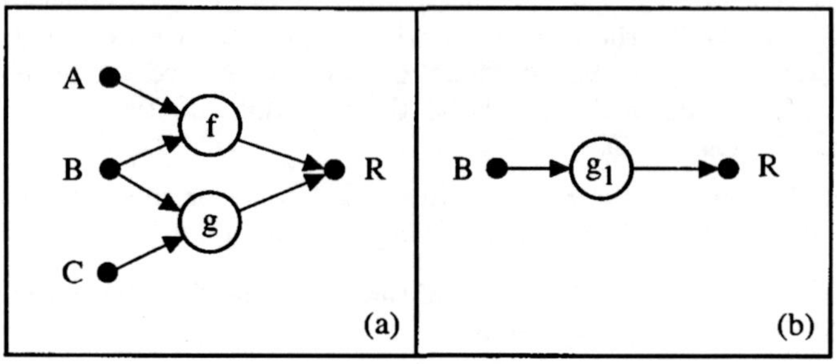

Figure 9.32: Two One-layer equivalent functional networks.

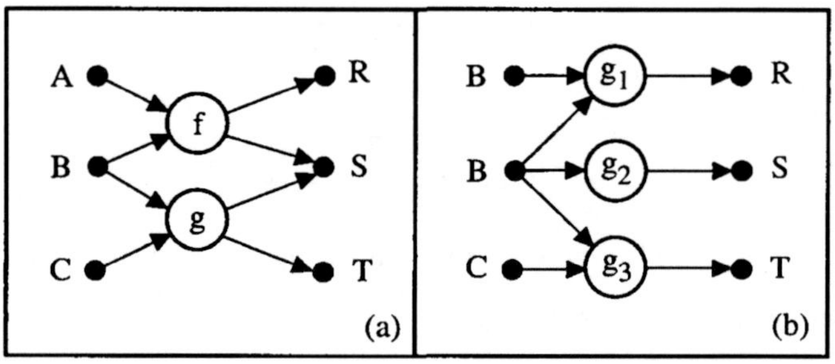

Figure 9.33: Two One-layer equivalent functional networks.

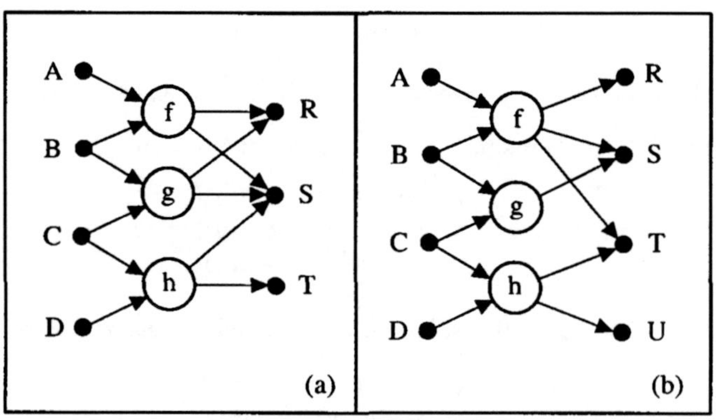

Figure 9.34: Two one-layer functional networks.

Table 9.16: A sample of size 40 of (x, y, u, v, w) data.

x	y	u	v	w	x	y	u	v	w
0.84	0.93	-0.97	2.90	4.98	0.54	0.31	-0.33	2.67	3.22
0.32	0.81	-0.84	2.06	5.83	0.85	0.93	-0.97	2.94	4.89
0.99	0.28	-0.67	3.66	2.28	0.54	0.44	-0.50	2.61	3.68
0.32	0.35	-0.16	2.31	3.66	0.72	0.54	-0.76	2.92	3.62
0.08	0.99	-0.94	1.63	7.31	0.80	0.27	-0.51	3.20	2.62
0.31	0.37	-0.19	2.29	3.77	0.97	0.34	-0.71	3.58	2.47
0.77	0.06	-0.25	3.15	2.19	0.34	0.22	-0.02	2.39	3.20
0.45	0.25	-0.16	2.54	3.16	0.50	0.30	-0.27	2.60	3.24
0.46	0.97	-1.00	2.14	6.49	0.96	0.85	-0.99	3.27	4.17
0.14	0.63	-0.43	1.96	5.03	0.24	0.32	-0.04	2.22	3.64
0.06	0.63	-0.36	1.87	5.11	0.44	0.05	0.08	2.55	2.60
0.75	0.26	-0.46	3.09	2.69	0.47	0.72	-0.81	2.35	4.99
0.99	0.20	-0.60	3.66	2.12	0.13	0.49	-0.19	2.01	4.42
0.54	0.95	-1.00	2.29	6.11	0.63	0.20	-0.27	2.86	2.74
0.08	0.98	-0.92	1.64	7.20	0.67	0.35	-0.49	2.89	3.10
0.94	0.35	-0.71	3.50	2.56	0.43	0.03	0.12	2.54	2.55
0.88	0.72	-0.98	3.17	3.87	1.00	0.98	-0.88	3.26	4.58
0.13	0.46	-0.14	2.04	4.26	0.52	0.26	-0.25	2.66	3.08
0.15	0.26	0.13	2.12	3.49	0.40	0.77	-0.83	2.21	5.40
0.61	0.31	-0.39	2.79	3.10	0.77	0.57	-0.82	3.00	3.61

9.11 Show that the functional networks shown in Figure 9.33 are equivalent.

9.12 Simplify the one-layer functional networks in Figure 9.34.

9.13 Using the Hénon series

$$x_n = 1 - 1.4x_{n-1}^2 + 0.3x_{n-2}, \tag{9.107}$$

do the following:

(a) Simulate 200 values of the series with added noise $N(0, 0.3^2)$.

(b) Fit a polynomial model by selecting the best degree of the polynomial using the minimum description length principle.

(c) Remove the noise to the series using the model.

(d) Repeat the above three steps but using added noise uniformly distributed with the same variance.

(e) Compare the results.

(f) Comment on the normality assumption.

9.14 The following are temperature measurements z made every minute on a chemical reactor:

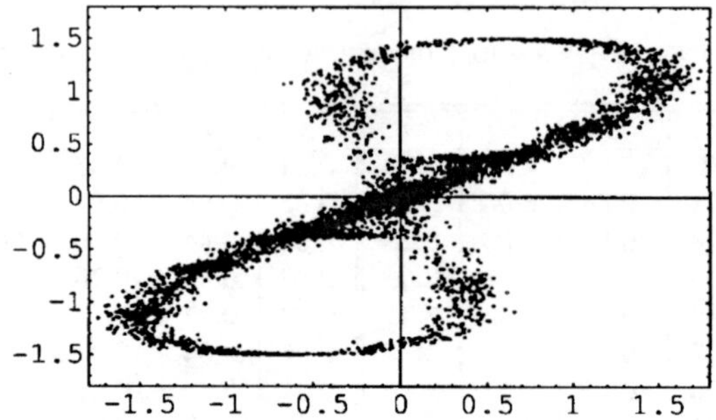

Figure 9.35: Embedding space (x_n, x_{n-1}) for a noisy time series of the Burger map with added normally distributed noise with $\sigma = 0.1$.

$$200, 202, 208, 204, 204, 207, 207, 204, 202, 199$$
$$201, 198, 200, 202, 203, 205, 207, 211, 204, 206$$
$$203, 203, 201, 198, 200, 200, 206, 207, 206, 200$$
$$203, 203, 200, 200, 195, 202, 204, 203, 204, 205.$$

Plot the series, fit a model and use it for predicting the temperature five minutes after the last measure.

9.15 Obtain a time series of the Burger map with added normally distributed noise with variance 0.1 (see Figure 9.35). Proceed as in Section 9.8.2 to obtain an approximate model using a polynomial and Fourier functional networks and clean the noise contained in the time series.

CHAPTER 10

Applications to Science and Engineering

10.1 Introduction

This chapter is devoted to describing some applications of functional equations to Science and Engineering. When designing a mathematical model to represent a physical reality it is a normal practice to choose functions that satisfy certain conditions. The aim of sections 10.3, 10.4 and 10.5 is to show how functional equations can be used to impose some constraints on the class of admissible functions. Each functional equation states a given property the mathematical model must satisfy. With the help of these examples, a new, powerful and rigorous methodology of model design is presented.

We start by using a motivating example in Section 10.2. In Section 10.3 we derive the general structure of the laws of Science and show that an arbitrary selection of these laws can lead to inconsistencies. Section 10.4 is devoted to a statistical model for lifetime analysis, which is based on a compatibility condition and some known results from the theory of extreme value distributions. Section 10.5 considers an interesting example in which we explain in detail all the steps followed by a three member team to design a fatigue life model, starting with the separate proposals of the three members and their physical or empirical bases and ending with a consensus solution. Finally, in Section 10.6 functional networks are used to predict values of magnitudes satisfying differential and/or difference equations and to obtain the differential/difference equation associated with a set of data. As shown in that section, the estimation of its coefficients is done by simply solving systems of linear equations.

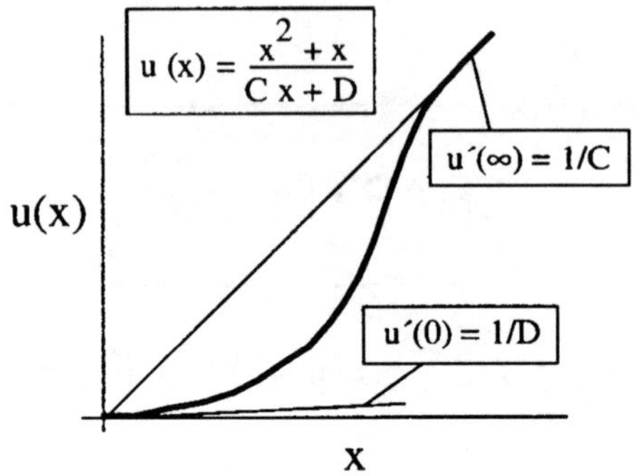

Figure 10.1: Graphical illustration of the tax function.

10.2 A motivating example

We start by giving an example where the need for careful selection of models is illustrated.

Example 10.1 (Families of tax functions). Let us assume that the person responsible for the tax policy of the European Community decides to establish a two-parameter family of tax functions given by

$$u(x) = \frac{x^2 + x}{Cx + D}, \tag{10.1}$$

where x is the income, $u(x)$ is the due tax and C and D are the two parameters (see Figure 10.1). Let us suppose, also, that this family of functions is to be utilized in all country members, using their corresponding monetary units.

Because the Expression (10.1) depends on two parameters, C and D, the legislator has only two degrees of freedom, that is, he/she is free to fix two points of the tax function. Let us assume that for country A a tax amount of 10 monetary units for an income of 100 units and 200 monetary units for an income of 1000 units are chosen. This leads to the following system of equations, values of constants C and D and tax function

$$\left.\begin{array}{l} 10 = \dfrac{100^2 + 100}{100C + D} \\ 200 = \dfrac{1000^2 + 1000}{1000C + D} \end{array}\right\} \Rightarrow \left\{\begin{array}{l} C = \dfrac{799}{180} \\ D = \dfrac{10190}{18} \end{array}\right\} \Rightarrow u(x) = \frac{180(x^2 + x)}{799x + 101900}, \tag{10.2}$$

where x and $u(x)$ are in monetary units of country A.

If the model (10.1) is to be used in another country, B, whose monetary unit is such that r units are equivalent to 1 unit of country A, and we desire to fix the same points as above, then the model should be fitted by the conditions: $100r$ units of income should pay $10r$ units of tax and $1000r$ units of income

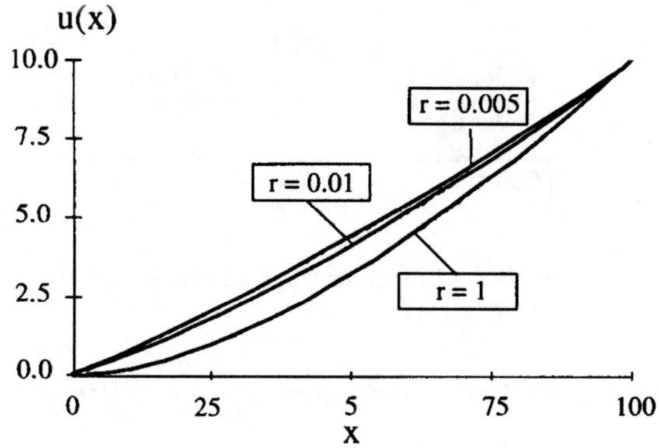

Figure 10.2: Tax function for several values of r.

should pay $200r$ units of tax. Thus, now the system of equations, the C and D constants and the tax function become

$$\left.\begin{array}{l} 10r = \dfrac{100^2r^2 + 100r}{100rC + D} \\ 200r = \dfrac{1000^2r^2 + 1000r}{1000rC + D} \end{array}\right\} \Rightarrow \left\{\begin{array}{l} C = \dfrac{800r - 1}{180r} \\ D = \dfrac{10000r + 190}{18} \end{array}\right\} \Rightarrow \tag{10.3}$$

$$\Rightarrow \quad v(y) = \frac{180r(y^2 + y)}{(800r - 1)y + 100000r^2 + 1900r},$$

where now, instead of x and $u(x)$, we use y and $v(y)$ because they are measured in the monetary units of country B.

In order to compare the tax functions in countries A and B, we now change y and $v(y)$ into the monetary units of country A ($v(y) = ru(x)$ and $y = rx$) to get

$$u(x) = \frac{180(rx^2 + x)}{(800r - 1)x + 100000r + 1900}, \tag{10.4}$$

which is not only different from the tax function in (10.2) but also depends on r. This implies that the citizens of countries A and B pay different taxes. Because of these facts, we say that the family of models (10.1) is inconsistent. Figure 10.2 shows tax functions (10.4) for several values of r (1, 0.01 and 0.005).

If, instead of (10.1), the following family of tax functions is selected

$$u(x) = \frac{x^2}{Cx + D}, \tag{10.5}$$

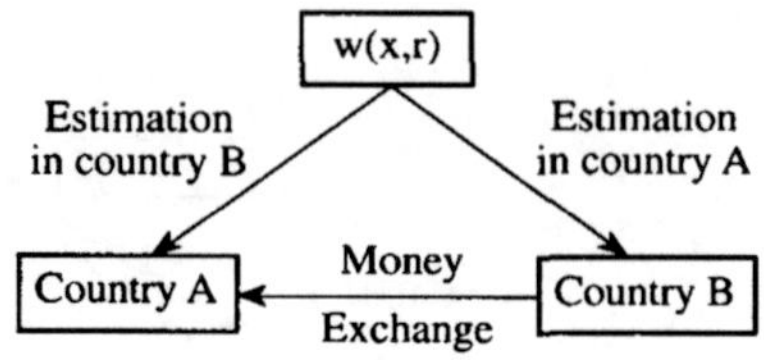

Figure 10.3: Two different ways of obtaining model A.

then, using the same conditions, instead of (10.3) and (10.4), we get

$$\left.\begin{array}{l} 10r = \dfrac{100^2r^2}{100rC+D} \\ 200r = \dfrac{1000^2r^2}{1000rC+D} \end{array}\right\} \Rightarrow \left\{\begin{array}{l} C = \dfrac{40}{9} \\ D = \dfrac{5000r}{9} \end{array}\right\} \Rightarrow v(y) = \frac{9y^2}{40y+5000r} \tag{10.6}$$

and

$$u(x) = \frac{9x^2}{40x+5000}. \tag{10.7}$$

Note that in this case (10.7) can be obtained from (10.6) replacing y by xt and $v(y)$ by $u(x)t$. Then, the family of models (10.5) is consistent.

Thus, we realize that the selection of the $u(x)$ family of tax functions, such as those in (10.1) and (10.5), cannot be arbitrary; on the contrary, it must satisfy some conditions in order to avoid the above problems.

Figure 10.3 shows a diagram in which we illustrate the two different processes we can follow in order to get model (10.7).

We want the diagram in Figure 10.3 to be commutative; that is, we want a family of functions $u(x, C, D)$ such that the same model A can be obtained either proceeding directly, by estimating the model in country A, or through model B, i.e., estimating the model in country B and then changing the monetary units.

Let the taxes associated with incomes x_1 and x_2 in country A be u_1 and u_2, respectively. By proceeding through Model B, we first solve the system of equations in the two unknowns $C(r)$ and $D(r)$:

$$\left.\begin{array}{l} u_1r = v[x_1r, C(r), D(r)] \\ u_2r = v[x_2r, C(r), D(r)] \end{array}\right\} \Rightarrow \left\{\begin{array}{l} C(r) \\ D(r) \end{array}\right. \Rightarrow v(y) = v[y, C(r), D(r)] \tag{10.8}$$

and then we change the monetary units to obtain the tax function for country A

$$u(x) = \frac{1}{r}u[rx, C(r), D(r)]. \tag{10.9}$$

If, on the contrary, we proceed directly to Model A, we solve the system of equations in the two unknowns $C(1)$ and $D(1)$:

$$\left.\begin{array}{l} u_1 = u[x_1, C(1), D(1)] \\ u_2 = u[x_2, C(1), D(1)] \end{array}\right\} \Rightarrow \left\{\begin{array}{l} C(1) \\ D(1) \end{array}\right. \Rightarrow u(x) = u[x, C(1), D(1))]. \tag{10.10}$$

For models (10.9) and (10.10) to coincide, we must have

$$u[rx, C(r), D(r)] = ru[x, C(1), D(1)]. \tag{10.11}$$

Equation (10.11) is a functional equation, which, by denoting

$$w(x,r) = u[x, C(r), D(r)],$$

can be written as

$$w(rx, r) = rw(x, 1). \tag{10.12}$$

We can give the following physical interpretation to functional Equation (10.12). Let $w(x,r)$, where x and w are in monetary units of country A, be the tax associated with an income x in country B. Then Equation (10.12), which is written in terms of monetary units of country B, says that a citizen of country B who earns x, in monetary units of A, that is, rx, in monetary units of B, pays the same taxes as a citizen of A earning x, in monetary units of A.

Equation (10.12) is the equation for homogeneous functions (5.35) and then its general solution is (see Theorem 5.9)

$$w(x,r) = u[x, C(r), D(r)] = r\phi\left(\frac{x}{r}\right), \tag{10.13}$$

where ϕ is an arbitrary function.

Expression (10.13) indicates that any consistent model can be written as a function of one single genuine parameter, r, and one arbitrary function, ϕ, and that if the tax function $\phi(x)$ is known for country A, then the tax function for any other country B must be given by (10.13). ∎

10.3 Laws of science

In a physical system there are some fundamental variables, such as length, time and space; from them, secondary or derived variables are obtained by certain, more or less complicated, formulas. In other cases, formulas relate to different variables, not necessarily fundamental. However, not every formula generates a valid variable, but only those satisfying some extra conditions (see Aczél (1987b), pp.35-70). It is necessary that a change of location or/and scale of the independent variables keep the same formula structure up to a change of location or/and scale of the derived or dependent variable. In other words, the formula should remain invariant under location and scale changes. This condition can be written as the following functional equation:

$$\begin{aligned} u(r_1x_1 + p_1, r_2x_2 + p_2, \ldots, r_nx_n + p_n) \\ = R(r_1, r_2, \ldots, r_n; p_1, p_2, \ldots, p_n)u(x_1, x_2, \ldots, x_n) \\ + P(r_1, r_2, \ldots, r_n; p_1, p_2, \ldots, p_n) \\ (r_i, x_i \in \mathbb{R}_{++};\ i = 1, 2, \ldots, n), \end{aligned} \tag{10.14}$$

where $u(x_1, x_2, \ldots, x_n)$ is the formula that gives the derived variable as a function of the fundamental variables and P and R are functions associated with the location and scale changes of the derived variable, respectively.

When a variable is allowed for location and scale changes we say that it has an *interval scale.* If a variable is allowed for scale changes only we say that it has a *ratio scale.* Examples of variables with interval scales are time, temperature and location. Examples of ratio scale variables are length, area, volume, speed and acceleration.

Equation (10.14) shows the more general situation, in that it includes a maximum of restrictions. However, some simpler cases can occur, depending on whether or not the location or the scale changes exist for the fundamental or the derived variables or whether they are homogeneous or heterogeneous for all the variables. In this manner, we can deal with very many different situations. The following theorems and corollaries give the solution of Equation (10.14) for six different cases.

Theorem 10.1 (General formula for the laws of Science I). *The general forms of dependent real variables with interval scale non-constant and continuous-at-a-point when all fundamental or independent variables have the same ratio scale, i.e.,the general solutions of the functional equation*

$$u(r\mathbf{x}) = R(r)u(\mathbf{x}) + P(r); \quad r, x_i \in \mathbb{R}_{++}, \quad (i = 1, 2, \ldots, n), \tag{10.15}$$

where scalars such as r must be interpreted as $(r, r, \ldots, r)$ when occupying a vector position, are

$$u(\mathbf{x}) = f\left(\frac{x_2}{x_1}, \frac{x_3}{x_1}, \ldots, \frac{x_n}{x_1}\right) + c\log(x_1); \quad R(r) = 1; \quad P(r) = c\log(r),$$

$$u(\mathbf{x}) = x_1^c f\left(\frac{x_2}{x_1}, \frac{x_3}{x_1}, \ldots, \frac{x_n}{x_1}\right) + b; \quad R(r) = r^c; \quad P(r) = b[1 - r^c]. \tag{10.16}$$

■

Proof: Setting $\mathbf{x} = \mathbf{1} = (1, 1, \ldots, 1)$ into (10.15) and subtracting this from (10.15) we get

$$\begin{aligned} v(r\mathbf{x}) &= R(r)v(\mathbf{x}) + v(r), \\ v(\mathbf{x}) &= u(\mathbf{x}) - u(\mathbf{1}). \end{aligned} \tag{10.17}$$

We distinguish two cases:

- (a) $R(r) = 1$. Then (10.17) becomes

$$v(r\mathbf{x}) = v(\mathbf{x}) + v(r).$$

If $\mathbf{x} = (s, s, \ldots, s)$ then

$$v(rs) = v(s) + v(r),$$

which is Equation (3.9). Thus, we have

$$v(r) = c\log(r),$$

and then

$$v(\mathbf{x}) = v\left(x_1\frac{1}{x_1}\mathbf{x}\right) = v\left(1, \frac{x_2}{x_1}, \frac{x_3}{x_1}, \dots, \frac{x_n}{x_1}\right) + c\log(x_1),$$

thus, we finally obtain

$$u(\mathbf{x}) = f\left(\frac{x_2}{x_1}, \frac{x_3}{x_1}, \dots, \frac{x_n}{x_1}\right) + c\log(x_1),$$

where

$$\begin{aligned} f\left(\frac{x_2}{x_1}, \frac{x_3}{x_1}, \dots, \frac{x_n}{x_1}\right) &= v\left(1, \frac{x_2}{x_1}, \frac{x_3}{x_1} \dots, \frac{x_n}{x_1}\right) + u(1) \\ R(r) &= 1, \\ P(r) &= c\log(r). \end{aligned}$$

- (b) $R(r)$ is not identically 1. If we set $\mathbf{x} = \mathbf{s}$ into (10.17) and take into account the symmetry between r and s we deduce

$$v(rs) = R(r)v(s) + v(r) \;\; \Rightarrow \;\; v(rs) = R(s)v(r) + v(s). \tag{10.18}$$

Because of the above condition on $R(r)$, there exists an r_0 such that $R(r_0) \neq 1$ and then

$$v(s) = a[R(s) - 1]; \;\; a = \frac{v(r_0)}{R(r_0) - 1}. \tag{10.19}$$

Since u, and thus v, is not constant, $a \neq 0$ and then the functional Equation (10.18) becomes

$$a[R(rs) - 1] = R(r)a[R(s) - 1] + a[R(r) - 1] \;\; \Rightarrow \;\; R(rs) = R(r)R(s),$$

which is Cauchy's Equation (3.10) and then, taking into account (10.19), we get

$$R(r) = r^c \;\; \Leftrightarrow \;\; v(r) = a(r^c - 1). \tag{10.20}$$

With this, Equation (10.17) transforms to

$$v(r\mathbf{x}) = r^c v(\mathbf{x}) + a(r^c - 1) \;\; \Rightarrow \;\; w(r\mathbf{x}) = r^c w(\mathbf{x}), \tag{10.21}$$

where $w(\mathbf{x}) = v(\mathbf{x}) + a$.

Thus, we have

$$\begin{aligned} w(\mathbf{x}) &= w\left(x_1\frac{1}{x_1}\mathbf{x}\right) = x_1^c w\left(1, \frac{x_2}{x_1}, \frac{x_3}{x_1} \dots, \frac{x_n}{x_1}\right) \\ &\Rightarrow \;\; v(\mathbf{x}) = x_1^c f\left(\frac{x_2}{x_1}, \frac{x_3}{x_1} \dots, \frac{x_n}{x_1}\right) - a, \end{aligned}$$

and then

$$u(\mathbf{x}) = x_1^c f\left(\frac{x_2}{x_1}, \frac{x_3}{x_1}, \dots, \frac{x_n}{x_1}\right) + b; \quad P(r) = b(1 - r^c).$$

■

Corollary 10.1 (Ratio scale variables). *The general form of dependent real-valued variables with ratio scale non-constant and continuous-at-a-point when all fundamental or independent variables have the same ratio scale, i.e., the general solution of the functional equation*

$$u(r\mathbf{x}) = R(r)u(\mathbf{x}) \; ; \quad r, x_i \in \mathbb{R}_{++} \, , \; (i = 1, 2, \dots, n) \tag{10.22}$$

is

$$u(\mathbf{x}) = x_1^c f\left(\frac{x_2}{x_1}, \frac{x_3}{x_1}, \dots, \frac{x_n}{x_1}\right), \quad R(r) = r^c. \tag{10.23}$$

■

Proof: Making $P(r) = 0$ in (10.16) we get either $c = 0$ or $b = 0$ and the resulting solution satisfies (10.22). Thus, (10.23) holds. ■

Example 10.2 (Areas of two-parameter families of surfaces). Let $u(x, y)$ be the area of a family of figures depending on two length parameters, as, for example, a family of ellipses with semi-axes x and y, or a family of rectangles with sides x and y. Then, we have two independent variables with the same dimensions. If we apply the ratio scale r to both variables we find that, according to Corollary 10.1, the only possibility for $u(x, y)$ is

$$u(x, y) = x^c f\left(\frac{y}{x}\right); \quad R(r) = r^c,$$

but we know (see Example 3.1) that $R(r) = r^2$. Thus, the formula, given the area of a two-parameter family, must be of the form

$$u(x, y) = x^2 f\left(\frac{y}{x}\right),$$

where f is an arbitrary non-negative function. Note that for the family of ellipses and rectangles we have

$$f(x) = \pi x \text{ and } f(x) = x,$$

respectively. ■

Theorem 10.2 (General formula for the laws of Science II). *The general forms of dependent real-valued variables with interval scale non-constant and continuous-at-a-point when all fundamental or independent variables have ratio scale, i.e., the general solutions of the functional equation*

$$u(\mathbf{r}\mathbf{x}) = R(\mathbf{r})u(\mathbf{x}) + P(\mathbf{r}); \quad \mathbf{r}, \mathbf{x} \in \mathbb{R}^n_{++} \tag{10.24}$$

are

$$\begin{aligned} u(\mathbf{x}) &= \sum_{i=1}^{n} c_i \log x_i + b, \\ R(\mathbf{r}) &= 1, \\ P(\mathbf{r}) &= \sum_{i=1}^{n} c_i \log r_i, \end{aligned} \tag{10.25}$$

and

$$\begin{aligned} u(\mathbf{x}) &= a \prod_{i=1}^{n} x_i^{c_i} + b, \\ R(\mathbf{r}) &= \prod_{i=1}^{n} r_i^{c_i}, \\ P(\mathbf{r}) &= b\left[1 - \prod_{i=1}^{n} r_i^{c_i}\right]. \end{aligned} \tag{10.26}$$

■

Proof: The proof is similar to that in Theorem 10.1. ■

Corollary 10.2 (Ratio scale variables). *The general form of dependent real-valued variables with ratio scale non-constant and continuous-at-a-point when all fundamental or independent variables have ratio scale, i.e., the general solution of the functional equation*

$$u(\mathbf{r}\mathbf{x}) = R(\mathbf{r})u(\mathbf{x}); \quad \mathbf{r}, \mathbf{x} \in \mathbb{R}^n_{++} \tag{10.27}$$

is

$$u(\mathbf{x}) = a \prod_{i=1}^{n} x_i^{c_i}; \quad R(\mathbf{r}) = \prod_{i=1}^{n} r_i^{c_i}, \tag{10.28}$$

with $a \neq 0$ *and* $\sum_{i=1}^{n} c_i^2 \neq 0$. ■

Proof: Making $P(r) = 0$ in (10.25) and (10.26) we get either $c_i = 0$ $(i = 1, 2, \ldots, n)$ or $b = 0$ and then (10.28) holds. ■

Theorem 10.3 (General formula for the laws of Science III). *The general form of dependent real-valued variables with interval scale non-constant and continuous-at-a-point when all fundamental or independent variables have interval scale with the same value of* r, *i.e., the general solution of the functional equation*

$$u(r\mathbf{x} + \mathbf{p}) = R(r, \mathbf{p})u(\mathbf{x}) + P(r, \mathbf{p}); \quad r, x_i, p_i \in \mathbb{R}_+, \quad (i = 1, 2, \ldots, n) \tag{10.29}$$

is

$$\begin{aligned} u(\mathbf{x}) &= \sum_{i=1}^{n} c_i x_i + b, \\ R(r, \mathbf{p}) &= r, \\ P(r, \mathbf{p}) &= b(1 - r) + \sum_{i=1}^{n} c_i p_i. \end{aligned} \tag{10.30}$$

■

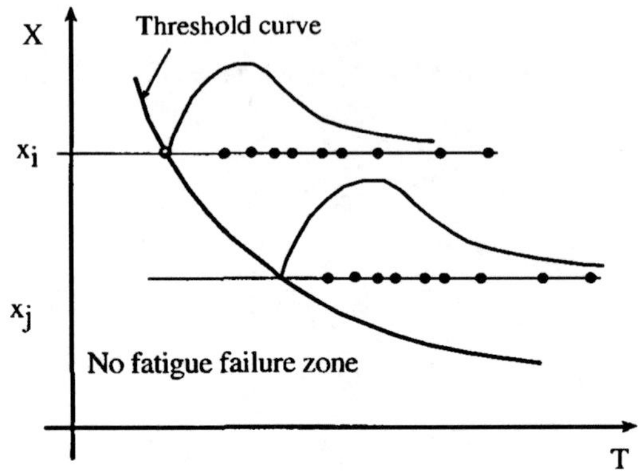

Figure 10.4: Regression model.

Proof: For the proof see Aczél (1987b), pp. 57-61. ∎

Corollary 10.1 *The only non-null solution of the functional equation*

$$u(r\mathbf{x}+\mathbf{p}) = R(r,\mathbf{p})u(\mathbf{x}); \quad r, x_i, p_i \in \mathbb{R}_+, \quad (i = 1, 2, \dots, n) \tag{10.31}$$

is

$$u(\mathbf{x}) = b; \quad R(r,\mathbf{p}) = 1, \tag{10.32}$$

where b is an arbitrary constant.

Proof: Making $P(r,\mathbf{p}) = 0$ in (10.30) we get $b = 0$ and $c_i = 0$ $(i = 1, 2, \dots, n)$. Thus, the only possible solution is the constant solution (10.32). ∎

10.4 A statistical model for lifetime analysis

In many practical engineering situations, the lifetime variable, T, appears as a random variable which depends on one regressor variable, X. This is, for example, the case of the fatigue life of wires, strands or tendons, the time up to breakdown in solid dielectrics or the time up to failure of marine breakwaters, which depend on the regressor variables stress range, voltage stress or wave height, respectively.

As a consequence, a regression model as shown in Figure 10.4, could be a convenient approach to the problem. The model is completely established as soon as the cumulative conditional distribution function of lifetime, $F(t, x)$, is defined for every value x of X.

Two different ways in which the engineer can tackle the problem are:

1. Using standard linear regression models in order to fit the experimental data.

2. Creating adequate models not only to fit the experimental data but also to satisfy physical and theoretical considerations.

By the first approach we mean the use of ready-made regression models, i.e., models not specially designed for the problem under consideration, but very well recognized by statisticians and experienced engineers. This is the most generally accepted approach because of its simplicity, its widespread use and the possibility of performing many standard and simple analyses, such as confidence limit analysis for example.

In the cases in which the application of the first approach is not satisfactory, the engineer tries to develop tailor-made regression models. These models can either reflect his experience and feeling about the problem or be based on physical and theoretical considerations.

In the following paragraphs, we derive a statistical model for lifetime analysis related to the weakest link principle with a wide applicability to engineering problems.

10.4.1 Derivation of the fatigue model

Castillo et al. (1985) justify the following assumptions for the fatigue model:

1. **Weakest link principle**: This principle establishes that the fatigue lifetime of a longitudinal element is the minimum fatigue life of its constituent pieces.

2. **Stability**: The selected distribution function type must hold (be valid) for different specimen lengths.

3. **Limit behavior**: To include the extreme case of the size of the supposed pieces constituting the element going to zero, or the number of pieces going to infinity, it is convenient for the distribution function family to be an asymptotic family (see Galambos (1987) and Castillo (1988)).

4. **Limited range**: Experience shows that the lifetime T and the stress range X, have a finite lower end, which must coincide with the theoretical lower end of the selected cdf. This implies that the Weibull distribution is the only one satisfying requirements 1 to 4.

5. **Compatibility**: In the X-T field, the cumulative distribution function of the lifetime given stress range, $F(t;x)$, should be compatible with the cumulative distribution function of the stress range given lifetime, $E(x;t)$.

These conditions lead to the following functional equation

$$\begin{aligned} F(t,x) &= 1-\exp\left\{-\left[\frac{t-\gamma_t(x)}{\alpha_t(x)}\right]^{\beta_t(x)}\right\} \\ &= 1-\exp\left\{-\left[\frac{x-\gamma_x(t)}{\alpha_x(t)}\right]^{\beta_x(t)}\right\} = E(x;t), \end{aligned} \tag{10.33}$$

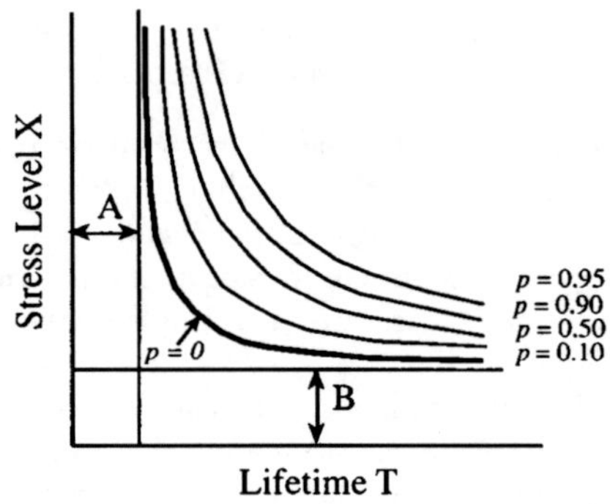

Figure 10.5: Wöhler field of model 1.

where $\gamma_t(x)$, $\alpha_t(x)$ and $\beta_t(x)$ are the location, scale and shape parameters of the Weibull laws for given x, and $\gamma_x(t)$, $\alpha_x(t)$ and $\beta_x(t)$ are the location, scale and shape parameters of the Weibull laws for given t.

Expression (10.33) is equivalent to the functional equation

$$\left[\frac{t-\gamma_t(x)}{\alpha_t(x)}\right]^{\beta_t(x)} = \left[\frac{x-\gamma_x(t)}{\alpha_x(t)}\right]^{\beta_x(t)}. \tag{10.34}$$

Gómez-Bayón (1984) and Castillo and Galambos (1987b) have shown that the only feasible solutions of Equation (10.33) are the three models:

- **Model 1** (see Figure 10.5):

$$F(t,x) = 1 - e^{-[(t-A)(x-B)/C+D]^E}. \tag{10.35}$$

- **Model 2** (see Figure 10.6):

$$F(t,x) = 1 - e^{-\left[C(t-A)^E(x-B)^D\right]}. \tag{10.36}$$

- **Model 3**:

$$F(t,x) = 1 - e^{-\left[C(t-A)^E(x-B)^D e^{F\log(t-A)\log(x-B)}\right]}, \tag{10.37}$$

where A,B, $C, E > 0$, D and F are arbitrary constants.

10.5 Statistical models for fatigue life of longitudinal elements

One of the most important problems when dealing with the statistical analysis of the fatigue life of longitudinal elements is the size effect; that is, the influence of length on the survivor function.

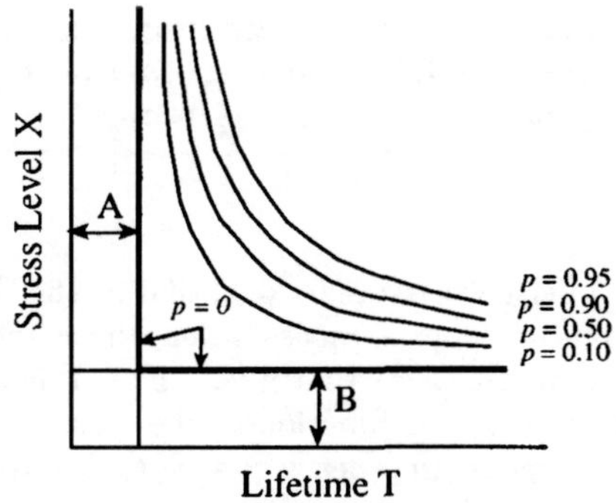

Figure 10.6: Wöhler field of model 2.

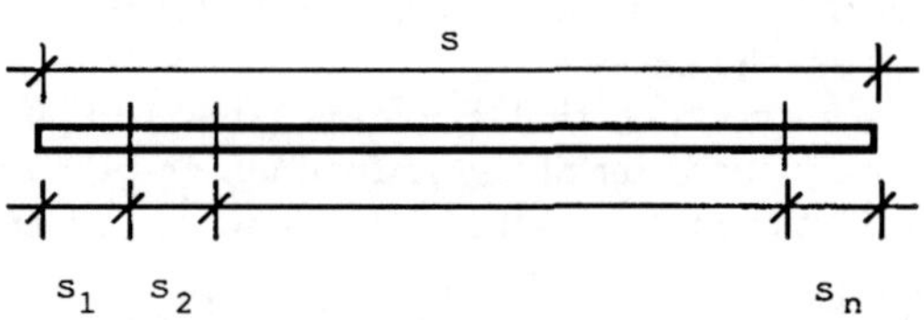

Figure 10.7: Illustration of the hypothesis of independence.

By longitudinal element we understand an element satisfying the following two conditions:

- only one dimension is important in the behavior of the element and
- if the element is longitudinally divided into imaginary pieces (see Figure 10.7) all pieces are subject to the same external action (stress, force, etc.)

Several models have been given in the past to solve this problem, but, unfortunately, most of them are based on the assumption of independence of the fatigue life of non-overlapping pieces. This assumption states that if an element of length s, such as that shown in Figure 10.7, is hypothetically divided into several pieces of lengths $s_1, s_2, ..., s_n$, then the survivor function of the element $S(s, z)$ must satisfy the equation

$$S(s, z) = \prod_{i=1}^{n} S(s_i, z).$$

Here we shall abandon the independence assumption and, making use of the functional equations theory, we shall state the problem in a very different way.

We shall assume here that a team of three members is required to design a consensus model for the analysis of the fatigue life of longitudinal elements.

However, they are required to give separate proposals before joining together and reaching a consensus. The three proposals associated with the three members will be denoted by models 1, 2 and 3, respectively.

Model 1

For the sake of simplicity we assume $n = 2$, that is, the element of length $x+y$ is divided into two non-overlapping pieces of lengths x and y. We also assume that there exists a function $S(x,z)$ that gives the survivor function of a piece of length x and that the survivor function of the element can be calculated in terms of that of the two pieces. In other words, $S(x,z)$ must satisfy the following functional equation

$$S(x+y,z) = H[S(x,z), S(y,z)], \tag{10.38}$$

where the function H indicates how the survivor function of the element can be obtained from those of the pieces.

It is worthwhile mentioning that Equation (10.38) implies the associativity and commutativity character of the H function and the dependence of the survivor function S on the total length of the element. In fact we can write

$$\begin{aligned} S(x+y+z,t) &= H[S(x+y,t), S(z,t)] \\ &= H[H[S(x,t), S(y,t)], S(z,t)] \\ &= H[S(x,t), S(y+z,t)] \\ &= H[S(x,t), H[S(y,t), S(z,t)]] \end{aligned}$$

and

$$S(x+y,t) = H[S(x,t), S(y,t)] = S(y+x,t) = H[S(y,t), S(x,t)].$$

Thus, the survivor function of an element of length s is independent of the number and size of the sub-elements into which it is divided in order to calculate it, using (10.38).

In the following paragraphs we solve functional Equation (10.38) in two different forms.

The functional Equation (10.38) is a particular case of the functional equation

$$S[G(x,y), z] = H[M(x,z), N(y,z)], \tag{10.39}$$

with $M = N = S$ and $G(x,y) = x + y$.

In this case we can easily satisfy all regularity conditions in Theorem 7.10, because $S(x,0) = 1$ and we can choose families of survivor functions such that $S_1(x,c) \neq 0$ and functions H such that $H_1 \neq 0$ and $H_2 \neq 0$. The general solution of (10.39) is (see Theorem 7.10 and Expression (7.80))

$$\begin{aligned} S(x,z) &= l[f(z)g^{-1}(x) + \alpha(z) + \beta(z)], \\ G(x,y) &= g[h(x) + k(y)], \\ H(x,y) &= l[m(x) + n(y)], \\ M(x,z) &= m^{-1}[f(z)h(x) + \alpha(z)] \\ N(x,z) &= n^{-1}[f(z)k(x) + \beta(z)] \end{aligned} \tag{10.40}$$

where g, h, k, l, m and n are arbitrary strictly monotonic continuously differentiable functions and f, α and β are arbitrary continuously differentiable functions.

Thus, for Equation (10.38) we have

$$\begin{aligned} S(x,z) &= l[f(z)g^{-1}(x)+\alpha(z)+\beta(z)] \\ &= m^{-1}[f(z)h(x)+\alpha(z)] \\ &= n^{-1}[f(z)k(x)+\beta(z)] \\ g[h(x)+k(y)] &= x+y, \end{aligned} \tag{10.41}$$

from which

$$g^{-1}(x+y)=h(x)+k(y),$$

which is Pexider's equation I with the general continuous-at-a-point solution (see Theorem 4.1)

$$g^{-1}(x)=Ax+B+C;\ \ h(x)=Ax+B;\ \ k(x)=Ax+C.$$

With this, expressions (10.41) become

$$\begin{aligned} S(x,z) &= l[f(z)(Ax+B+C)+\alpha(z)+\beta(z)] \\ &= m^{-1}[f(z)[Ax+B]+\alpha(z)] \\ &= n^{-1}[f(z)[Ax+C]+\beta(z)], \end{aligned}$$

and making $Af(z)x=u$ we obtain

$$\begin{aligned} S(x,z) &= l[u+(B+C)f(z)+\alpha(z)+\beta(z)] \\ &= m^{-1}[u+Bf(z)+\alpha(z)] \\ &= n^{-1}[u+Cf(z)+\beta(z)], \end{aligned} \tag{10.42}$$

and, using Lemma 6.1, we get

$$\begin{aligned} u &= cu+a \ \Rightarrow\ c=1,\ a=0, \\ l(x) &= m^{-1}\left(\frac{x-a-b}{c}\right) \\ &= m^{-1}(x-b) \\ (B+C)f(z)+\alpha(z)+\beta(z) &= Bf(z)+\alpha(z)+b \Rightarrow \beta(z)=b-Cf(z). \end{aligned}$$

Then, (10.42) becomes

$$S(x,z)=m^{-1}[u+Bf(z)+\alpha(z)]=n^{-1}(u+b),$$

which implies

$$\begin{aligned} u &= c_1u+a_1 \ \Rightarrow\ c_1=1,\ a_1=0, \\ m^{-1}(x) &= n^{-1}\left(\frac{x-a_1-b_1}{c_1}\right)=n^{-1}(x-b_1), \\ Bf(z)+\alpha(z) &= c_1b+b_1=b+b_1 \ \Rightarrow\ \alpha(z)=b+b_1-Bf(z), \end{aligned}$$

and finally we get the desired solution

Model 1: The general solution of (10.38) is:

$$S(x,z) = w[f(z)x]; \;\; H(x,y) = w[w^{-1}(x) + w^{-1}(y)],$$

where we have made $w^{-1}(x) = \dfrac{[n(x) - b]}{A}$.

Due to the weakest link principle and because $S(x,z)$ is a survivor function, it must be non-increasing in z and x. Then, in addition, we must have

$$S(x,0) = 1 \;\Rightarrow\; w[f(0)x] = 1 \;\Rightarrow\; \begin{cases} [f(0) = 0\,;\; w(0) = 1], \text{ or} \\ [f(0) = \infty\,;\; w(\infty) = 1], \text{ or}\,. \\ [f(0) = -\infty\,;\; w(-\infty) = 1] \end{cases}$$

$$S(x,\infty) = 0 \;\Rightarrow\; w[f(\infty)x] = 0 \;\Rightarrow\; \begin{cases} [f(\infty) = 0\,;\; w(0) = 0], \text{ or} \\ [f(\infty) = \infty\,;\; w(\infty) = 0], \text{ or}\,. \\ [f(\infty) = -\infty\,;\; w(-\infty) = 0] \end{cases}$$

If $w(x) = \exp(Dx)$ we get the model of independence.

The structure of the function H reveals its above mentioned associative and commutative character.

We can solve (10.38) in a much easier way if we observe that the variable z plays the role of one parameter, i.e., for any fixed value of z, Equation (10.38) can be written in the form

$$S(x+y) = H[S(x), S(y)],$$

and due to the associative character of H we can write (see Theorem 6.6)

$$H(x,y) = w[w^{-1}(x) + w^{-1}(y)],$$

and its substitution into (10.38) leads to

$$\begin{aligned} S(x+y,z) &= w\{w^{-1}[S(x,z)] + w^{-1}[S(y,z)]\} \Rightarrow \\ &\Rightarrow G(x+y,z) = G(x,z) + G(y,z), \end{aligned}$$

with $G(x,z) = w^{-1}S(x,z)$, which, for z held constant, is Cauchy's Equation (3.7) and then (see Theorem 3.3):

$$G(x,z) = f(z)x \;\Rightarrow\; S(x,z) = w[f(z)x].$$

Model 2

Member 2 in the team wants to start from the following result : Bogdanoff and Kozin (1987) based on some experimental results of Picciotto (1970), suggest the following model for the survivor function

$$S(x,z) = S(y,z)^{N(y,x)}, \tag{10.43}$$

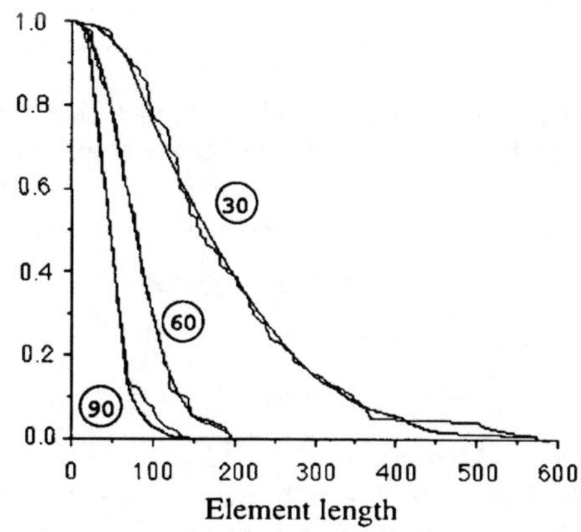

Figure 10.8: Experimental and theoretical survivor functions for lengths 30, 60 and 90 cm. (from Bogdanoff and Kozin (1987)).

where $S(x,z)$ and $S(y,z)$ are the survivor functions associated with two elements of lengths x and y, respectively, and $N(y,x)$ is an unknown function (see Example 6.2).

Figure 10.8 shows the experimental survivor functions and those obtained using model (10.43) (see Bogdanoff and Kozin (1987)).

Note that (10.43) is an implicit function of $S(x,z)$, or in other words, it is a functional equation. Thus, it must be solved to know what the Bogdanoff and Kozin proposal is. Castillo et al. (1990a) showed that the only compatible functions for $N(y,x)$ are those of the form $N(y,x) = \dfrac{q(x)}{q(y)}$.

From Example 6.2, we get

$$\textbf{Model 2}: \; S(x,z) = p(z)^{q(x)}; \; N(y,x) = \frac{q(x)}{q(y)}. \tag{10.44}$$

For $S(x,z)$ to be a survivor function it must be non-increasing in z and we must have

$$\begin{array}{lllll} S(x,0) = 1 & \Rightarrow & p(0)^{q(x)} = 1 & \Rightarrow & p(0) = 1, \\ S(x,\infty) = 0 & \Rightarrow & p(\infty)^{q(x)} = 0 & \Rightarrow & p(\infty) = 0. \end{array}$$

If $q(x) = x$ we get the model of independence.

The hazard function associated with $S(x,z)$ is

$$h(x,z) = \frac{-p'(z)}{p(z)} q(x) = -[\log p(z)]' q(x) = s(z) q(x),$$

which shows that Model 2 is the Cox–proportional hazards model (see Cox (1972)). Thus, functional Equation (10.43) characterizes the proportional hazards Cox-model.

Model 3

Member 3, based on expression (10.43), assumes that the survivor function of one element of length x can be obtained from the survivor function of one element of length y and a given, but unknown, function of x and y. In other words he assumes that the survivor function must satisfy the functional equation

$$S(x,z)=K[S(y,z),N(x,y)], \tag{10.45}$$

which is a particular case of (6.36)

$$G(x,y)=K[M(x,z),N(y,z)],$$

with $S(x,y)=G(y,x)=M(y,x)$.

If we choose F and N to be invertible with respect to their first argument and K invertible with respect to its first argument for a fixed value of the second, then the regularity conditions in Theorem 6.5 hold and the general solution of the last equation is (see Expression (6.37)):

$$\begin{array}{rclrcl} G(x,y) &=& f^{-1}[p(x)+q(y)]; & K(x,y) &=& f^{-1}[l(x)+n(y)]; \\ M(x,y) &=& l^{-1}[p(x)+r(y)]; & N(x,y) &=& n^{-1}[q(x)-r(y)], \end{array} \tag{10.46}$$

and then, for (10.45) we must have

$$\begin{array}{rcl} S(x,z) &=& f^{-1}[p(z)+q(x)] \\ &=& l^{-1}[p(z)+r(x)], \\ K(x,y) &=& f^{-1}[l(x)+n(y)], \\ N(y,z) &=& n^{-1}[q(y)-r(z)], \end{array} \tag{10.47}$$

which, by Lemma 6.1, implies

$$\begin{array}{rcl} p(z) &=& cp(z)+a \;\Rightarrow\; c=1,\;\; a=0, \\ f^{-1}(x) &=& l^{-1}\left(\dfrac{x-a-b}{c}\right)=l^{-1}(x-b), \\ q(x) &=& cr(x)+b=r(x)+b, \end{array}$$

and then, from Expression (10.47), model 3 becomes

$$\begin{array}{rrcl} \textbf{Model 3}: & S(x,z) &=& l^{-1}[p(z)+r(x)], \\ & K(x,y) &=& l^{-1}[l(x)+m(y)], \\ & N(x,y) &=& m^{-1}[r(x)-r(y)], \end{array} \tag{10.48}$$

where we have made $m(x)=n(x)-b$.

For $S(x,z)$ to be a survivor function it must be non-increasing in z and we must have

$$S(x,0)=l^{-1}[p(0)+r(x)]=1 \;\Rightarrow\; \left\{\begin{array}{l} [l(1)=p(0)=-\infty],\;\; \text{or} \\ [l(1)=p(0)=\infty],\;\; \text{or} \\ [r(x)=l(1)-p(0)] \end{array}\right. .$$

$$S(x,\infty)=l^{-1}[p(\infty)+r(x)]=0 \;\Rightarrow\; \left\{\begin{array}{l} [l(0)=p(\infty)=\infty],\;\; \text{or} \\ [l(0)=p(\infty)=-\infty],\;\; \text{or}\,. \\ [r(x)=l(0)-p(\infty)] \end{array}\right.$$

If $l^{-1}(x) = \exp[D\exp(Cx)]$ we get the model of independence.

One important aspect to point out here is that Equation (10.45) is more than a simple generalization of Equation (10.43). In fact, it includes some extra compatibility conditions, in the sense that no arbitrary $N(y,x)$ is admissible in Model 3, even though, initially, the function $N(y,x)$ seems to be arbitrary. In order to prove this, we show that the only admissible N functions are those appearing in Model 2 (Expression (10.44)).

Let us assume that function K is that implied from Equation (10.48), that is

$$K(x,y) = x^y = l^{-1}[l(x) + m(y)] \;\Rightarrow\; l(x^y) = l(x) + m(y),$$

which, by making the change of variable $u = \log(x)$, can be written

$$l[\exp(uy)] = l[\exp(u)] + m(y).$$

This is a Pexider functional equation with solution

$$l(x) = c\log[\alpha\log(x)]; \quad m(y) = c\log(y).$$

Thus, we finally get

$$\begin{aligned} N(x,y) &= m^{-1}[r(x) - r(y)] = \exp\left[\frac{r(x)-r(y)}{c}\right] = \frac{q(x)}{q(y)}, \\ q(x) &= \exp\left[\frac{r(x)}{c}\right], \end{aligned}$$

which is model 2.

Reaching a consensus

In the second and final step the team is required to join and reach a consensus.

Normally, a consensus solution is understood as a linear combination of the quantitative judgments of several individuals. However, in many cases the consensual solution reached does not satisfy many of the properties that were satisfied by the solutions in the proposals given by the different individuals (see Genest and Zidek (1986)). This fact is irrelevant when one tries to use the consensus model to make some evaluations, such as to calculate some probabilities, for example, but becomes a very serious inconvenience when one tries to model a physical system. In fact, the functional equations (10.38), (10.43) and (10.45) state some properties, which the different members understand the physical system must satisfy. Thus, any member would not accept models violating his/her associated functional equation. Thus, in the following, we shall understand consensus as the intersection of the three families of models, if it exists, i.e., as models satisfying all the requirements. We start by analyzing the common part of Models 1 and 2 (see the corresponding equations).

$$S(x,z) \;=\; w[f(z)x] = p(z)^{q(x)},$$

which implies

$$\log[S(x,z)] = \log\{w[f(z)x]\} = q(x)\log[p(z)],$$

and making the change of variable $u = f(z)$ we get

$$\log\{w[ux]\} = q(x)\log\{p[f^{-1}(u)]\},$$

which is Pexider's functional Equation (4.4) with the general continuous-at-a-point solution (see Theorem 4.4)

$$\log\{[w(x)]\} = ABx^C, \;\; q(x) = Ax^C, \;\; \log\{p[f^{-1}(x)]\} = Bx^C.$$

Thus,

$$\begin{array}{lllllll} S(x,z) & = & \exp\left\{AB[f(z)x]^C\right\}; & w(x) & = & \exp\left(ABx^C\right); \\ p(z) & = & \exp[Bf^C(z)]; & q(x) & = & Ax^C, \end{array} \tag{10.49}$$

which shows that Models 1 and 2 are not coincident but they share the common model

$$S(x,z) = \exp\left\{AB[f(z)x]^C\right\} \equiv S(x,z) = \exp[f(z)x]^C = \beta(z)^{x^C}, \tag{10.50}$$

where $\beta(z)$ is an arbitrary positive function.

Model (10.50) for $C = 1$ becomes the model of independence.

The hazard function for this model is

$$h(x,z) = \frac{\dfrac{d\beta(z)}{dz}}{\beta(z)}x^c = -\frac{d\{\log[\beta(z)]\}}{dz}x^c = s(z)x^c.$$

If now we look for the common part of Models 1 and 3 we get the functional equation

$$S(x,z) = l^{-1}[p(z) + r(x)] = w[f(z)x],$$

which, by making the change of variable $u = f(z)$, becomes Pexider's Equation (4.3)

$$l[w(ux)] = p[f^{-1}(u)] + r(x),$$

with the general continuous-at-a-point solution (see Theorem 4.3)

$$l[w(x)] = A\log(BCx); \;\; p[f^{-1}(x)] = A\log(Bx); \;\; r(x) = A\log(Cx),$$

and then we finally get

$$w(x) = l^{-1}[A\log(BCx)]; \;\; f^{-1}(x) = p^{-1}[A\log(Bx)]; \;\; r(x) = A\log(Cx),$$

which shows that Model 1 is a particular case of Model 3.

Finally, we compare Models 2 and 3. For the coincidence we must have

$$S(x,z) = l^{-1}[p_1(z) + r(x)] = p(z)^{q(x)},$$

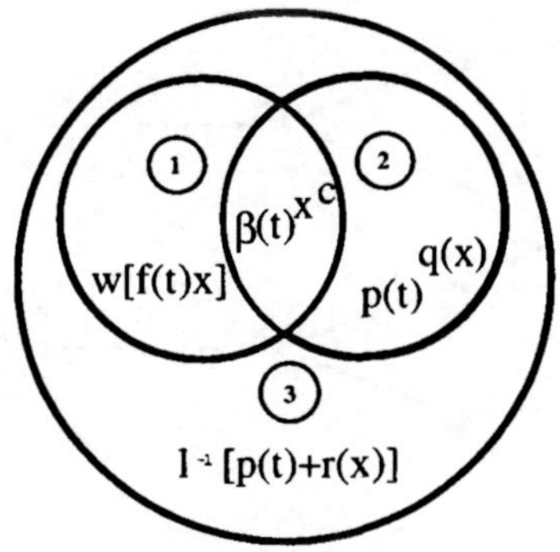

Figure 10.9: Illustration of separate and consensus proposals.

and taking logarithms we get

$$q(x)\log p(z) = \log\{l^{-1}[p_1(z) + r(x)]\}$$

which implies

$$l\{\exp[q(x)\log p(z)]\} = p_1(z) + r(x),$$

which is Pexider's Equation (4.3). Thus, we have

$$\begin{array}{rcl} l[\exp(x)] & = & A\log(BCx), \\ r[q^{-1}(x)] & = & A\log(Bx), \\ p_1\left\{p^{-1}[\exp(x)]\right\} & = & A\log(Cx), \end{array}$$

and then

$$\begin{array}{rcl} l^{-1}(x) & = & \exp\left[\exp(x/A)/(BC)\right], \\ q(x) & = & \exp\left(r(x)/A\right)/B, \\ p(z) & = & \exp\left[\exp\left(p_1(z)/A\right)/C\right], \end{array}$$

which shows that Model 2 is a particular case of Model 3.

Thus, we can conclude that a consensus model could be model (10.50), which is the family of models common to all three members of the team.

Figure 10.9 shows the required separate and consensus proposals as well as the common proposals associated with all three groups of only two members.

As a final conclusion, we can add that functional equations can prove themselves to be a very powerful tool to be used in model design. As a matter of fact, the engineer can state all the conditions to be satisfied by the desired model in terms of functional equations. Then, by first solving the resulting system and then in terms of its general solution, one can make the selection by playing with the remaining degrees of freedom.

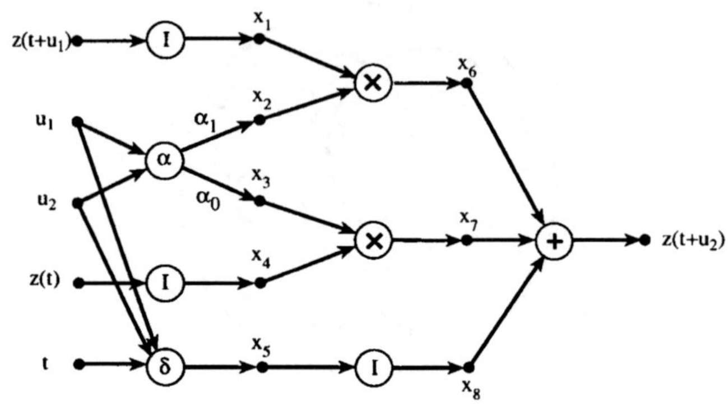

Figure 10.10: Functional network associated with functional Equation (7.2).

10.6 Differential, functional and difference equations

In this section we use the functional network methodology described in Chapter 9 for predicting values of magnitudes satisfying differential, functional and/or difference equations, and for obtaining the difference and differential equation associated with a set of data. As we shall show, the estimation of the differential or difference equation coefficients is carried out simply solving systems of linear equations, in the cases of equally or unequally spaced or missing data points.

10.6.1 A motivating example

In this section we use the example in Section 7.2.

Equation (7.2) can be represented by the network in Figure 10.10, where I is used to refer to the identity function. Similarly, Equation (7.4) can be represented by the network in Figure 10.11. Both cases correspond to functional networks, such as those introduced in Chapter 9. Therefore, we proceed by applying the functional networks formalism, as shown in Section 9.5.

Step 1 (Statement of the problem): Understanding of the problem to be solved, which was done in Chapter 7.

Step 2 (Initial topology): Based on the knowledge of the problem, the topology of the initial functional network is selected. For example, the vibrating functional, (7.2), and difference, (7.4), equations of the mass problem have led to the two functional networks in Figures 10.10 and 10.11, respectively.

Step 3 (Simplification): In this step, the initial functional network is simplified using functional equations. It must be noted that when there are coincident

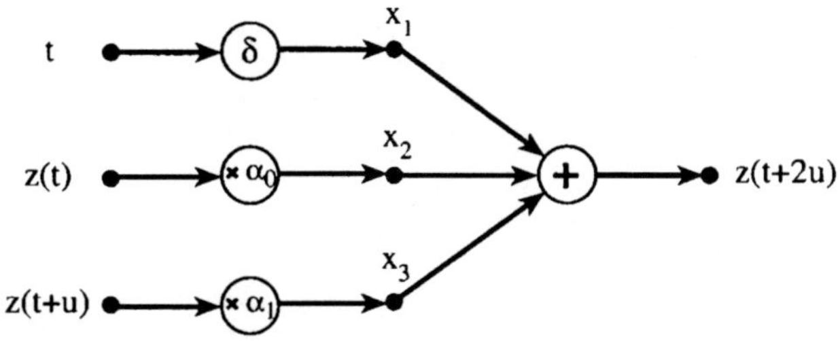

Figure 10.11: Functional network associated with difference Equation (7.4).

neural outputs, they must coincide in values, and this leads to a functional equation which allows the initial topology of the functional network to be simplified. This is not the case of the functional networks in Figures 10.10 and 10.11. Thus, no simplification is possible here. For some illustrative examples on this step, the reader is referred to Section 9.6.

Step 4 (Uniqueness of representation): In this step, uniqueness conditions for the neural functions to be unique must be found. For example, in the case of the network in Figure 10.10, we can consider the possibility of the existence of two sets of functions $\{\alpha_0, \alpha_1, \delta\}$ and $\{\alpha_0^*, \alpha_1^*, \delta^*\}$ such that

$$\begin{aligned} z(t+u_2) &= \alpha_0(u_1,u_2)z(t) + \alpha_1(u_1,u_2)z(t+u_1) + \delta(t;u_1,u_2) \\ &= \alpha_0^*(u_1,u_2)z(t) + \alpha_1^*(u_1,u_2)z(t+u_1) + \delta^*(t;u_1,u_2); \end{aligned} \tag{10.51}$$

that is, about the existence of two different functional networks with the same structure leading to the same outputs for the same inputs.

Step 5 (Data collection): For the learning to be possible we need some data. In this step, we consider two different cases: (a) equally spaced data, and (b) unequally spaced data. For the sake of clarity, the following steps of the functional network approach will be described for each of the two cases.

Equally spaced data

We start by analyzing the case with equally spaced data. Let us assume that we have available the data given in Table 10.1, which consists of the vibrating mass displacements z corresponding to different times t.

In the case of equally spaced data (constant u), we use Equation (7.4), where $\alpha_0(u)$ and $\alpha_1(u)$ for constant u are constants, and function $\delta(t)$ can be approximated by a linear combination of a set of linearly independent functions $\{\phi_i(t) | i = 1, \ldots, m\}$.

If $z(t_j)$ for $j = 0, \ldots, n$ are the observed data for equally spaced times t_j, the solution of a differential equation of order k with constant coefficients in

Table 10.1: Observed displacements z of system in Figure 7.2 for different times t.

t	z	t	z	t	z	t	z	t	z
0.0	0.100	0.04	0.177	0.08	0.241	0.12	0.277	0.16	0.278
0.2	0.238	0.24	0.155	0.28	0.036	0.32	-0.113	0.36	-0.277
0.4	-0.441	0.44	-0.590	0.48	-0.708	0.52	-0.781	0.56	-0.800
0.6	-0.760	0.64	-0.660	0.68	-0.507	0.72	-0.31	0.76	-0.086
0.8	0.151	0.84	0.381	0.88	0.586	0.92	0.752	0.96	0.867
1.0	0.924	1.04	0.921	1.08	0.861	1.12	0.754	1.16	0.612
1.2	0.450	1.24	0.284	1.28	0.130	1.32	0.001	1.36	-0.093
1.4	-0.145	1.44	-0.156	1.48	-0.130	1.52	-0.074	1.56	-0.000
1.6	0.077	1.64	0.144	1.68	0.188	1.72	0.199	1.76	0.170
1.8	0.098	1.84	-0.012	1.88	-0.153	1.92	-0.313	1.96	-0.478
2.0	-0.632	2.04	-0.758	2.08	-0.843	2.12	-0.876	2.16	-0.850
2.2	-0.764	2.24	-0.623	2.28	-0.435	2.32	-0.214	2.36	0.023
2.4	0.258	2.44	0.474	2.48	0.654	2.52	0.786	2.56	0.861
2.6	0.876	2.64	0.833	2.68	0.740	2.72	0.608	2.76	0.452
2.8	0.287	2.84	0.130	2.88	-0.005	2.92	-0.107	2.96	-0.169
3.0	-0.189	3.04	-0.170	3.08	-0.119	3.12	-0.047	3.16	0.032
3.2	0.105	3.24	0.158	3.28	0.180	3.32	0.163	3.36	0.105
3.4	0.006	3.44	-0.126	3.48	-0.282	3.52	-0.446	3.56	-0.603
3.6	-0.737	3.64	-0.834	3.68	-0.880	3.72	-0.870	3.76	-0.799
3.8	-0.671	3.84	-0.494	3.88	-0.280	3.92	-0.045	3.96	0.194
4.0	0.417								

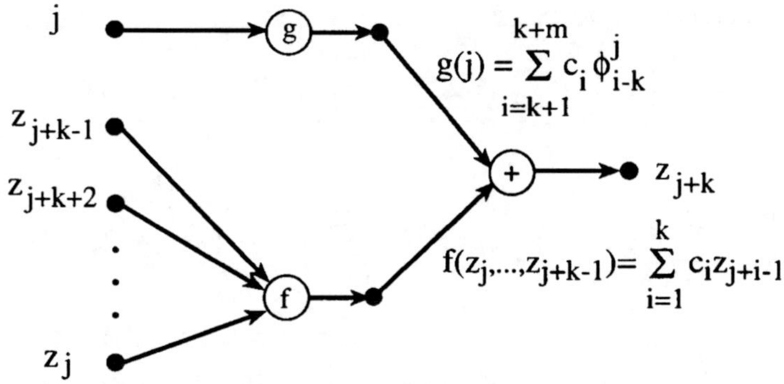

Figure 10.12: Functional network associated with Equation (10.52).

$z(t)$ can be approximated using the model

$$z_{j+k} = \sum_{i=1}^{k} c_i \; z_{j+i-1} + \sum_{i=k+1}^{k+m} c_i \; \phi^j_{i-k}; \;\; j = 0, \ldots, n-k, \tag{10.52}$$

where $u = t_{j+1} - t_j$ for $j = 0, \ldots, n-1$, $z_j = z(t_0 + ju)$, $\phi^j_i = \phi_i(t_0 + ju)$ and $c_1, \ldots, c_{k+m}$ are constant coefficients.

The functional network associated with (10.52) is given in Figure 10.12.

Step 6 (Learning): At this point, the neural functions are estimated (learned) by using some minimization methods. In our example, the error e_{j+k} at the point $t_{j+k} = t_0 + (j+k)u$ using the approximation given by (10.52) becomes

$$e_{j+k} = z_{j+k} - \sum_{i=1}^{k} c_i \; z_{j+i-1} - \sum_{i=k+1}^{k+m} c_i \phi^j_{i-k}; \;\; j = 0, \ldots, n-k. \tag{10.53}$$

Thus, the parameters $c_1, \ldots, c_{k+m}$ can be estimated by minimizing

$$Q = \sum_{j=0}^{n-k} e^2_{j+k} = \sum_{j=0}^{n-k} \left(z_{j+k} - \sum_{i=1}^{k} c_i z_{j+i-1} - \sum_{i=k+1}^{k+m} c_i \phi^j_{i-k} \right)^2. \tag{10.54}$$

The minimum is obtained for

$$-\frac{1}{2}\frac{\partial Q}{\partial c_r} = \sum_{j=0}^{n-k}\left(z_{j+k} - \sum_{i=1}^{k} c_i z_{j+i-1} - \sum_{i=k+1}^{k+m} c_i \phi_{i-k}^j\right) z_{j+r-1} = 0,$$

$$r = 1, \ldots, k, \tag{10.55}$$

$$-\frac{1}{2}\frac{\partial Q}{\partial c_r} = \sum_{j=0}^{n-k}\left(z_{j+k} - \sum_{i=1}^{k} c_i z_{j+i-1} - \sum_{i=k+1}^{k+m} c_i \phi_{i-k}^j\right) \phi_{r-k}^j = 0,$$

$$r = k+1, \ldots, k+m.$$

This leads to the system of linear equations with $k+m$ unknowns

$$\mathbf{Ac} = \mathbf{b} \Leftrightarrow \begin{pmatrix} \mathbf{A}_{11} & | & \mathbf{A}_{12} \\ -- & + & -- \\ \mathbf{A}_{21} & | & \mathbf{A}_{22} \end{pmatrix} \begin{pmatrix} \mathbf{c}_1 \\ -- \\ \mathbf{c}_2 \end{pmatrix} = \begin{pmatrix} \mathbf{b}_1 \\ -- \\ \mathbf{b}_2 \end{pmatrix}. \tag{10.56}$$

From (10.55) we can write the expressions for each element a_{rs} of $\mathbf{A}$ and b_r of $\mathbf{b}$:

$$\begin{array}{rcll} a_{rs} & = & \sum_{j=0}^{n-k} z_{j+s-1} z_{j+r-1} & \text{if} \quad r = 1, \ldots, k, \;\; s = 1, \ldots, k, \\ a_{rs} & = & \sum_{j=0}^{n-k} \phi_{s-k}^j z_{j+r-1} & \text{if} \quad r = 1, \ldots, k, \;\; s = k+1, \ldots, k+m, \\ a_{rs} & = & \sum_{j=0}^{n-k} z_{j+s-1} \phi_{r-k}^j & \text{if} \quad r = k+1, \ldots, k+m, \;\; s = 1, \ldots, k, \\ a_{rs} & = & \sum_{j=0}^{n-k} \phi_{s-k}^j \phi_{r-k}^j & \text{if} \quad r = k+1, \ldots, k+m, \;\; s = k+1, \ldots, k+m, \\ b_r & = & \sum_{j=0}^{n-k} z_{j+k} z_{j+r-1} & \text{if} \quad r = 1, \ldots, k, \\ b_r & = & \sum_{j=0}^{n-k} z_{j+k} \phi_{r-k}^j & \text{if} \quad r = k+1, \ldots, k+m. \end{array} \tag{10.57}$$

Finally, from (10.56) we get

$$\mathbf{c} = \mathbf{A}^{-1}\mathbf{b}, \tag{10.58}$$

which gives the solution.

Returning to the system in Figure 7.2, if we use the functions

$$\{\phi_1(t), \phi_2(t), \phi_3(t), \phi_4(t), \phi_5(t)\} = \{1, \sin(t), \cos(t), \sin(2t), \cos(2t)\}.$$

and the equally spaced observed displacements for different times shown in Table 10.1 or Figure 10.13, we get:

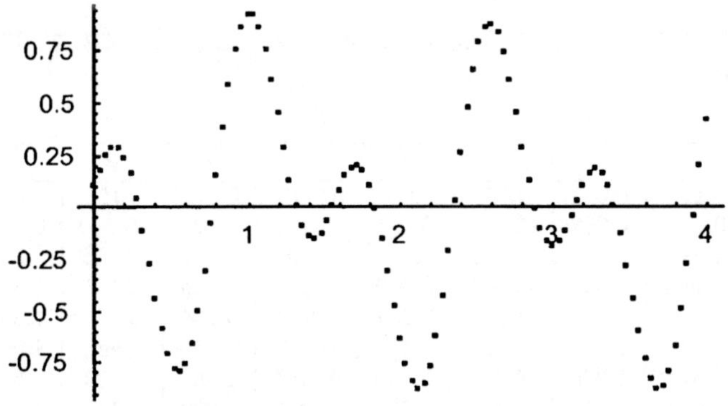

Figure 10.13: Observed data z for the displacement of system in Figure 7.2.

$$\mathbf{A} = \begin{pmatrix} 24.40 & 23.60 & -3.67 & 5.08 & 3.56 & -4.46 & -1.63 \\ 23.60 & 24.40 & -3.58 & 4.80 & 3.52 & -4.12 & -2.10 \\ -3.67 & -3.58 & 99.00 & 41.70 & -19.10 & 13.80 & 11.90 \\ 5.08 & 4.80 & 41.70 & 43.0 & 6.91 & -6.85 & -20.10 \\ 3.56 & 3.52 & -19.10 & 6.91 & 55.50 & 21.60 & -12.20 \\ -4.46 & -4.12 & 13.80 & -6.85 & 21.60 & 50.40 & 6.18 \\ -1.63 & -2.10 & 11.90 & -20.10 & -12.0 & 6.18 & 48.60 \end{pmatrix} \tag{10.59}$$

and

$$\mathbf{b} = \begin{pmatrix} 21.30 \\ 23.70 \\ -3.34 \\ 4.35 \\ 3.25 \\ -3.52 \\ -2.62 \end{pmatrix}, \tag{10.60}$$

which leads to

$$\mathbf{c} = \begin{pmatrix} -1.0130 \\ 1.9442 \\ -0.0214 \\ 0.0335 \\ -0.0155 \\ 0.0154 \\ 0.0091 \end{pmatrix}. \tag{10.61}$$

Step 7 (Model validation): Finally, using (10.52) with the values in (10.61) we can predict displacements which are visually indistinguishable from those in

Table 10.2: Observed displacements z of system in Figure 7.2 for different random times t.

t	z	t	z	t	z	t	z	t	z
0.000	0.100	0.001	0.102	0.009	0.119	0.028	0.155	0.040	0.177
0.258	0.105	0.308	-0.067	0.321	-0.117	0.327	-0.140	0.331	-0.155
0.413	-0.491	0.494	-0.739	0.534	-0.794	0.561	-0.800	0.588	-0.777
0.639	-0.664	0.640	-0.661	0.660	-0.588	0.666	-0.565	0.695	-0.436
0.741	-0.195	0.781	0.038	0.781	0.039	0.784	0.056	0.813	0.230
0.831	0.333	1.050	0.915	1.100	0.826	1.110	0.786	1.200	0.459
1.220	0.377	1.370	-0.109	1.450	-0.154	1.580	0.036	1.580	0.048
1.590	0.054	1.610	0.094	1.610	0.097	1.620	0.110	1.660	0.166
1.780	0.135	1.790	0.125	1.820	0.057	1.820	0.052	1.890	-0.188
1.950	-0.437	1.960	-0.487	1.970	-0.520	1.980	-0.542	2.040	-0.747
2.120	-0.876	2.140	-0.871	2.220	-0.707	2.230	-0.646	2.330	-0.178
2.340	-0.099	2.410	0.315	2.420	0.385	2.430	0.434	2.450	0.517
2.490	0.682	2.600	0.877	2.610	0.867	2.670	0.767	2.710	0.640
2.740	0.545	2.760	0.432	2.820	0.223	2.830	0.164	2.840	0.115
3.000	-0.189	3.001	-0.189	3.040	-0.173	3.090	-0.109	3.110	-0.066
3.110	-0.065	3.200	0.103	3.220	0.136	3.260	0.173	3.310	0.172
3.310	0.169	3.340	0.133	3.420	-0.047	3.480	-0.300	3.500	-0.345
3.520	-0.437	3.520	-0.453	3.630	-0.817	3.690	-0.884	3.700	-0.883
3.750	-0.823	3.750	-0.821	3.770	-0.769	3.770	-0.758	3.780	-0.750
3.850	-0.454	3.880	-0.294	3.940	0.057	3.950	0.127	3.970	0.238

Figure 10.13. In fact, we get a maximum absolute prediction error of 0.0334 and a medium absolute prediction error of 0.0132.

To test the possibility of over-fitting, we have obtained the RMSE (root mean squared error) for the training data and a set of 1000 test data points, obtaining the following results

$$RMSE_{training} = 0.018; \quad RMSE_{testing} = 0.042,$$

which shows that the error increase is not very high.

Step 8 (Use of the model): At this step, the model is ready to be used.

Unequally spaced data

Table 10.2 shows 100 observed displacements of the system in Figure 7.2 for random times. In this case we use the Expression (10.58) to predict the behavior of the system using these observed displacements and two different models approximating the functions α_0, α_1 and δ involved in it. Note that this approach is also valid for the case of missing data.

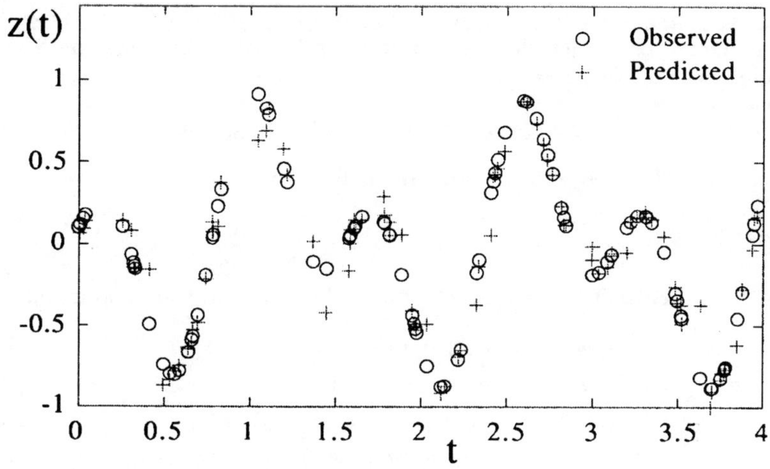

Figure 10.14: Observed and predicted displacements.

Model 1:

Step 6 (Learning): Suppose that functions α_0, α_1 and δ are approximated by

$$\alpha_0(u_1, u_2) = a_1 + a_2 u_1 + a_3 u_2, \quad (10.62)$$

$$\alpha_1(u_1, u_2) = b_1 + b_2 u_1 + b_3 u_2, \quad (10.63)$$

$$\delta(t, u_1, u_2) = c_1 + c_2 \sin(t) + c_3 \cos(t) + c_4 \sin(2t) + c_5 \cos(2t), \quad (10.64)$$

where a_i, b_i and c_i are parameters to be estimated. To this aim, we define the function

$$F(\mathbf{a}, \mathbf{b}, \mathbf{c}) = \sum_{i=3}^{100} [z^*(t_i) - z(t_i)]^2, \quad (10.65)$$

where $z^*(t_i)$ is the predicted displacement for time t_i using Expression (10.58). The minimum of this function is attained at:

$$\begin{array}{llllll} a_1 = & 1.603, & a_2 = & -23.663, & a_3 = & 15.514, \\ b_1 = & -0.591, & b_2 = & 26.331, & b_3 = & -18.058, \\ c_1 = & -0.007, & c_2 = & 0.030, & c_3 = & -0.013, \\ c_4 = & 0.014, & c_5 = & -0.015. & & \end{array}$$

Step 7 (Model validation): Using these parameters, the medium absolute prediction error is $E = 0.064$. Figure 10.14 shows the observed and predicted displacements of the system in Figure 7.2.

To test the possibility of over-fitting, we have obtained the RMSE (root mean squared error) for the training data and a set of 1000 test data points, obtaining the following results

$$RMSE_{training} = 0.11; \quad RMSE_{testing} = 0.15,$$

which shows that the error increase is small.

Model 2:

Step 6 (Learning): Suppose that functions α_0, α_1 and δ are approximated by

$$\begin{aligned}
\alpha_0(u_1, u_2) &= a_1 + a_2 u_1 + a_3 u_2, &(10.66)\\
\alpha_1(u_1, u_2) &= b_1 + b_2 u_1 + b_3 u_2, &(10.67)\\
\delta(t, u_1, u_2) &= c_1(u_1, u_2) + c_2(u_1, u_2)\sin(t) &(10.68)\\
&\quad + c_3(u_1, u_2)\cos(t) + c_4(u_1, u_2)\sin(2t) &(10.69)\\
&\quad + c_5(u_1, u_2)\cos(2t), &(10.70)
\end{aligned}$$

where, $c_i(u_1, u_2) = c_{i1} + c_{i2}u_1 + c_{i3}u_2$, for $i = 1, \dots, 5$, and a_i, b_i and c_{ij} are parameters to be estimated. To this purpose, we define the function

$$F(\mathbf{a}, \mathbf{b}, \mathbf{c}) = \sum_{i=3}^{100} [z^*(t_i) - z(t_i)]^2, \tag{10.71}$$

where $z^*(t_i)$ is the predicted displacement for time t_i using Expression (10.58). The minimum of this function is attained for:

$$\begin{array}{llllll}
a_1 = & 1.749, & a_2 = & -27.282, & a_3 = & 16.893,\\
b_1 = & -0.694, & b_2 = & 30.703, & b_3 = & -20.628,\\
c_{11} = & 0.042, & c_{12} = & 1.706, & c_{13} = & -1.999,\\
c_{14} = & -0.077, & c_{15} = & -2.620, & c_{21} = & 3.415,\\
c_{22} = & 0.006, & c_{23} = & 0.302, & c_{24} = & -0.717,\\
c_{25} = & -0.049, & c_{31} = & -0.189, & c_{32} = & 1.196,\\
c_{33} = & 0.014, & c_{34} = & -1.870, & c_{35} = & 0.929.
\end{array}$$

Step 7 (Model validation): Using these parameters, the medium absolute prediction error is $E = 0.060$. This shows that it is not worthwhile including non-constant $c_i(u_1, u_2)$, $i = 1, 2, 3, 4, 5$ functions.

Exercises

10.1 To solve the problem in Section 10.5 a fourth member proposes a model of the type

$$S(x, u + v) = H[S(x, u), S(x, v)].$$

Obtain its general solution, discuss the new consensus solution and give it a physical interpretation.

10.2 Obtain a general equation for hydraulic problems knowing that in the most general case you can have the following magnitudes:

- Four geometric length parameters: length (a), width (b), depth (c) and roughness (r)
- One cinematic parameter: velocity (v)
- One parameter related to internal forces: pressure (p)
- Density (d)
- Specific weight (g)
- Viscosity (m)
- Surface stress (s)
- Elasticity (E)

10.3 Consider the physical relations:

$$e = log(vt).$$

and

$$e = c_1 + c_2 t.$$

where e is the space, v the velocity and t the time. Are they valid physical relations?

Hint: perform the change of variables

$$e \to r_e e + e_0; \quad v \to r_v v + v_0; \quad t \to r_t t + t_0$$

and see what you get.

10.4 Consider the formula

$$V_T = \beta U^\gamma (S^{1/2}/K)^{\gamma-1}(q_s - q).$$

where

- V_T evaporation under turbulent conditions $[LT^{-1}]$
- $q_s - q$ humidity difference (saturation deficit) where q is the specific humidity of air and q_s is the specific humidity of saturation at the temperature of the water surface (both are dimensionless)
- K mean turbulent diffusion coefficient of water vapor in air $[L^2T^{-1}]$
- S water surface area $[L^2]$
- U wind speed (mean value) $[LT^{-1}]$
- β and γ are dimensionless constants.

Is it a valid physical relation?

10.5 Build a fatigue model with a Weibull type survival function and check which of the three models proposed in Section 10.5 it belongs to.

10.6 Add a new proposal for a model of the statistical analysis of the fatigue life of longitudinal elements (size effect) described in Section 10.5 such that the resulting model after consensus with the three other proposals becomes more restrictive than the model in (10.50).

10.7 Design a functional network to solve the differential equation

$$z''(x) + (a + b)z'(x) + abz(x) = 0.$$

Hint: Use the results in Example 7.12.

10.8 Design two functional networks to solve the differential equations

$$xf'(x) - kf(x) = 0$$

$$f'(x) - \frac{k}{x}f(x) = x^2$$

and

$$f''(x) - f(x) = 0.$$

discussed in Examples 7.14, 7.15 and 7.16, respectively.

CHAPTER 11

Applications to Geometry and CAGD

11.1 Introduction

The aim of this chapter is to show some interesting applications of functional equations to Geometry and CAGD (*Computer Aided Geometric Design*). The chapter starts, in Section 11.2, with a characterization of the well known Euler formula for polyhedra by means of several sets of functional equations which arise when some perturbations are performed on polyhedra. Section 11.3 deals with some functional equations used to characterize two functions (the absolute value and the integral part functions) with many applications in computer graphics. Section 11.4 analyzes the problem of finding the functions preserving a geometric invariant (such as the distance between two points, angles between two intersecting lines, tangential distance between two spheres, etc.) through the functional equations to be satisfied by these functions. Then, some applications of functional equations to CAGD are considered in Section 11.5. First, after introducing some preliminary results in Section 11.5.1, we apply functional equations to characterize the most general surfaces in implicit form, such that their intersections with the planes $z = z_0$, $y = y_0$ and $x = x_0$ are linear combinations of sets of given functions of the other two variables (Section 11.5.2). In Section 11.5.3 we find the most general surfaces in explicit form such that their intersections with planes parallel to the planes $y = 0$ and $x = 0$ belong to given parametric families of curves. In Section 11.5.4 we use functional equations to analyze the uniqueness of representation of Gordon-Coons-surfaces. In Section 11.5.5 we discuss the tensor product surfaces. Finally, in Section 11.6 we apply functional networks to solve the problem of fitting surfaces, showing its main advantages with respect to neural networks. The performance of this new method is illustrated by revisiting the previously analyzed cases of implicit and explicit surfaces.

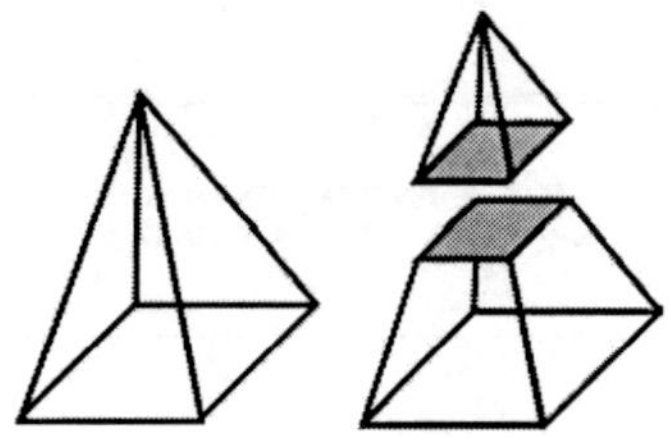

Figure 11.1: Truncation of one vertex with 4 intersecting edges.

11.2 Fundamental formula for polyhedra

It is well known that every polyhedron satisfies the following formula

$$F = E - V + 2, \tag{11.1}$$

where F is the number of faces, E the number of edges and V the number of vertices of the polyhedron.

In this example, we characterize this formula by means of equations or systems of functional equations. We assume that there is an unknown expression $F = f(E, V)$ that gives the number of faces as a function of the number of edges and the number of vertices of the polyhedron.

To derive some functional equations to be satisfied by the Euler formula, we perform some perturbations of a polyhedron such that it is transformed into another polyhedron with different numbers of faces, vertices and edges.

Perturbation 1 : *Truncation of one vertex with x intersecting edges.*
Figure 11.1 illustrates how a polyhedron can be obtained from another polyhedron by truncating one vertex with 4 intersecting edges.

In this case we have

$$F_2 = F_1 + 1; \;\; V_2 = V_1 + x - 1; \;\; E_2 = E_1 + x, \tag{11.2}$$

where E_1, F_1, V_1 and E_2, F_2, V_2 are the numbers of edges, faces and vertices of polyhedra 1 and 2, respectively, and x is the number of edges intersecting at the given vertex.

Thus, we have

$$F_1 + 1 = f(E_1, V_1) + 1 = F_2 = f(E_2, V_2) = f(E_1 + x, V_1 + x - 1), \tag{11.3}$$

which leads to the functional equation

$$f(e, v) + 1 = f(e + x, v + x - 1); \;\; \forall\ x, e \text{ and } v. \tag{11.4}$$

Figure 11.2: Splitting a x-polygon face of a polyhedron.

Making $e = e_0$ and calling

$$g(v) = f(e_0, v) + 1; \quad r = x + e_0; \quad s = v + x - 1,$$

we get

$$h(s - r) = f(r, s) \quad \text{where} \quad h(s - r) = g(s - r + e_0 + 1), \tag{11.5}$$

and substitution into (11.4) leads to

$$h(v - e) + 1 = h(v - e - 1) \Rightarrow h(y) - h(y - 1) + 1 = 0, \tag{11.6}$$

which is a difference equation with the general solution

$$h(y) = K - y, \tag{11.7}$$

where K is an arbitrary constant.

With this, (11.5) becomes

$$f(e, v) = e - v + K,$$

which is the general solution of (11.4).

Perturbation 2: *Splitting an x-polygon face into x faces.*
In Figure 11.2 we split an x-polygon face of a polyhedron into x faces to get a new polyhedron. According to this figure we have

$$F_2 = F_1 + x - 1; \quad V_2 = V_1 + 1; \quad E_2 = E_1 + x,$$

and then

$$F_2 = F_1 + x - 1 = f(E_1, V_1) + x - 1 = f(E_2, V_2) = f(E_1 + x, V_1 + 1).$$

Thus, we have the functional equation

$$f(e + x, v + 1) = f(e, v) + x - 1. \tag{11.8}$$

To solve (11.8) we make $x = y$ and $x = y + 1$ and we get

$$f(e + y, v + 1) = f(e, v) + y - 1,$$
$$f(e + y + 1, v + 1) = f(e, v) + y,$$

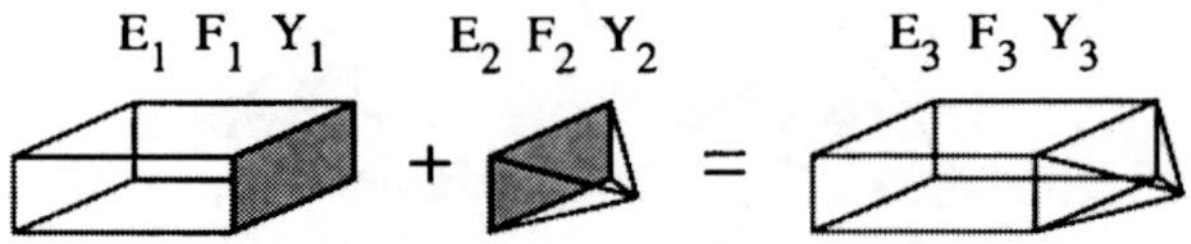

Figure 11.3: Joining two polyhedra with a common face ($z = 0$).

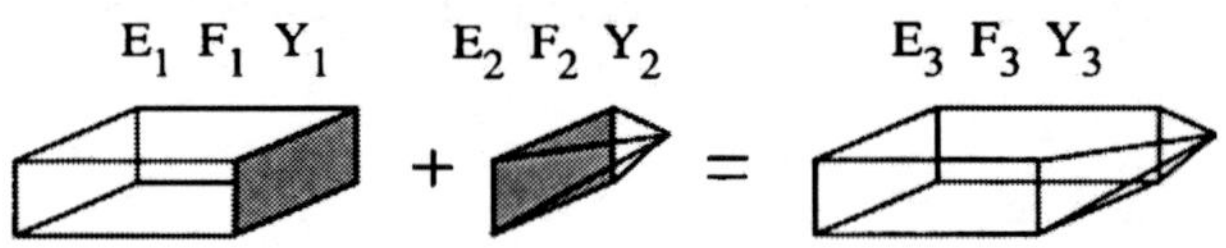

Figure 11.4: Joining two polyhedra with a common face ($z = 1$).

and subtracting both equations and calling $p = e + y$ and $q = v + 1$ we get

$$f(p + 1, q) - f(p, q) = 1, \tag{11.9}$$

which for any q is a difference equation with general solution

$$f(e, v) = C(v) + e, \tag{11.10}$$

where $C(v)$ is an arbitrary function.

Substitution of (11.10) into (11.8) leads to

$$C(v + 1) - C(v) + 1 = 0, \tag{11.11}$$

which is a difference equation with general solution

$$C(v) = K - v, \tag{11.12}$$

where K is an arbitrary constant.

Thus, we finally get

$$f(e, v) = e - v + K, \tag{11.13}$$

which is the general solution of (11.8).

Perturbation 3: *Joining two polyhedra with a common face.* When joining two polyhedra with a common face (see Figures 11.3 to 11.6), the number of faces of the new generated polyhedron depends on whether or not the two faces sharing one edge of the initially common face are coplanary. This dilemma exists for every edge of the common face. So, if we call z the number of coplanary

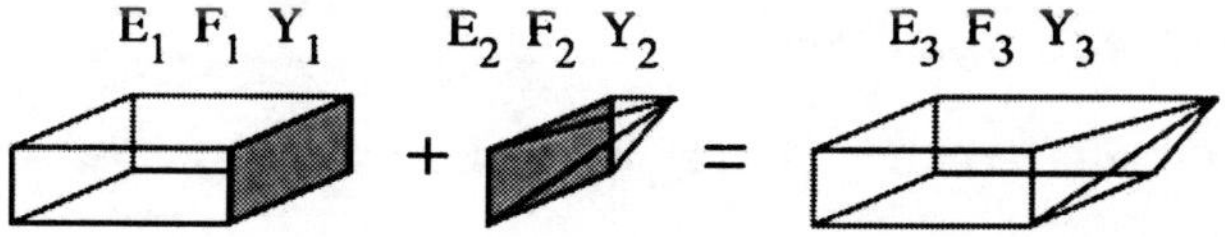

Figure 11.5: Joining two polyhedra with a common face ($z = 2$).

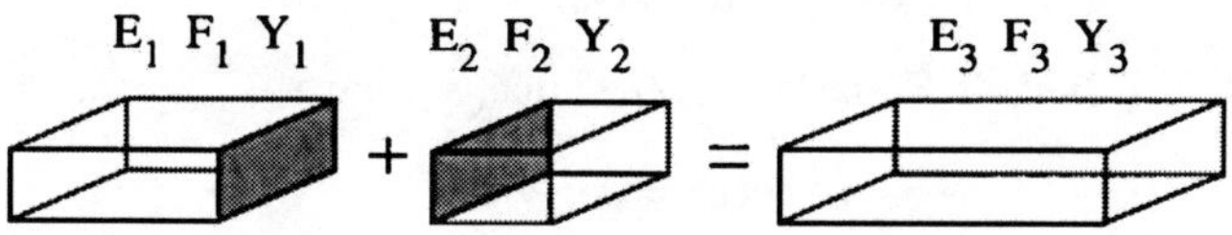

Figure 11.6: Joining two polyhedra with a common face ($z = 4$).

occurrences and x the number of sides of the common face, we get

$$\begin{array}{l} F_3 = F_1 + F_2 - 2 - z, \\ V_3 = \left\{ \begin{array}{lll} V_1 + V_2 - x - z + 1 & \text{if} & 0 < z < x, \\ V_1 + V_2 - x - z & \text{if ,} & z = x \text{ or } z = 0, \end{array} \right. \\ E_3 = \left\{ \begin{array}{lll} E_1 + E_2 - x - 2z + 1 & \text{if} & 0 < z < x, \\ E_1 + E_2 - x - 2z & \text{if} & z = x \text{ o } z = 0. \end{array} \right. \end{array} \tag{11.14}$$

Thus, we get the following functional equations

$$\begin{array}{l} f(E_1, V_1) + f(E_2, V_2) - 2 - z \\ = \left\{ \begin{array}{lll} f(E_1 + E_2 - x - 2z + 1, V_1 + V_2 - x - z + 1) & \text{if} & 0 < z < x, \\ f(E_1 + E_2 - x - 2z, V_1 + V_2 - x - z) & \text{if} & z = 0 \text{ or } x. \end{array} \right. \end{array} \tag{11.15}$$

Both right hand sides of (11.15) are equivalent; i.e., they imply the same solutions. Note that the first comes from the second by substituting x by $x - 1$.

To solve (11.15) we make $z = z_0$ and $z = z_1$ and we subtract both to get

$$\begin{array}{l} f(E_1 + E_2 - x - 2z_0 + 1, V_1 + V_2 - x - z_0 + 1) - \\ \quad -f(E_1 + E_2 - x - 2z_1 + 1, V_1 + V_2 - x - z_1 + 1) = z_1 - z_0, \end{array}$$

and if we now call

$$p = E_1 + E_2 - x - 2z_0 + 1, \quad q = V_1 + V_2 - x - z_0 + 1, \quad h_0 = z_0 - z_1,$$

the result is

$$f(p, q) - f(p + 2h_0, q + h_0) = -h_0, \tag{11.16}$$

the general solution of which is

$$f(p, q) = g(2q - p) + \frac{p}{2}, \tag{11.17}$$

where g is an arbitrary function.

Now substituting (11.17) into (11.15) and making $x = x_0$ we get

$$h(u) + h(v) = h(u + v), \tag{11.18}$$

where we have made

$$h(u) = g(u + x_0) - 2 + \frac{x_0}{2}.$$

But (11.18) is Cauchy's Equation (3.7). Thus, we have

$$g(u) = \alpha u + \beta,$$

and then (11.17) becomes

$$f(p, q) = 2\alpha q + (\frac{1}{2} - \alpha)p + \beta. \tag{11.19}$$

Substituting this into (11.15) we finally get

$$\beta - 2 = -x(\alpha + \frac{1}{2}) \quad \Rightarrow \quad \alpha = \frac{-1}{2}; \ \beta = 2,$$

which implies

$$f(e, v) = e - v + 2, \tag{11.20}$$

which is the general solution of (11.15).

Perturbation 3a : *Joining two polyhedra with a common face and no coplanary faces.* This is the case of perturbation 3 with $z = 0$, which leads to the functional equation

$$f(E_1 + E_2 - x, V_1 + V_2 - x) = f(E_1, V_1) + f(E_2, V_2) - 2; \ \forall x. \tag{11.21}$$

Making $x = V_1 + V_2 - K$ and $x = K$ with $K = constant$, we get

$$\begin{aligned} f(E_1 + E_2 - V_1 - V_2 + K, K) &= f(E_1, V_1) + f(E_2, V_2) - 2, \\ f(E_1 + E_2 - K, V_1 + V_2 - K) &= f(E_1, V_1) + f(E_2, V_2) - 2, \end{aligned}$$

from which, by making

$$E = E_1 + E_2 - K, \ \ V = V_1 + V_2 - K,$$

we get

$$f(E - V + K, K) = f(E, V) \Rightarrow f(E, V) = h(E - V). \tag{11.22}$$

Substitution of (11.22) into (11.21) and calling $x = E_1 - V_1$ and $y = E_2 - V_2$ leads to

$$h(x + y) = h(x) + h(y) - 2,$$

the general solution of which is

$$h(y) = \alpha y + 2.$$

Thus, (11.22) becomes

$$f(e, v) = \alpha(e - v) + 2. \tag{11.23}$$

Now we shall prove that equations (11.4) and (11.21) or equations (11.8) and (11.21) characterize the Euler formula (11.1). Because equations (11.4) and (11.8) have the same general solution they are equivalent. Thus, we only demonstrate the first case.

Taking into account that the general solutions of the functional equations (11.4) and (11.21) are (11.13) and (11.23), the general solution of the system (11.4)-(11.21) is given by the solution of the functional equation

$$e - v + K = \alpha(e - v) + 2 \quad \Rightarrow \quad e(\alpha - 1) + v(1 - \alpha) + 2 - K = 0,$$

which implies $\alpha = 1$ and $K = 2$. Thus, the general solution of the system (11.4)-(11.21) becomes

$$f(e, v) = e - v + 2.$$

Then, we have characterized Euler's formula by three different forms, i.e., Equation (11.15) and systems (11.4)–(11.21) and (11.8)–(11.21).

11.3 Two interesting functions in computer graphics

This section is devoted to describing an interesting application of functional equations to computer graphics. In a recent paper, Monreal and Santos (1998) pointed out that, despite the most usual curves and surfaces in the computer graphics field being based on piecewise polynomial functions (Bézier curves, B-splines, etc.), some other functions may also give a natural and nice looking solution for modelling certain shapes. In particular, in the context of architectural design a particular need for non-smooth functions is often found. Among them, they considered the absolute value function for surfaces with sharp edges (roofs, staircases, etc.), the integral part function for surfaces with jump discontinuities, or the Max and Min functions for truncations. These functions, though readily available in most computer packages, have not been used as frequently as their smooth counterparts (splines, etc.). In this section, following Monreal and Santos (1998), we proceed to derive and solve some functional equations that, under some assumptions, characterize some non–smooth functions.

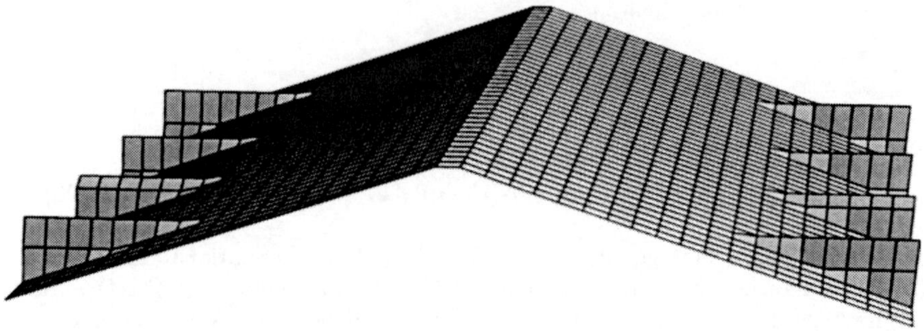

Figure 11.7: A graphical representation of a roof obtained by the simple use of the absolute value and Max functions.

11.3.1 Absolute value function

The problem to be solved can be better understood by looking at Figure 11.7, which is given by the following expression:

$$z = \text{Max}\left(1 - \left|x - 1 - 2\lfloor\frac{x}{2}\rfloor\right|, \frac{3}{2} - \left|\frac{2y}{3}\right|\right), \tag{11.24}$$

and shows a roof consisting of two parts: the central part depends only on y and the dormer windows come from the absolute value function and the integral part function (indicated by "$\lfloor . \rfloor$" in expression 11.24) depending on x. The Max function provides the upper part of the intersection.

In the generalization of this roof, expressions of the Max, Min and combinations of Max and Min of n elements are required.

For $n = 2$ it is obvious that

$$Max(x, y) = \frac{x + y + |x - y|}{2}. \tag{11.25}$$

Using this expression recursively leads to

$$Max(x, y, z) = Max(x, Max(y, z)) = Max\left(x, \frac{y + z + |y - z|}{2}\right), \tag{11.26}$$

where the last term includes the absolute value of differences and sums of absolute values. Such an operation is too difficult for practical calculations. This fact compels the authors to obtain formulas for $||x| - |y||$ or $||x| - y|$ in terms of $|x + y|$, $|x - y|$, $|x|$ and $|y|$.

Considering the relation

$$||x| + |y|| = |x| + |y|,$$

jointly with

$$||x| + |y|| + ||x| - |y|| = |x + y| + |x - y|,$$

they obtain the expression

$$||x| - |y|| = |x + y| + |x - y| - |x| - |y|,$$

which reduces the nesting level of the absolute value.

The last equation shows that the absolute value function is a solution of the functional equation

$$f(x + y) = f(x) + f(y) + f(f(x) - f(y)) - f(x - y). \tag{11.27}$$

The next theorem solves this equation.

Theorem 11.1 (An absolute value functional equation). *A function $f : \mathbb{R} \to \mathbb{R}$ which is continuous in $\mathbb{R}^+$, is a solution of Equation (11.27) if and only if $f(x) = x$ or $f(x) = -|x|$ or $f(x) = 0$ or $f(x) = |x|$.* ∎

Proof: If we substitute in (11.27):

1. $x = y = 0$, we obtain $f(0) = 0$.

2. $y = 0$, we have $f(f(x)) = f(x)$, $\forall x$ in $\mathbb{R}$.

3. $x = y$, we obtain $f(2x) = 2f(x)$, $\forall x$ in $\mathbb{R}$.

4. $x = 2y$, and taking into account (b) and (c), we have $f(3x) = 3f(x)$, $\forall x$ in $\mathbb{R}$.

By induction, for all n, m in $\mathbb{Z}$, $f(2^n x) = 2^n f(x)$ and $f(3^m x) = 3^m f(x)$ and consequently $f(2^n 3^m x) = 2^n 3^m f(x)$. Since the set $\{2^n 3^m / n, m \in \mathbb{Z}\}$ is dense in $\mathbb{R}^+$, we obtain $f(xy) = yf(x)$ for all x in $\mathbb{R}$ and y in $\mathbb{R}^+$. In particular, we have $f(x) = xf(1)$ for all x in $\mathbb{R}^+$ and $f(x) = -xf(-1)$ for all x in $\mathbb{R}^-$. Thus

$$f(x) = \begin{cases} ax & \text{if} \quad x > 0 \\ bx & \text{if} \quad x < 0 \end{cases}; \quad \text{for some } a,\ b \text{ in } \mathbb{R}.$$

Now, Equation (11.27) with $x > 0$, $y < 0$, $x + y > 0$, yields $f(ax - by) = ax - by$ and either $f \equiv 0$, $a = 1$ or $b = 1$. If $a = 1$, for $x < 0$, $y > 0$, $x + y > 0$ in (11.27), $x - by = f(bx - y)$. There are two possibilities

1. if $bx - y > 0$, then $f(bx - y) = bx - y = x - by$ and we deduce $b = 1$. Thus, $x - y > 0$, which is a contradiction.

2. if $bx - y < 0$, then $b^2 = 1$.

Therefore, if $a = 1$ then $f(x) = x$ for all x in $\mathbb{R}$ or f is the absolute value function. Analogously, if $b = 1$ we deduce $a^2 = 1$ and in this case the solution is the minus absolute value function. ∎

Let us consider now the expression for $|x - |y||$. First of all, note that:

$$\text{Max}(0, y - |x|) = \frac{|x + \text{Max}(y, 0)| - 2|x| + |x - \text{Max}(y, 0)|}{2},$$

which jointly with

$$\operatorname{Max}(0, y - |x|) = \frac{y + ||x| - y| - |x|}{2},$$

yields

$$|y - |x|| = |x + \operatorname{Max}(y,0)| + |\operatorname{Max}(y,0) - x| - y - |x|.$$

Hence, the absolute value function is a solution of

$$y + f(y - f(x)) + f(x) = f(\operatorname{Max}(y,0) + x) + f(\operatorname{Max}(y,0) - x), \quad (11.28)$$

the solution of which is given by the following theorem.

Theorem 11.2 (Characterization of the absolute value function). *The only solution of Equation (11.28) is the absolute value function.* ■

Proof: For $y \le 0$ and $x = 0$ in (11.28) we obtain $f(y - f(0)) = f(0) - y$, thus $f(z) = -z$ if $z + f(0) \le 0$.

For $y \le 0$ and $x + f(0) \le 0$, we deduce, using (11.28), that $f(-x) = f(x) = -x$.

Now, if $z \le -f(0)$, we have $f(z) = -z$ and if $z \ge f(0)$, then $f(z) = -z$ and we have that $f(0) \ge 0$.

But, for $x = 0$ and $y = f(0)$ in (11.28), we obtain $f(0) = 0$ and hence, f is the absolute value function. ■

In a similar way, we can take the relation

$$2|y||x| - y| = (y + |y|)(|x + y| + |y - x|) - 2y(|x| + |y|),$$

that leads to the functional equation

$$2f(x)f(f(x) - y) = (y + f(y)(f(x + y) + f(x - y) - 2y(f(x) + f(y)), \quad (11.29)$$

whose solution is given by the following theorem.

Theorem 11.3 (Another absolute value functional equation). *The unique solution $f : \mathbb{R} \to \mathbb{R}$ of Equation (11.29), injective on $\mathbb{R}^+$, or surjective, is the absolute value function.* ■

See Monreal and Santos (1998) for a proof.

11.3.2 Integral part function

This section characterizes another interesting non-smooth function in the field of computer graphics: the integral part function. As an illustration, Figure 11.8 shows a staircase, which has been modelled combining an integral part parametric function given by:

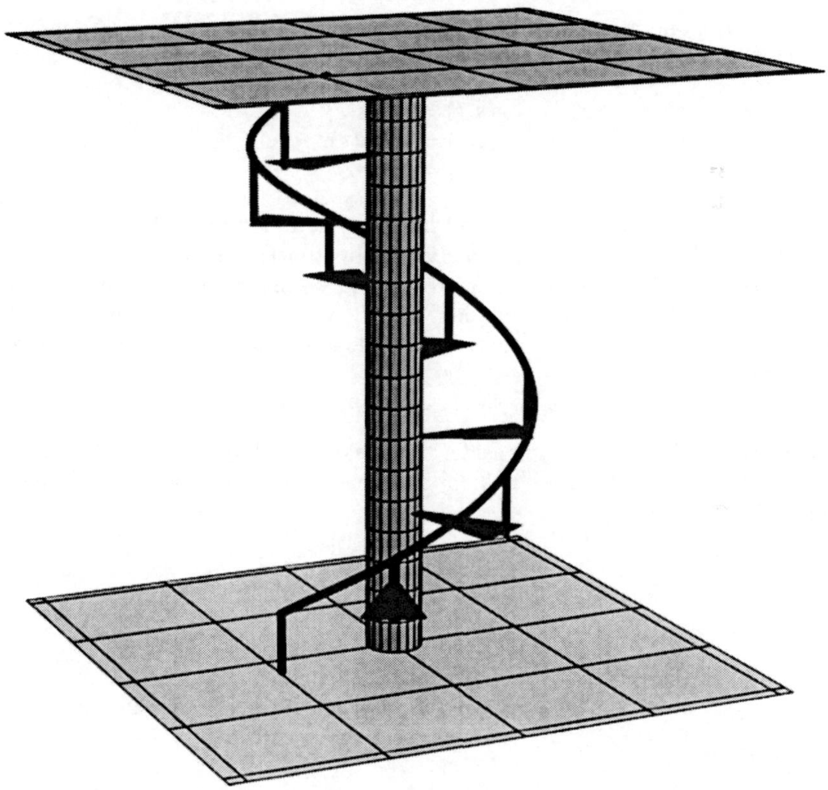

Figure 11.8: A graphical representation of a staircase obtained from the integral part function.

$$
\begin{aligned}
x(s,t) &= t\ Cos\left(\frac{s}{m}\right),\\
y(s,t) &= t\ Sin\left(\frac{s}{m}\right),\\
z(s,t) &= \frac{h}{n}\lfloor n\ \frac{s}{\alpha}\rfloor,
\end{aligned}
\tag{11.30}
$$

where m, n, h and α are fixed parameters, and the symbol $\lfloor . \rfloor$ is used to indicate the integral part function, with the parametric function describing the column:

$$
\begin{aligned}
x(s,t) &= rCos(s),\\
y(s,t) &= rSin(s),\\
z(s,t) &= t,
\end{aligned}
$$

where r takes the value 0.2. The rail has been obtained from (11.30) by taking $t = 1$ and additively increasing the s values by a factor of 0.5.

Following again Monreal and Santos' work, we include here some characterizations of the integral part function. First, we realize that the integral part does not satisfy the Cauchy equation. So, the aim is to find some expression of the form

$$[x+y] = [x] + [y] + h(x,y).$$

For (u,v) in $\mathbb{R}^+ \times \mathbb{R}^+$, they consider the square $R = [[u],[u]+1) \times [[v],[v]+1)$ and analyze the behavior of $[x]+[y]$ and $[x+y]$ in this square. According to Figure 11.9, we have

$$[x] + [y] = [u] + [v]; \quad \forall x, y \in R,$$

and

$$
[x+y] = \begin{cases} [u]+[v] & \text{if} \quad 0 \le x - [u] + y - [v] < 1,\\ [u]+[v]+1 & \text{if} \quad 1 \le x - [u] + y - [v] < 2. \end{cases}
$$

Therefore,

$$[x+y] = [x] + [y] + [x - [x] + y - [y]]; \quad \forall x, y \in \mathbb{R}.$$

Thus, the integral part function satisfies the functional equation

$$f(x+y) = f(x) + f(y) + f(x - f(x) + y - f(y)). \tag{11.31}$$

It is easy to verify that the function f is a solution of (11.31) if and only if the function h defined by $h(x) = x - f(x)$, $\forall x \in \mathbb{R}$ is a solution of

$$h(x+y) = h\left(h(x) + h(y)\right). \tag{11.32}$$

This leads to the theorem:

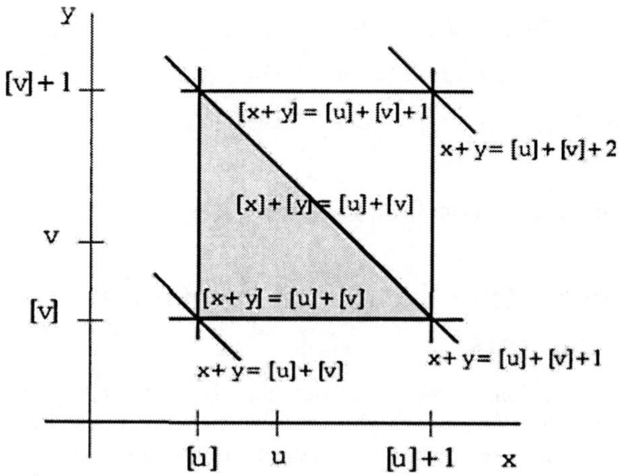

Figure 11.9: Relation between $[x+y]$ and $[x]+[y]$ functions in a certain square.

Theorem 11.4 (A characterization of the integral part function). *If $h : \mathbb{R} \to [0,1]$ is surjective, with $h(0) = h(1) = 0$ then h satisfies Equation (11.32) if and only if $h(x) = x - [x]$ for all x in $\mathbb{R}$.* ■

Proof: Taking $y = 0$ in (11.32) we get: $h(h(x)) = h(x)$, $\forall x \in \mathbb{R}$ so $h(p) = p$, $\forall p \in [0,1)$.

Taking now $y = 1$ in (11.32) we get: $h(x+1) = h(x)$, $\forall x \in \mathbb{R}$. Furthermore, since $x - [x] \in [0,1)$, $h(x) = h(x - [x]) = x - [x]$, $\forall x \in \mathbb{R}$. ■

Finally, some functional equations in a single variable can be considered in the problem of characterization of the integral part function. For example, the integral part function obviously holds:

$$f(f(x)) = f(x), \tag{11.33}$$

and

$$f(x+1) = f(x) + 1, \tag{11.34}$$

Moreover, the integral part function also verifies the functional equation:

$$f(x + \lambda(f(x+1) - x)) = f(x); \quad \lambda \in [0,1). \tag{11.35}$$

For the last equation they observed that $[x] = [u]$ for all x in $[u, [u]+1]$, i.e., for all λ in $[0,1)$, $[u + \lambda([u] + 1 - u)] = [u]$.

The next theorem gives some new characterizations of the integral part function:

Theorem 11.5 (Other characterizations of the integral part function). *If $f : \mathbb{R} \to \mathbb{Z}$ is a surjective function, then the following statements are equivalent:*

1. *f is the integral part function,*

2. *f satisfies Equation (11.33) and (11.35),*

3. *f satisfies Equation (11.34) and (11.35) and $f(1) > 0$.*

∎

See Monreal and Santos (1998) for the proofs.

11.4 Geometric invariants given by functional equations

Functional equations have been traditionally applied to many interesting geometric problems. Among them, we may cite the problem of finding the functions preserving a geometric invariant through functional equations to be satisfied by these functions. By geometric invariants we mean, for example, the distance between two points, angles between two intersecting lines, tangential distance between two spheres, etc. This problem was already analyzed in Benz (1993). This section is devoted merely to describing some of his interesting results. For more information, the reader is referred to the original paper.

11.4.1 The preserve distance functional equation

Suppose that $M \neq \emptyset$ and W are sets and that $d : M \times M \to W$ is a mapping. The structure (M, W, d) is called a *metric space* and $d(x, y)$ the *distance* of $x, y \in \mathbb{R}$.

Let now S be a fixed subset of $M \times M$. The problem of distance preserve consists of finding all functions $f : M \to M$ such that the the functional equation

$$d(f(x), f(y)) = d(x, y) \tag{11.36}$$

holds for all $(x, y) \in S$. In case $S = M \times M$ we call (11.36) universal.

We include here some examples from Benz's paper.

Example 11.1 (Example of a distance preserve problem). Let us suppose that $M = \mathbb{R}^2$, $W = \mathbb{R}$, and

$$d(x, y) = \sqrt{(x_1 - y_1)^2 + (x_2 - y_2)^2},$$

where we assume $x = (x_1, x_2)$ and $y = (y_1, y_2) \in \mathbb{R}^2$. We also take

$$S = \{(x, y) \in M \times M | d(x, y) = 1\}.$$

Putting $f(x) = (\varphi(x_1, x_2), \psi(x_1, x_2))$, Equation (11.36) becomes

$$\begin{aligned}&[\varphi(x_1 + \cos x_3, x_2 + \sin x_3) - \varphi(x_1, x_2)]^2\\&+ [\psi(x_1 + \cos x_3, x_2 + \sin x_3) - \psi(x_1, x_2)]^2 = 1; \quad \forall x_1, x_2, x_3 \in \mathbb{R},\end{aligned}$$

where we have selected two points $(x_1 + \cos x_3, x_2 + \sin x_3)$ and (x_1, x_2), at a unit distance. The solutions of this functional equation are given by

$$\begin{aligned}\varphi(x_1, x_2) &= x_1 \cos(t) - x_2 \sin(t) + a,\\ \psi(x_1, x_2) &= x_1 \sin(t) + x_2 \cos(t) + b,\end{aligned} \tag{11.37}$$

and

$$\begin{aligned}\varphi(x_1, x_2) &= x_1 \cos(t) + x_2 \sin(t) + a,\\ \psi(x_1, x_2) &= x_1 \sin(t) - x_2 \cos(t) + b,\end{aligned} \tag{11.38}$$

where $a, b, t \in \mathbb{R}$ are constants.

Note that both solutions can be expressed in terms of matrices and vectors, as

$$\begin{pmatrix} \cos(t) & \mp\sin(t) \\ \sin(t) & \pm\cos(t) \end{pmatrix} \cdot \begin{pmatrix} x_1 \\ x_2 \end{pmatrix} + \begin{pmatrix} a \\ b \end{pmatrix}, \tag{11.39}$$

corresponding to rotations (given by a certain angle t), reflections and translations by the vector $(a, b)^T$, where $(.)^T$ means the transpose of vector (a, b), which are *affine transformations*. This implies that, given a subset of $\mathbb{R}^2$, the only two–dimensional transformations preserving distances between their points are affine transformations. ■

Note that the preserve distance property corresponds to a limit case of contractions, Hutchinson (1981), which have great applications in fractal images, Barnsley (1990) and fractal compression, Barnsley and Hurd (1993). Suppose, for example, that in the same conditions of Example 11.1 we consider the condition (11.36) to be universal, i.e., $S = \mathbb{R}^2 \times \mathbb{R}^2$. In addition, we impose the pair (M, d) to be a metric space. In this case, let $\mathcal{F}$ be the set of compact subsets of M:

$$\mathcal{F} = \{T \subset \mathbb{R}^2 | T \;\text{ compact}\},$$

we can define the *distance from a point x to the set $B \in \mathcal{F}$* as:

$$d(x, B) = min\{d(x, y)/y \in B\}. \tag{11.40}$$

This definition can be extended to the *distance from the set $A \in \mathcal{F}$ to the set $B \in \mathcal{F}$* as:

$$d(A, B) = max\{d(x, B)/x \in A\}. \tag{11.41}$$

Note that, since the sets A and B are compact, this definition is meaningful. In particular, there are points $\bar{x} \in A$ and $\bar{y} \in B$ such that

$$d(A, B) = d(\bar{x}, \bar{y}).$$

In this situation, let w be a contraction mapping on the metric space (M, d). This means that there is a constant $0 \le s < 1$ such that

$$d(w(x), w(y)) \le s d(x, y). \tag{11.42}$$

Any such number is called a *contractivity factor for w*.

Note the differences between expressions (11.42) and (11.36). The last one is the limit case of the first one, in the sense that Expression (11.42) goes to (11.36) as $s \to 1$. However, the contractivity property is the key point to get an effective image rendering algorithm. Indeed, if w is a contraction, it induces another contraction mapping $W : \mathcal{F} \to \mathcal{F}$ defined by:

$$w(B) = \{w(x)|x \in B\}; \ \forall B \in \mathcal{F}, \tag{11.43}$$

with the same contractivity factor. Now, by the fixed point theorem, W poses exactly one fixed point P and, moreover, for any point B (compact subset of $\mathbb{R}^2$) of $\mathcal{F}$, the sequence $W^{\circ n}(B)$, where $W^{\circ n}$ indicates that W is composed n times with itself, converges to P.

The last step is to construct an iterated function system. It consists of a complete metric space (M, d) together a finite set of contraction mappings $W_i : \mathcal{F} \to \mathcal{F}$. Usually, the abbreviation "IFS" is used for "iterated function systems". Given an IFS, the transformation $W : \mathcal{F} \to \mathcal{F}$ defined by:

$$W(B) = \cup_{i=1}^{n} W_i(B) \tag{11.44}$$

is a contraction mapping too. Its unique fixed point satisfies:

$$W(A) = \cup_{i=1}^{n} W_i(A) = A, \tag{11.45}$$

and is given by

$$A = \lim_{n \to \infty} W^{\circ n}(B), \tag{11.46}$$

for any $B \in \mathcal{F}$.

This fixed point is then called the "attractor" of the IFS, because it attracts any trajectory, no matter what $B \in \mathcal{F}$ is taken as the initial point. The natural question now is: how can we render the fractal image for a given IFS? Equation (11.44) provides the simplest rendering algorithm, called the deterministic algorithm. It works by iterating the operator W, starting with an arbitrary subset of the plane. The iterates form a sequence of sets that converge to the attractor of the IFS. Figure 11.10 illustrates the convergence of this algorithm, starting with three different initial sets: a square, a triangle, and a line segment.

Figure 11.11 shows some examples of fractals obtained by using this algorithm. These pictures have been obtained by using a Mathematica package, `IFS.m`, described in Gutiérrez et al. (1997). For more information on fractal rendering we also refer the reader to Gutiérrez et al. (1996).

Example 11.2 (Example of a preserve distance problem). Now we take $M = \mathbb{R}^2$, $W = \mathbb{R}$,

$$S = \{(x, y) \in M \times M | d(x, y) = 1\}$$

and

$$d(x, y) = (x_1 - y_1)(x_2 - y_2),$$

Figure 11.10: The convergence of the deterministic algorithm for the Sierpinsky fractal, starting with three different initial sets: a square, a triangle, and a line segment.

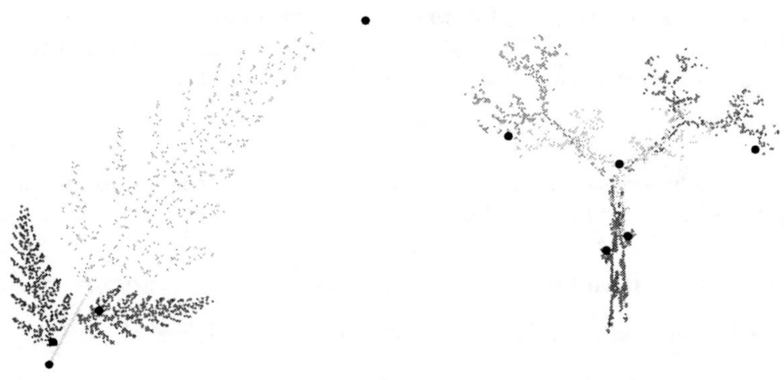

Figure 11.11: The fixed points and affine copies of the fern and tree fractals.

where once again $x = (x_1, x_2), y = (y_1, y_2) \in \mathbb{R}^2$. The notion of distance here is the Lorentz-Minkowski metric. In this case, (11.36) leads to

$$\left[\varphi\left(x_1 + z, x_2 + \frac{1}{z}\right) - \varphi(x_1, x_2)\right]\left[\psi\left(x_1 + z, x_2 + \frac{1}{z}\right) - \psi(x_1, x_2)\right] = 1,$$

for all $x_1, x_2, z \in \mathbb{R}$ with $z \neq 0$.

Its solutions are given by

$$\begin{aligned} \varphi(x_1, x_2) &= ax_1 + b, \\ \psi(x_1, x_2) &= \frac{1}{a}x_2 + c, \end{aligned}$$

and

$$\begin{aligned} \varphi(x_1, x_2) &= ax_2 + c, \\ \psi(x_1, x_2) &= \frac{1}{a}x_1 + c, \end{aligned}$$

where $a, b, c \in \mathbb{R}$ are constants with $a \neq 0$. ■

11.4.2 The functional equation of preserve area

Suppose that $M \neq \emptyset$ and W are sets and let $\mathcal{F} \neq \emptyset$ be a set of non-empty subsets of M. Suppose that $\alpha : \mathcal{F} \to W$ is a mapping. We then call $(M, \mathcal{F}, W, \alpha)$ an *area space.* The elements of $\mathcal{F}$ are called *figures* and $\alpha(F)$ is called the *area* of $F \in \mathcal{F}$. Let S be a fixed subset of $\mathcal{F}$.

The problem of *area preserve* consists in finding all functions $f : M \to M$ such that $f(F) \in \mathcal{F}$ and the functional equation

$$\alpha(f(F)) = \alpha(F) \tag{11.47}$$

holds for all $F \in S$.

Example 11.3 (Example of a preserve area problem). Consider $M = \mathbb{R}^3$, $W = \mathbb{R}$, $\mathcal{F} = \{T \subset \mathbb{R}^3 | 1 \leq \#T \leq 3\}$, where $\#T$ is the cardinal of the set T, and $\alpha(T) =$ area of the triangle $\{x, y, z\}$. For $T = \{x, y, z\} \in \mathcal{F}$ and

$$S = \{T \in \mathcal{F} | \alpha(T) = 1\}.$$

The set of solutions of (11.47) in this case is given by the group of congruent mappings of $\mathbb{R}^3$ (see Lester (1986)). ■

11.4.3 The functional equation of angle preserve

In this case, suppose that $M \neq \emptyset$ and W are sets and define

$$\mathcal{A} = \{(x, \{y, z\}) |\ x, y, z \in M \text{ with } x \notin \{y, z\}\}.$$

Let $\omega : \mathcal{A} \to W$ be a mapping. We then call $(M, \mathcal{A}, W, \omega)$ an *angle space.* Note that this name comes from the fact that if $M = \mathbb{R}^2$, we think of ω as being the angle between the lines xy and xz.

Let now S be a fixed subset of $\mathcal{A}$. Then the problem of angle preserve consists of finding all functions $f : M \to M$ such that $f(A) \in \mathcal{A}$ and the functional equation

$$\omega(f(A)) = \omega(A) \tag{11.48}$$

holds for all $A \in S$.

Example 11.4 (Example of angle preserve problem). For the universal case $S = \mathcal{A}$ and for $M = \mathbb{R}^2$, $W = \mathbb{R}$,

$$\omega(x, \{y, z\}) = \frac{[(y-x)\cdot(z-x)]^2}{(y-x)^2(z-x)^2},$$

where "." denotes the dot product. In this case, (11.48) becomes

$$\frac{[(f(y)-f(x))\cdot(f(z)-f(x))]^2}{(f(y)-f(x))^2(f(z)-f(x))^2} = \frac{[(y-x)\cdot(z-x)]^2}{(y-x)^2(z-x)^2},$$

for all $x, y, z \in \mathbf{R}^2$ such that $x \notin \{y, z\}$. The assumption $f(A) \in \mathcal{A}$ for $A \in S$ guarantees that $f(x) \notin \{f(y), f(z)\}$. ∎

11.5 Using functional equations for CAGD

Properties of cross sections of surfaces or intersections with planes parallel to the coordinate planes are very important in many applied fields (see Castillo and Iglesias (1997)). For example, some methods of representing an algebraic curve require the intersection of a surface $z = F(x, y)$ with the plane $z = 0$ (see, for example, Hoschek and Lasser (1993), page 496 and references therein).

As shown in Farin (1987) (page 304), in scientific computing, the contour lines (obtained through intersections between a number of parallel planes and a given surface) are often of great importance. For example, Figure 11.12 shows the bathymetry (contour lines) of the Santa Marina beach (Spain). Some algorithms for obtaining such contour lines have been studied in Petersen (1983) and even generalized to the case of an adaptive contouring of a trivariate interpolant. Efficient procedures for tracing completely the intersection of a plane with rational parametric surfaces have been analyzed by Chandru and Kochar (1987), and along the same lines, Farouki (1986) considers a generalization of this problem using algebraic surfaces of low degree instead of a plane. Other computational methods for sectioning can be found in Hoitsma and Roche (1983) and in Lee and Fredericks (1984). Some other applications of cross sections for the medical area can be found in *et. al.* Ekoule et al. (1991), and in Boissonat (1985) for pattern recognition and computer vision. In general, computing planar intersections of geometric objects is a very important capability of CAD/CAM and many other geometric-modelling systems (see Mortenson (1985)).

In addition to this, in several areas of Engineering, contouring techniques are used to represent a three dimensional surface in two dimensions (see Anand (1993b)). Contour plots are created as intersections of the surface with planes

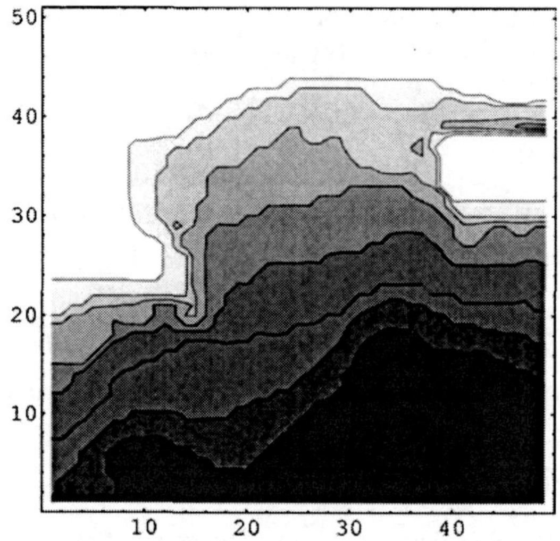

Figure 11.12: Bathymetry (contour lines) of the Santa Marina beach in Spain.

$z = z_0$ for different values of z_0. Similarly, for non-parametric, i.e., implicit or explicit, surfaces this is usually done by drawing its intersections with planes $x = x_0$ and $y = y_0$, for selected values of x_0 and y_0. From these intersections, areas or volumes can be easily approximated. As a first example, Figure 11.13 shows a perspective of the Santa Marina beach sea bottom, corresponding to the contour lines in Figure 11.12, obtained by this method.

In CAGD surfaces are dealt with in several forms: parametric, explicit and implicit equations. The most common representation in the commercial software and research fields are parametric equations. This representation allows a quick computation of the coordinates of all points on a curve or surface. Moreover, the parametric form can be used to define a curve segment or surface patch constraining the parameters to intervals. Because curves and surfaces are usually bounded in computer graphics, this characteristic is of considerable importance.

Nevertheless, in the last few years, explicit and implicit representations are being used more frequently in CAGD, allowing a better treatment of several problems. As one example, the point classification problem is easily solved with the implicit representation: it consists of a simple evaluation of the implicit functions. This is useful in many applications, such as solid modelling, for example, where points must be defined inside or outside the boundaries of an object (see Hoffmann (1989)). Through implicit representation, the problem is reduced to a trivial sign test. Furthermore, the implicit representation offers surfaces of desired smoothness with the lowest possible degree. Finally, when we restrict it to polynomial functions, the implicit representation is more general than the

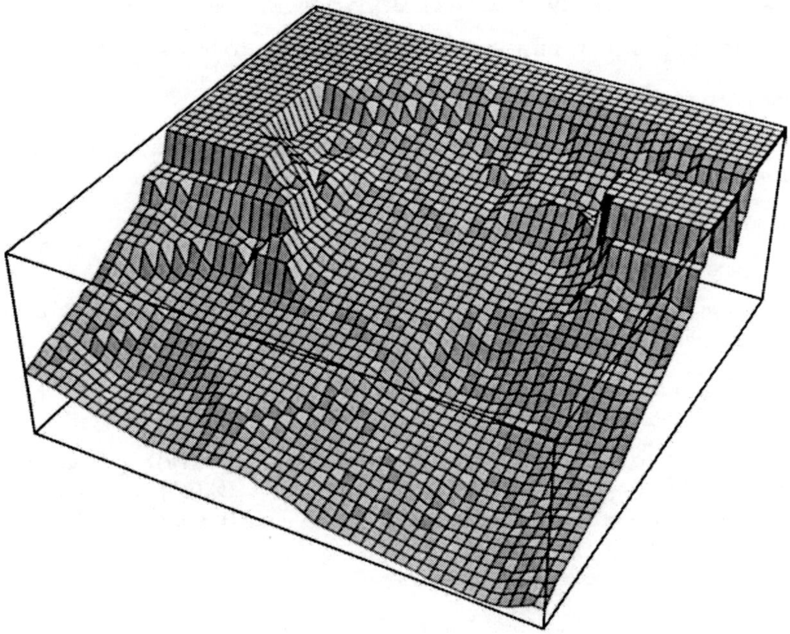

Figure 11.13: Bathymetry (perspective) of the Santa Marina beach in Spain.

parametric representation (but probably not if we allow arbitrarily complicated functions) and several methods have been described to solve the problem of *implicitation*, that is, obtaining the implicit representation of rational surfaces (see, for example Sederberg (1983), Sederberg et al. (1984), Farin (1990), or Hoffmann (1989)). Although this conversion is always possible, difficulties arise when base points are present, but these problems are not insurmountable (see Chionh and Goldman (1992)).

From the above considerations, it is clear that identification of the most general families of surfaces with arbitrary planar cross sections parallel to the coordinate axes belonging to given parametric and non-parametric families is an important problem. Our aim in the three next sections is twofold: to obtain such characterizations and to show the power of functional equations in this process of characterization.

11.5.1 Preliminary Results

Before starting with the main results of the above introduced problem, we introduce three preliminary results to be used later in Sections 11.5.2, 11.5.3 and 11.5.4.

From Chapter 4 we already know the solutions of the functional equation $\sum_{k=1}^{n} f_k(x)g_k(y) = 0$. For consistency with the notation to be used in the following sections, we rewrite Theorem 4.5 in vectorial form:

Theorem 11.6 (Vectorial form of the sum of products equation). *All solutions of the equation*

$$\mathbf{f}(x) \bullet \mathbf{g}(y) = \sum_{k=1}^{n} f_k(x)g_k(y) = 0, \tag{11.49}$$

where $\mathbf{f}(x) = (f_1(x), \ldots, f_n(x))$, $\mathbf{g}(y) = (g_1(y), \ldots, g_n(y))$ *and* $\bullet$ *is used to denote the dot product of two vectors, can be written in the form*

$$\begin{aligned} \mathbf{f}(x) &= \boldsymbol{\varphi}(x)\mathbf{A}, \\ \mathbf{g}(y) &= \boldsymbol{\psi}(y)\mathbf{B}, \end{aligned} \tag{11.50}$$

where $\boldsymbol{\varphi}(x) = (\varphi_1(x), \ldots, \varphi_r(x))$, $\boldsymbol{\psi}(y) = (\psi_{r+1}(y), \ldots, \psi_n(y))$, r *is an integer between 0 and* n, $\{\varphi_1(x), \ldots, \varphi_r(x)\}$ *and* $\{\psi_{r+1}(x), \ldots, \psi_n(x)\}$ *are two arbitrary systems of linearly independent functions, and* $\mathbf{A}$ *and* $\mathbf{B}$ *are constant matrices, which satisfy*

$$\mathbf{A}\mathbf{B}^T = \mathbf{0}. \tag{11.51}$$

■

As a consequence of this result we obtain the following corollary:

Corollary 11.1 (Sum of products). *Let* $\{u_1(x), \ldots, u_I(x)\}$ *and* $\{v_1(y), \ldots, v_J(y)\}$ *be two linearly independent sets of known functions, then the solution of the functional equation*

$$\boldsymbol{\alpha}(y) \bullet \mathbf{u}(x) = \boldsymbol{\beta}(x) \bullet \mathbf{v}(y), \tag{11.52}$$

where $\{\alpha_1(y), \ldots, \alpha_I(y)\}$ *and* $\{\beta_1(x), \ldots, \beta_J(x)\}$ *are the unknown functions, is*

$$\boldsymbol{\alpha}(y) = \mathbf{v}(y)\mathbf{D}; \;\; \boldsymbol{\beta}(x) = \mathbf{u}(x)\mathbf{D}^T, \tag{11.53}$$

where $\mathbf{D}$ *is an arbitrary constant matrix.* ■

Proof: Expression (11.52) is equivalent to

$$(\boldsymbol{\alpha}(y)| - \mathbf{v}(y)) \bullet (\mathbf{u}(x)|\boldsymbol{\beta}(x)) = 0.$$

Thus, according to Theorem 11.6, we have

$$\begin{aligned} (\mathbf{u}(x)|\boldsymbol{\beta}(x)) &= \mathbf{u}(x)(\mathbf{I}_I|\mathbf{C}), \\ (\boldsymbol{\alpha}(y)| - \mathbf{v}(y)) &= \mathbf{v}(y)(\mathbf{D}| - \mathbf{I}_J), \end{aligned}$$

where $\mathbf{C}$ and $\mathbf{D}$ are $I \times J$ and $J \times I$ constant matrices, respectively, such that

$$(\mathbf{I}_I|\mathbf{C})(\mathbf{D}| - \mathbf{I}_J)^T = \mathbf{0} \Leftrightarrow \mathbf{D}^T = \mathbf{C}.$$

■

The following theorem will be used later.

Theorem 11.7 (An exponential equation). *The general non-negative and integrable solution of the functional equation*

$$z(x,y) = (\alpha_1(y)x + \alpha_2(y))^{\alpha_3(y)} = (\beta_1(x)y + \beta_2(x))^{\beta_3(x)}, \tag{11.54}$$

where $\alpha_1(.), \alpha_2(.), \alpha_3(.)$ *and* $\beta_1(.), \beta_2(.), \beta_3(.)$ *are unknown functions, is*

$$\begin{aligned} z(x,y) &= [C(x-A)(y-B)+D]^E, \\ z(x,y) &= E(x-A)^C(y-B)^D \exp[M\log(x-A)\log(y-B)], \end{aligned}$$

which depend on 5 *and* 6 *parameters, respectively.* ■

Proof: See Castillo and Galambos (1987a) for a proof. ■

11.5.2 Families of implicit surfaces

In this section we deal with the case of surfaces in implicit form. We look for the most general surfaces in implicit form, such that their planar cross sections parallel to the coordinate planes satisfy some conditions. This allows the designer to choose families of surfaces with cross sections at convenience. The following theorem gives these conditions in a precise form and the corresponding solution.

Theorem 11.8 (Implicit surfaces). *The most general family of implicit surfaces* $f(x,y,z) = 0$, *such that their intersections with the planes* $z = z_0$, $y = y_0$, *and* $x = x_0$ *are linear combinations of sets* $\mathcal{U} = \{u_1(y,z), u_2(y,z), \ldots, u_I(y,z)\}$, $\mathcal{V} = \{v_1(z,x), v_2(z,x), \ldots, v_J(z,x)\}$ *and* $\mathcal{W} = \{w_1(x,y), w_2(x,y), \ldots, w_K(x,y)\}$, *of functions of the other two variables, is of the form*

$$f(x,y,z) = \sum_{i=1}^{I}\sum_{j=1}^{J}\sum_{k=1}^{K} [C_{ijk}\alpha_i(x)\beta_j(y)\gamma_k(z)], \tag{11.55}$$

where $\boldsymbol{\alpha}(x), \boldsymbol{\beta}(y)$ *and* $\boldsymbol{\gamma}(z)$ *are vectors of arbitrary functions and* C_{ijk} *are the elements of an arbitrary constant matrix* $\mathbf{C}$.

In addition, the $\mathcal{U}, \mathcal{V}$ *and* $\mathcal{W}$ *functions cannot be arbitrary, but of the form*

$$u_i(y,z) = \sum_{j=1}^{J}\sum_{k=1}^{K} [C_{ijk}\beta_j(y)\gamma_k(z)]; \quad i = 1, \ldots, I, \tag{11.56}$$

$$v_j(z,x) = \sum_{i=1}^{I}\sum_{k=1}^{K} [C_{ijk}\alpha_i(x)\gamma_k(z)]; \quad j = 1, \ldots, J, \tag{11.57}$$

$$w_k(x,y) = \sum_{i=1}^{I}\sum_{j=1}^{J} [C_{ijk}\alpha_i(x)\beta_j(y)]; \quad k = 1, \ldots, K. \tag{11.58}$$

■

Proof: According to the above assumptions, we look for surfaces $f(x,y,z) = 0$ such that they satisfy the system of functional equations

$$\begin{aligned} f(x,y,z) &\equiv \sum_{i=1}^{I} \alpha_i(x) u_i(y,z) = \sum_{j=1}^{J} \beta_j(y) v_j(z,x), \\ f(x,y,z) &\equiv \sum_{j=1}^{J} \beta_j(y) v_j(z,x) = \sum_{k=1}^{K} \gamma_k(z) w_k(x,y), \end{aligned} \tag{11.59}$$

where the sets $\{\alpha_i(x);\ i = 1,\ldots,I\}$, $\{\beta_j(y);\ j = 1,\ldots,J\}$ and $\{\gamma_k(z); k = 1,\ldots,K\}$ can be assumed, without loss of generality, as sets of linearly independent functions. Note that if they are not, we can rewrite equations in (11.59) in the same form but with linearly independent sets.

The system of equations (11.59) state the conditions for the sets $\mathcal{U}$, $\mathcal{V}$ and $\mathcal{W}$ to be *compatible* with the sets $\{\alpha_i(x);\ i = 1,2,\ldots,I\}$, $\{\beta_j(y);\ j = 1,2,\ldots,J\}$ and $\{\gamma_k(z);\ k = 1,2,\ldots,K\}$.

For any fixed z, the first equation in (11.59) is of the form (11.52) and, according to Corollary 11.1 we have (see (11.53)):

$$\mathbf{u}(y,z) = \boldsymbol{\beta}(y)\mathbf{A}^T(z), \tag{11.60}$$

$$\mathbf{v}(z,x) = \boldsymbol{\alpha}(x)\mathbf{A}(z), \tag{11.61}$$

where $\mathbf{A}(z)$ is a matrix the elements of which are functions of z.

If we now replace (11.60) or (11.61) in (11.59) we get

$$f(x,y,z) = \boldsymbol{\alpha}(x)\mathbf{A}(z)\boldsymbol{\beta}^T(y) = \sum_{i=1}^{I}\sum_{j=1}^{J} A_{ij}(z)\alpha_i(x)\beta_j(y).$$

Similarly, for any fixed x, the second equation in (11.59) leads to

$$\begin{aligned} \mathbf{v}(z,x) &= \boldsymbol{\gamma}(z)\mathbf{B}^T(x), \\ \mathbf{w}(x,y) &= \boldsymbol{\beta}(y)\mathbf{B}(x). \end{aligned} \tag{11.62}$$

Now, from equations (11.61) and (11.62) we obtain for each $j = 1,\ldots,J$,

$$v_j(z,x) = \sum_{i=1}^{I} A_{ij}(z)\alpha_i(x) = \sum_{k=1}^{K} B_{jk}(x)\gamma_k(z),$$

which is also of the form given in Corollary 11.1, and then we can write

$$v_j(z,x) = \sum_{i=1}^{I}\sum_{k=1}^{K} C_{ijk}^{(2)}\alpha_i(x)\gamma_k(z),$$

where $C_{ijk}^{(2)}$ are the elements of a constant matrix.

It is clear that similar expressions can be obtained for $u_i(y,z)$ and $w_k(x,y)$.

If we now replace this into equations (11.59), we get

$$\begin{aligned} f(x,y,z) &= \sum_{i=1}^{I}\sum_{j=1}^{J}\sum_{k=1}^{K}\left[C_{ijk}^{(2)}\alpha_i(x)\beta_j(y)\gamma_k(z)\right] \\ &= \sum_{i=1}^{I}\sum_{j=1}^{J}\sum_{k=1}^{K}\left[C_{ijk}^{(3)}\alpha_i(x)\beta_j(y)\gamma_k(z)\right], \\ f(x,y,z) &= \sum_{i=1}^{I}\sum_{j=1}^{J}\sum_{k=1}^{K}\left[C_{ijk}^{(3)}\alpha_i(x)\beta_j(y)\gamma_k(z)\right] \\ &= \sum_{i=1}^{I}\sum_{j=1}^{J}\sum_{k=1}^{K}\left[C_{ijk}^{(1)}\alpha_i(x)\beta_j(y)\gamma_k(z)\right], \end{aligned} \tag{11.63}$$

but, taking into account that the above sets of functions are linearly independent, we get

$$C_{ijk}^{(1)} = C_{ijk}^{(2)} = C_{ijk}^{(3)} = C_{ijk}.$$

This, together with (11.63), leads to (11.55). ∎

Remarks :

1. Note that no constraints have been imposed on the $f(x,y,z)$ functions. Thus, an arbitrary class of functions has been assumed.

2. The implicit equation of the surface (11.55) is a linear combination of the tensor product of the sets of functions in $\boldsymbol{\alpha}(x)$, $\boldsymbol{\beta}(y)$ and $\boldsymbol{\gamma}(z)$.

3. Algebraic surfaces are particular cases of this family.

4. The only functions $u_i(y,z)$, $v_j(z,x)$ and $w_k(x,y)$ satisfying condition (11.59) are of the form (11.56)–(11.58), where $\alpha(.), \beta(.)$ and $\gamma(.)$ are the function coefficients in (11.59).

5. Note that the functional equations are imposed not only on a fixed number of given cross sections, but on an arbitrary planar section parallel to the coordinate planes.

6. Functional equations allow the functional form of the solution to be characterized from a simple compatibility condition.

11.5.3 Families of explicit surfaces

In this section we look for the most general surface in explicit form, $z = z(x,y)$, such that their intersections with planes parallel to the planes $y = 0$ and $x = 0$ belong to given (not necessarily equal) parametric families of curves, i.e.,

$$
\begin{aligned}
z = z(x,y) &= h\left(x, \alpha_1(y), \alpha_2(y), \ldots, \alpha_k(y)\right), \\
z = z(x,y) &= r\left(y, \beta_1(x), \beta_2(x), \ldots, \beta_m(x)\right).
\end{aligned} \tag{11.64}
$$

This leads to the functional equation

$$
h\left(x, \alpha_1(y), \alpha_2(y), \ldots, \alpha_k(y)\right) = r\left(y, \beta_1(x), \beta_2(x), \ldots, \beta_m(x)\right), \tag{11.65}
$$

where we assume that $\alpha_1(.), \alpha_2(.), \ldots, \alpha_k(.)$ and $\beta_1(.), \beta_2(.), \ldots, \beta_m(.)$ are unknown functions to be determined and $h(.)$ and $r(.)$ are known functions, which are selected for having convenient cross sections.

Note that a given surface is completely defined by one of the two equations in (11.64). For two different representations (the two equations in (11.64)) to correspond to the same surface they must satisfy some conditions. Thus, the functional Equation (11.65) plays the role of a *compatibility condition.*

In the following paragraphs we analyze different cases of $h(.)$ and $r(.)$ parametric families of curves.

Explicit surfaces of the form $z = s\left(\sum\limits_{i=1}^{n} p_i(x)q_i(y)\right)$

In this case we take

$$
h(x, \alpha_1, \alpha_2, \ldots, \alpha_k) = \sum_{i=1}^{k} \alpha_i u_i(x),
$$

and

$$
r(y, \beta_1, \beta_2, \ldots, \beta_m) = \sum_{i=1}^{m} \beta_i v_i(y),
$$

and we solve the associated problem by means of the following theorem.

Theorem 11.9 (Explicit surfaces). *The most general surface in explicit form, $z = z(x,y)$, such that all sections with planes parallel to the planes $y = 0$ and $x = 0$ are linear combinations of given sets of linearly independent functions $\mathcal{U} = \{u_1(.), u_2(.), \ldots, u_k(.)\}$ and $\mathcal{V} = \{v_1(.), v_2(.), \ldots, v_m(.)\}$, respectively, is*

$$
z = z(x,y) = \mathbf{v}(y)\mathbf{A}\mathbf{u}^T(x), \tag{11.66}
$$

where $\mathbf{A}$ is an arbitrary constant matrix. Thus, we obtain a linear combination of the tensor products of the vectors of functions $\mathbf{u}(x)$ and $\mathbf{v}(y)$. ∎

Proof: According to the above assumptions, we must have

$$
\begin{aligned}
z(x,y) &= \boldsymbol{\alpha}(y) \bullet \mathbf{u}(x), \\
z(x,y) &= \boldsymbol{\beta}(x) \bullet \mathbf{v}(y),
\end{aligned} \tag{11.67}
$$

where $\mathbf{u}(x) = (u_1(x), \ldots, u_k(x))$ and $\mathbf{v}(y) = (v_1(y), \ldots, v_m(y))$ are known and $\boldsymbol{\alpha}(y) = (\alpha_1(y), \ldots, \alpha_k(y))$ and $\boldsymbol{\beta}(x) = (\beta_1(x), \ldots, \beta_m(x))$ are to be determined.

Expressions (11.67) imply

$$\boldsymbol{\alpha}(y) \bullet \mathbf{u}(x) = \boldsymbol{\beta}(x) \bullet \mathbf{v}(y),$$

which is a functional equation of the form given in Corollary 11.1. Thus, its general solution is

$$\begin{aligned} \boldsymbol{\alpha}(y) &= \mathbf{v}(y)\mathbf{A}, \\ \boldsymbol{\beta}(x) &= \mathbf{u}(x)\mathbf{A}^T, \end{aligned}$$

where $\mathbf{A}$ is an arbitrary $(m \times k)$ constant matrix.

Thus, the explicit equation of the parametric family of surfaces becomes (11.66). ■

Since any explicit surface $z = z(x, y)$ can be written as an equivalent implicit surface $f(x, y, z) = z(x, y) - z$, the above theorem can be obtained as a particular case of Theorem 11.9, as follows. According to (11.55) we can write

$$f(x,y,z) = \sum_{i=1}^{I}\sum_{j=1}^{J}\sum_{k=1}^{K}[C_{ijk}\alpha_i(x)\beta_j(y)\gamma_k(z)] = z(x,y) - z,$$

and making $z = z_0$ we obtain

$$\sum_{i=1}^{I}\sum_{j=1}^{J}\alpha_i(x)\beta_j(y)\sum_{k=1}^{K}[C_{ijk}\gamma_k(z_0)] = z(x,y) - z_0,$$

and then we get

$$z(x,y) = z_0 + \sum_{i=1}^{I}\sum_{j=1}^{J}C'_{ij}\alpha_i(x)\beta_j(y),$$

where

$$C'_{ij} = \sum_{k=1}^{K}[C_{ijk}\gamma_k(z_0)],$$

which is equivalent to (11.66).

Corollary 11.2 (I). *f, instead of (11.67), we consider*

$$\begin{aligned} z(x,y) &= s\left(\boldsymbol{\alpha}(y) \bullet \mathbf{u}(x)\right), \\ z(x,y) &= s\left(\boldsymbol{\beta}(x) \bullet \mathbf{v}(y)\right), \end{aligned}$$

where $s(.)$ is an invertible function, the general solution for $z(x, y)$ becomes

$$z(x,y) = s\left(\mathbf{v}(y)\mathbf{A}\mathbf{u}^T(x)\right). \tag{11.68}$$

■

This corollary can be directly proved by applying Theorem 11.9 to the explicit function $z = s^{-1}(z(x, y))$.

So, given $\mathbf{u}(x)$ and $\mathbf{v}(y)$, Equation (11.68) defines its associated parametric family of surfaces and any surface from this family has an associated matrix $\mathbf{A}$.

The intersections of two surfaces of this family, with parameters defined by matrices $\mathbf{A}_1$ and $\mathbf{A}_2$, have as a projection on the plane $z = 0$ the curve

$$\mathbf{v}(y)(\mathbf{A}_1 - \mathbf{A}_2)\mathbf{u}^T(x) = 0, \tag{11.69}$$

which together with (11.68) allows the curve to be drawn.

The intersections of these surfaces with planes $z = z_0$ have as projections on the plane $z = 0$, the curve

$$\mathbf{v}(y)\mathbf{A}\mathbf{u}^T(x) = z_0. \tag{11.70}$$

Example 11.5 (Non-parametric bicubic tensor product surfaces). If the surface $z = z(x, y)$ is such that all sections with planes parallel to the planes $y = 0$ and $x = 0$ are third degree polynomials, we have $s(w) = w$,

$$(u_1(x), u_2(x), u_3(x), u_4(x)) = \left(1, x, x^2, x^3\right),$$

and

$$(v_1(x), v_2(x), v_3(x), v_4(x)) = \left(1, y, y^2, y^3\right).$$

Then, according to (11.68), the explicit equation of the surface becomes

$$z = z(x, y) = \left(1, y, y^2, y^3\right) \mathbf{A} \left(1, x, x^2, x^3\right)^T. \tag{11.71}$$

It is obvious that intersections of the above surface by planes $z = z_0$ are algebraic curves (see (11.70)). Similarly, (11.69) shows that the projection on the plane $z = 0$ of the intersections of two surfaces of this family are algebraic curves too.

The family (11.71) depends on 16 parameters and then one surface can be forced to pass through 16 given points. Figure 11.14 shows one example, where the 16 points have been selected in such a way that certain sets of 4 points belonging to the planes $x = i$ and $y = j$ with $i = 1, \ldots, 4$; $j = 1, \ldots, 4$. The figure also shows the polynomial curves passing through every 4 points in each of these sets. Figure 11.15 shows the interpolating surface.

Now we force all vertical plane intersections to be third degree polynomials in X or Y. Making $y = px + q$ or $x = ry + t$ in (11.71) and cancelling coefficients of x and y of degree larger than 3, we get

$$a_{44} = a_{43} = a_{34} = a_{42} = a_{33} = a_{24} = 0.$$

Thus, the new family becomes

$$z = z(x, y) = \begin{pmatrix} 1, y, y^2, y^3 \end{pmatrix} \begin{pmatrix} a_{11} & a_{12} & a_{13} & a_{14} \\ a_{21} & a_{22} & a_{23} & 0 \\ a_{31} & a_{32} & 0 & 0 \\ a_{41} & 0 & 0 & 0 \end{pmatrix} \begin{pmatrix} 1 \\ x \\ x^2 \\ x^3 \end{pmatrix}.$$

■

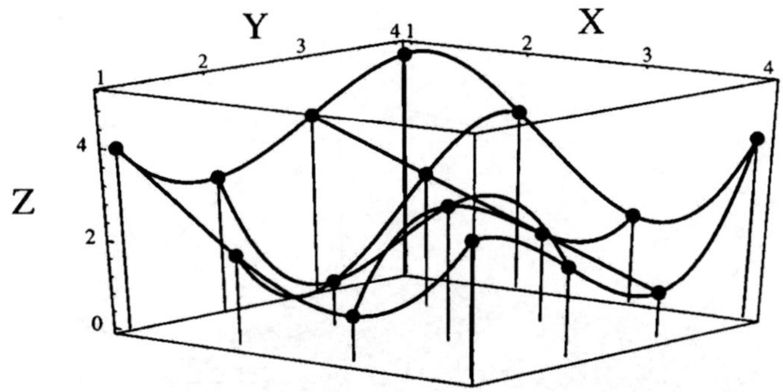

Figure 11.14: Set of points defining the surface.

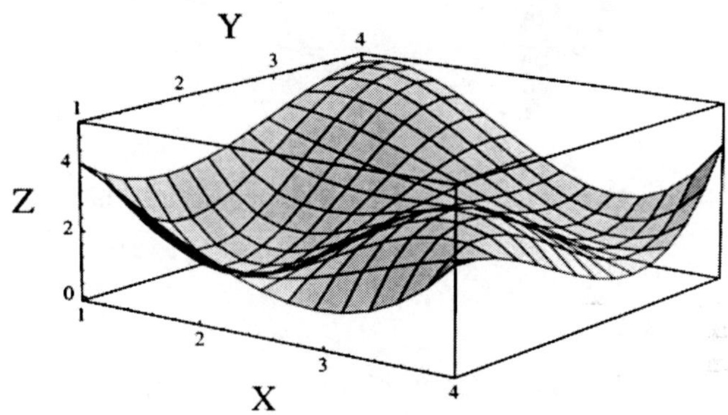

Figure 11.15: Interpolating surface.

Example 11.6 (Some exponential families). Now we consider the case $s(w) = exp(w)$, $(u_1(x), u_2(x), u_3(x)) = (1, x, log(x))$ and $(v_1(y), v_2(y), v_3(y)) = (1, y, y^2)$. Then we get the family of surfaces

$$z(x,y) = \exp\left[\left(1, y, y^2\right) \mathbf{A} \left(1, x, \log(x)\right)^T\right],$$

which depends on 9 parameters.

The intersections of two surfaces of this family, with associated parameters $\mathbf{A}_1$ and $\mathbf{A}_2$, have the following projection on the plane $z = 0$:

$$\left(1, y, y^2\right) \left(\mathbf{A}_1 - \mathbf{A}_2\right) \left(1, x, \log(x)\right)^T = 0.$$

As one example, in Figure 11.16 we show the surface with equation

$$z(x,y) = \exp\left[-x(1 + y + y^2) + \log(x)(1 + y/2)\right].$$

■

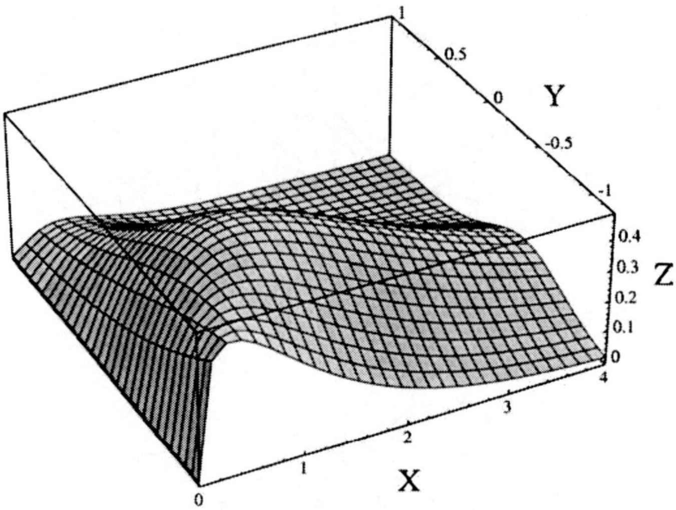

Figure 11.16: One surface from the exponential family.

Some other surfaces

In this section we illustrate the power of the method giving two more particular examples of (11.64). We take

$$\begin{aligned} h(x,\alpha_1,\alpha_2,\alpha_3) &= s\left((\alpha_1 x+\alpha_2)^{\alpha_3}\right), \\ r(y,\beta_1,\beta_2,\beta_3) &= s\left((\beta_1 y+\beta_2)^{\beta_3}\right), \end{aligned}$$

i.e.,

$$\begin{aligned} z(x,y) &= s\left[(\alpha_1(y)x+\alpha_2(y))^{\alpha_3(y)}\right], \\ z(x,y) &= s\left[(\beta_1(x)y+\beta_2(x))^{\beta_3(x)}\right], \end{aligned}$$

which leads to the functional Equation (11.54).

Theorem 11.7 gives the two families of solutions of (11.54) and figures 11.17 and 11.18 show the two surfaces associated with the explicit equations $z = 1-\exp\left(-(xy-1)^2\right)$ and $z=\exp\left(-(xy-1)^2\right)$, respectively. Their cross planar sections by planes of the form $x=x_0$ and $y=y_0$ are obvious.

It is worthwhile pointing out that, in spite of the general form of the functional Equation (11.54), the only feasible solutions are parametric families; that is, the functional form of the solutions is determined by the compatibility condition (11.54).

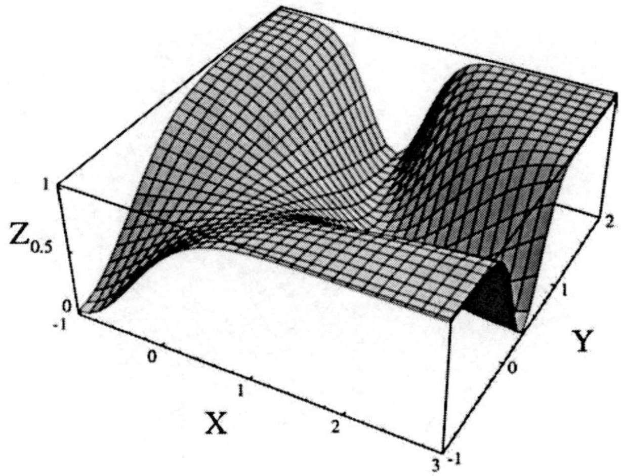

Figure 11.17: Surface $z = 1 - \exp\left(-(xy-1)^2\right)$.

11.5.4 Gordon–Coons-surfaces

Functional equations are equally useful to prove certain properties of surfaces in parametric form. In this section we prove a uniqueness theorem for this type of surfaces. To start we introduce them and the basic notation using Gordon's approach (see Gordon (1993)). Consider the following problem: assume two families of parametric curves

$$\{\mathbf{g}_i(t)|i = 1, 2, \ldots, M\} \text{ and } \{\mathbf{f}_j(s)|j = 1, 2, \ldots, N\},$$

where

$$\mathbf{g}_i(t) = \begin{pmatrix} g_i^{(1)}(t) \\ g_i^{(2)}(t) \\ g_i^{(3)}(t) \end{pmatrix}; \quad \mathbf{f}_j(s) = \begin{pmatrix} f_j^{(1)}(s) \\ f_j^{(2)}(s) \\ f_j^{(3)}(s) \end{pmatrix},$$

and the superindices 1, 2 and 3 of the g and f functions refer to the x, y and z components, respectively. All functions $\mathbf{g}_i(t)$ and $\mathbf{f}_j(s)$ have to be defined on common parameter intervals $[\bar{t}_0, \bar{t}_1]$ and $[\bar{s}_0, \bar{s}_1]$, respectively.

These curves intersect in a set of space points, the coordinates of which are obtained for some values of the parameters $t_1 < t_2 \ldots < t_N$ and $s_1 < s_2 < \ldots < s_M$. For these two families to define a surface they must satisfy the following *compatibility conditions*

$$\mathbf{u}_{ij} = \mathbf{g}_i(t_j) = \mathbf{f}_j(s_i).$$

Then, it is possible to build a surface $\mathbf{v}(s, t)$ interpolating the $M + N$ given curves; that is, satisfying the system of vector equations

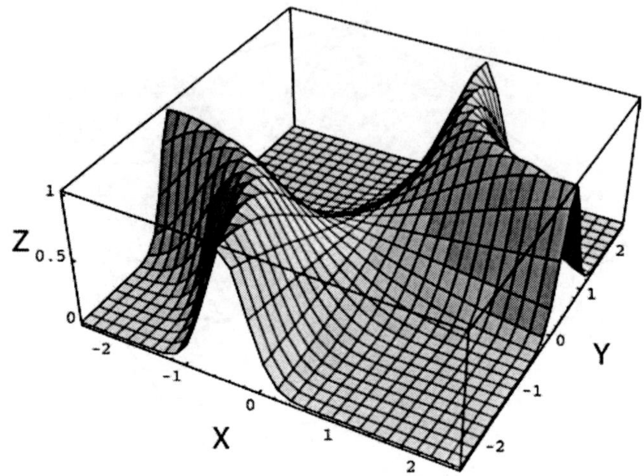

Figure 11.18: Surface $z = \exp\left(-(xy-1)^2\right)$.

$$\mathbf{v}(s_i, t) = \mathbf{g}_i(t); \quad i = 1, 2, \ldots, M,$$

$$\mathbf{v}(s, t_j) = \mathbf{f}_j(s); \quad j = 1, 2, \ldots, N.$$

Thus, the problem of interpolating a surface through an intersecting skeletal network of three-dimensional curves can be reduced to the following scalar-value problem:

Construct a bivariate function $\mathbf{v}(s,t)$, which interpolates the $M+N$ univariate functions $\{\mathbf{g}_i(t)|i=1,\ldots,M\}$, $\{\mathbf{f}_j(s)|j=1,\ldots,N\}$. To solve this problem, Gordon at the end of the 60' and 70's proposed the solution given by the following theorem:

Theorem 11.10 (Gordon's theorem). *If*

$$\{\phi_i(s)|i=1,2,\ldots,M\}$$

and

$$\{\psi_j(t)|j=1,2,\ldots,N\}$$

are any two sets of functions such that they satisfy the conditions:

$$\phi_i(s_k) = \delta_{ik}, \tag{11.72}$$

$$\psi_j(t_k) = \delta_{jk}, \tag{11.73}$$

where δ_{ij} *is the Kronecker* δs*, and if*

$$\{\mathbf{g}_i(t) | i = 1, 2, \ldots, M\}$$

and

$$\{\mathbf{f}_j(s) | j = 1, 2, \ldots, N\}$$

are any two sets of functions such that the following compatibility conditions are satisfied:

$$\mathbf{g}_i(t_j) = \mathbf{f}_j(s_i), \quad \forall i, j. \tag{11.74}$$

Then, the bivariate function

$$\mathbf{v}(s,t) = \sum_{i=1}^{M} \mathbf{g}_i(t)\phi_i(s) + \sum_{j=1}^{N} \mathbf{f}_j(s)\psi_j(t) - \sum_{i=1}^{M}\sum_{j=1}^{N} \mathbf{u}_{ij}\phi_i(s)\psi_j(t) \tag{11.75}$$

is one solution of the interpolation problem

$$\begin{array}{rcl} \mathbf{v}(s_k, t) & = & \mathbf{g}_k(t), (k = 1, 2, \ldots, M), \\ \mathbf{v}(s, t_p) & = & \mathbf{f}_p(t), (p = 1, 2, \ldots, N). \end{array}$$

■

Note that (11.75) gives $\mathbf{v}(s,t)$ as a function of the three surfaces

$$\mathbf{v}_1(s,t) = \sum_{i=1}^{M} \mathbf{g}_i(t)\phi_i(s),$$

which interpolates the family of curves $\{\mathbf{g}_i(t) | i = 1, 2, \ldots, M\}$,

$$\mathbf{v}_2(s,t) = \sum_{j=1}^{N} \mathbf{f}_j(s)\psi_j(t),$$

which interpolates the family of curves $\{\mathbf{f}_j(s) | j = 1, 2, \ldots, N\}$, and

$$\mathbf{v}_3(s,t) = \sum_{i=1}^{M}\sum_{j=1}^{N} \mathbf{u}_{ij}\phi_i(s)\psi_j(t),$$

which interpolates the set of points $\{\mathbf{u}_{ij} | i = 1, 2, \ldots, M; j = 1, 2, \ldots, N\}$. Note that it is again a tensor product.

Gordon–Coons surfaces (see Gordon (1993) and Coons (1964)) are frequently referred to as *transfinite interpolants* (Farin (1990), page 381, Hoschek and Hoschek and Lasser (1993), page 371), since they interpolate a continuous data, i.e., all points along the prescribed boundary curves.

Uniqueness theorem

In this section, the uniqueness of representation for the surface family defined by (11.75), will be analyzed; i.e., we try to find out whether there is only one representation for such a surface. To be more precise we give the following theorem.

Theorem 11.11 (Uniqueness Theorem). *Given the sets of linearly independent functions*

$$\{\phi_i(s)|i = 1, 2, \dots, M\} \text{ and } \{\psi_j(t)|j = 1, 2, \dots, N\},$$

then there exist unique sets of functions

$$\{\mathbf{g}_i(t)|i = 1, 2, \dots, M\} \text{ and } \{\mathbf{f}_j(s)|j = 1, 2, \dots, N\},$$

satisfying the compatibility conditions

$$\mathbf{u}_{ij} = \mathbf{g}_i(t_j) = \mathbf{f}_j(s_i),$$

such that the surface $\mathbf{v}(s,t)$ *is given by (11.75).* ∎

Proof: For this proof we use functional equations. Let $\{\mathbf{g}_i(t), \mathbf{f}_j(s)\}$ and $\{\mathbf{g}_i^*(t), \mathbf{f}_j^*(s)\}$ be two such representations satisfying the same compatibility conditions

$$\mathbf{u}_{ij} = \mathbf{g}_i(t_j) = \mathbf{f}_j(s_i) = \mathbf{g}_i^*(t_j) = \mathbf{f}_j^*(s_i), \tag{11.76}$$

then, we have

$$\sum_{i=1}^{M} \mathbf{g}_i(t)\phi_i(s) + \sum_{j=1}^{N} \mathbf{f}_j(s)\psi_j(t) = \sum_{i=1}^{M} \mathbf{g}_i^*(t)\phi_i(s) + \sum_{j=1}^{N} \mathbf{f}_j^*(s)\psi_j(t),$$

so, the following functional equation holds

$$\sum_{i=1}^{M} \left[\mathbf{g}_i(t) - \mathbf{g}_i^*(t)\right] \phi_i(s) + \sum_{j=1}^{N} \left[\mathbf{f}_j(s) - \mathbf{f}_j^*(s)\right] \psi_j(t) \equiv 0.$$

Its solution is given by (see Corollary 11.1):

$$\mathbf{g}_i(t) - \mathbf{g}_i^*(t) = \sum_{j=1}^{N} \mathbf{A}_{ij}\psi_j(t) \Rightarrow \mathbf{g}_i(t) = \sum_{j=1}^{N} \mathbf{A}_{ij}\psi_j(t) + \mathbf{g}_i^*(t), \ \forall i,$$
$$\mathbf{f}_j(s) - \mathbf{f}_j^*(s) = -\sum_{i=1}^{M} \mathbf{A}_{ij}\phi_i(s) \Rightarrow \mathbf{f}_j(s) = -\sum_{i=1}^{M} \mathbf{A}_{ij}\phi_i(s) + \mathbf{f}_j^*(s), \ \forall j,$$

where $\mathbf{A}$ is a matrix of constants. Returning now to the compatibility conditions (11.76), and using definition (11.73), we get

$$\mathbf{u}_{ij} = \mathbf{g}_i(t_j) = \sum_{K=1}^{N} \mathbf{A}_{ik}\psi_k(t_j) + \mathbf{g}_i^*(t_j)$$

which implies

$$\mathbf{u}_{ij} = \mathbf{A}_{ij}\psi_j(t_j) + \mathbf{u}_{ij} = \mathbf{A}_{ij} + \mathbf{u}_{ij}, \ \forall i, j,$$

which is the desired uniqueness. ∎

A particular case

In this subsection we analyze the particular case of surfaces in which the sets of functions $\{\mathbf{g}_i(t)\}$ and $\{\mathbf{f}_j(s)\}$ are linear combinations of the sets $\{\psi_j(t)\}$ and $\{\phi_i(s)\}$, respectively. In this case, the surface $\mathbf{v}(s,t)$ appears in a very simple form, as indicated by the following theorem.

Theorem 11.12 (A particular case). *If the conditions of the above theorem hold, and the two sets of functions*

$$\{\mathbf{g}_i(t)|i=1,2,\ldots,M\} \text{ and } \{\mathbf{f}_j(s)|j=1,2,\ldots,N\}$$

are linear combinations of the sets of functions

$$\{\psi_j(t)|j=1,2,\ldots,N\} \text{ and } \{\phi_i(s)|i=1,2,\ldots,M\},$$

that is,

$$\mathbf{g}=\mathbf{A}\boldsymbol{\psi};\ \ \mathbf{f}=\mathbf{B}\boldsymbol{\phi}, \tag{11.77}$$

then $\mathbf{A}=\mathbf{B}^T=\mathbf{U}=\{\mathbf{u}_{ij}|i=1,\ldots,M;j=1,\ldots,N\}$. ■

Proof: The surface $\mathbf{v}(s,t)$ satisfies (11.75), which can be written, in matrix form, as:

$$\mathbf{v}=\boldsymbol{\phi}^T\mathbf{g}+\mathbf{f}^T\boldsymbol{\psi}-\boldsymbol{\phi}^T\mathbf{U}\boldsymbol{\psi}. \tag{11.78}$$

Substitution of (11.77) into (11.78) leads to

$$\mathbf{v}=\boldsymbol{\phi}^T\mathbf{A}\boldsymbol{\psi}+\boldsymbol{\phi}^T\mathbf{B}^T\boldsymbol{\psi}-\boldsymbol{\phi}^T\mathbf{U}\boldsymbol{\psi}=\boldsymbol{\phi}^T(\mathbf{A}+\mathbf{B}^T-\mathbf{U})\boldsymbol{\psi},$$

and then

$$\mathbf{v}(s_i,t_j)=\mathbf{g}_i(t_j)=\mathbf{f}_j(t_i)=\boldsymbol{\phi}^T(s_i)(\mathbf{A}+\mathbf{B}^T-\mathbf{U})\boldsymbol{\psi}(t_j). \tag{11.79}$$

Using now the definitions (11.72) and (11.73) we get

$$\mathbf{u}_{ij}=\mathbf{v}(s_i,t_j)=\mathbf{A}_{ij}+\mathbf{B}_{ji}-\mathbf{U}_{ij}. \tag{11.80}$$

Applying the same definitions (11.72) and (11.73) to equations (11.77) we get

$$\mathbf{g}_i(t_j)=(\mathbf{g}(t_j))_i=(\mathbf{A}\boldsymbol{\psi})_i=\mathbf{A}_{ij} \tag{11.81}$$

and

$$\mathbf{f}_j(s_i)=(\mathbf{f}(s_i))_j=(\mathbf{B}\boldsymbol{\phi})_j=\mathbf{B}_{ji}. \tag{11.82}$$

Now, substitution of (11.80), (11.81) and (11.82) into the compatibility conditions (11.74) we immediately have

$$\mathbf{U}_{ij}=\mathbf{A}_{ij}=\mathbf{B}_{ji},$$

which proves the theorem. ■

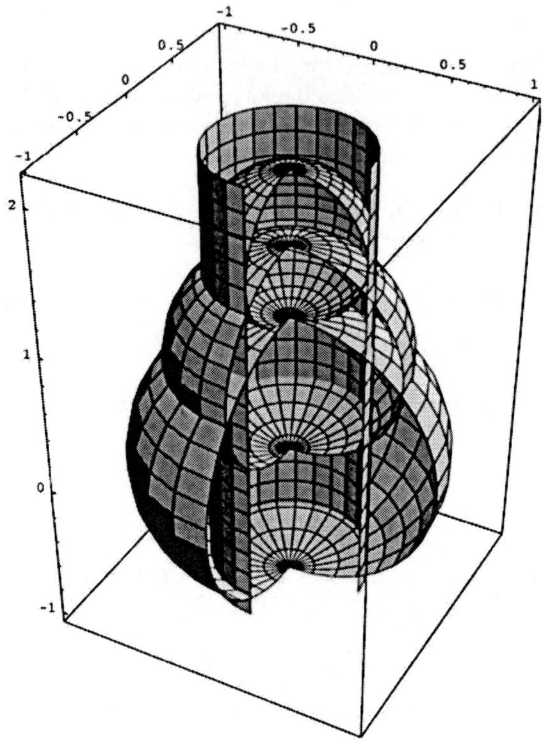

Figure 11.19: Four parametric surfaces and their intersections.

Note that in this case we have not only uniqueness of representation, which holds under more general conditions, but we also know the form of matrix $\mathbf{U}$. Note also that the form of the surface $\mathbf{v}(s,t)$ is extremely simple. In fact, for this particular case, the three terms in Expression (11.75) coincide; that is,

$$\mathbf{v}(s,t) = \mathbf{v}_1(s,t) = \mathbf{v}_2(s,t) = \mathbf{v}_3(s,t).$$

These particular surfaces are usually called tensor product surfaces in CAGD. Consequently, for interpolating the surface $\mathbf{v}(s,t)$ it is enough to interpolate any one of the families of curves $\{\mathbf{g}_i(t)\}$ or $\{\mathbf{f}_j(s)\}$, or to interpolate the set of points $\{\mathbf{u}_{ij}\}$.

Adequate selection of the sets of functions $\{\mathbf{g}_i(t)\}$ and $\{\mathbf{f}_j(s)\}$ can simplify the problem of determining the intersection curves of two different surfaces, as is illustrated in the following example.

Example 11.7 (Parametric form). In this example we build four parametric surfaces in parametric form. For the first three surfaces (spheres) we have used the base curves

$$\mathbf{g}_i(t) = \begin{pmatrix} asin(t)cos(s_i) \\ asin(t)sin(s_i) \\ acos(t) + b \end{pmatrix}; \quad \mathbf{f}_i(s) = \begin{pmatrix} asin(t_j)cos(s) \\ asin(t_j)sin(s) \\ acos(t_j) + b \end{pmatrix},$$

with $a = 1, b = 0$, $a = 0.75, b = 0.75$ and $a = 0.5, b = 1.5$, respectively. For the fourth surface (cylinder) we have taken the base curves

$$\mathbf{g}_i(t) = \begin{pmatrix} cos(s_i)/2 \\ sin(s_i)/2 \\ t - 1 \end{pmatrix}; \quad \mathbf{f}_i(s) = \begin{pmatrix} cos(s)/2 \\ sin(s)/2 \\ t_j - 1 \end{pmatrix},$$

The parameters s and t belong to the intervals $[0, 3\pi/2]$ and $[0, \pi]$, respectively. The surfaces and their intersections are shown in Figure 11.19.

Note that intersections of these surfaces can be easily determined, since the sets $\{\mathbf{f}_j(s)\}$ coincide for all of them and they correspond to $t = constant$. ∎

11.5.5 Tensor-product surfaces

A number of interesting problems in computer aided design (CAD) can be solved by using the so-called free-form curves and surfaces. Roughly speaking, they are parametric functions governed by a set of points (called control points) that more or less determine the shape of the curve or surface and many of its geometric properties. Free-form curves and surfaces are essential tools (among others) in the automotive, aircraft and ship building industries (see Bu-Qing and Ding-Yuan (1989)). For a soft introduction to the field, the reader is referred to Faux and Pratt (1985) and Anand (1993a). A more detailed mathematical description can also be found in Farin (1993), Rogers and Adams (1990) and Hoscheck and Lasser (1993).

A free-form parametric curve $\mathbf{c}(u)$ is given by:

$$\mathbf{C}(u) = \sum_{i=0}^{n} \mathbf{P}_i f_i(u), \tag{11.83}$$

where $\{\mathbf{P}_i, i = 0, 1, \ldots, n\} \subset \mathbb{R}^k$ are the control points in a two- or three-dimensional space ($k = 2, 3$ respectively) and $f_i(u)$, $i = 0, \ldots, n$ a family of basis functions, which determine the type of curve we are dealing with. For instance, a natural family in many cases is the monomial one, where $f_i(u) = u^i$, $\forall i$. However, in this case we are not able to give a geometric interpretation to the coefficients of the resulting curve. Clearly, this question is of great importance for interactive work, one of the main requirements in CAD.

There are, however, other polynomial basis functions for which the coefficients have geometric significance. The most usual among them are the Bézier and B-spline functions.

Given a set of $m + 1$ points $\{\mathbf{P}_i, i = 0, 1, \ldots, m\}$, we can define a Bézier curve of degree m as

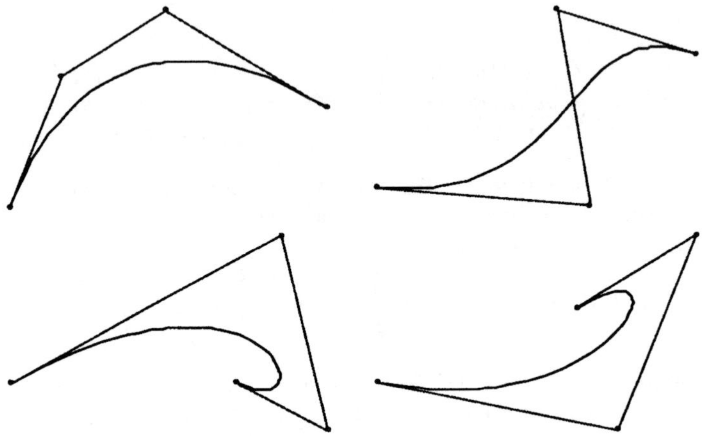

Figure 11.20: Four planar Bézier curves.

$$\mathbf{C}(u) = \sum_{i=0}^{m} \mathbf{P}_i B_i^m(u) \qquad u \in [0,1] \tag{11.84}$$

where the $B_i^m(u)$ are the so-called Bernstein polynomials, defined as

$$B_i^m(u) = \binom{m}{i} u^i \, (1-u)^{m-i}$$

with

$$\binom{m}{i} = \frac{m!}{i!\,(m-i)!}$$

where by convention $0^0 \equiv 1$ and $0! \equiv 1$.

Note that the curve generally follows the shape of the control polygon, which consists of the segments joining the control points. This fact is illustrated in Figure 11.20, which displays four examples of planar Bézier curves.

Note that if we impose that each $\mathbf{P}_i$ in (11.83) traverses a curve of the form

$$\mathbf{P}_i = \mathbf{P}_i(v) = \mathbf{C}(u) = \sum_{j=0}^{m} \mathbf{P}_{ij} g_j(v), \tag{11.85}$$

combining Equations (11.83) and (11.85), we obtain the surface given by

$$\mathbf{S}(u,v) = \sum_{i=0}^{n} \mathbf{P}_i(v) f_i(u) = \sum_{i=0}^{n} \sum_{j=0}^{m} \mathbf{P}_{ij} g_j(v) f_i(u), \tag{11.86}$$

where the parameters u and v are assumed to be valued on the rectangle $[a,b] \times [c,d]$. For instance, a *tensor product Bézier surface* $\mathbf{P}(u,v)$ *of degree* (m,n) is given by

$$\mathbf{P}(u,v) = \sum_{i=0}^{n}\sum_{j=0}^{m} \mathbf{P}_{ij} B_i^n(u) B_j^m(v) \tag{11.87}$$

where $\{\mathbf{P}_{ij} | \; i = 0,\dots,n; \; j = 0,\dots,m\}$ are also control points and $B_i^n(u)$, $B_j^m(v)$ are the Bernstein polynomials of degrees n and m respectively. Once again, the variables u and v are to be valued onto the square $[0,1] \times [0,1]$.

Note that if we fix $u = u_0$ in (11.86), then

$$\mathbf{S}(u_0,v) = \sum_{i=0}^{n}\sum_{j=0}^{m} \mathbf{P}_{ij} g_j(v) f_i(u_0) = \sum_{j=0}^{m}\left(\sum_{i=0}^{n} \mathbf{P}_{ij} f_i(u_0)\right) g_j(v) = \sum_{j=0}^{m} d_j g_j(v) \tag{11.88}$$

where

$$d_j = \sum_{i=0}^{n} \mathbf{P}_{ij} f_i(u_0). \tag{11.89}$$

Similarly, $\mathbf{S}(u,v_0)$ is a curve lying on $\mathbf{S}(u,v)$ and the curves $\mathbf{S}(u_0,v)$ and $\mathbf{S}(u,v_0)$ intersect at the surface point $\mathbf{S}(u_0,v_0)$. These curves are called *isoparametric curves*, $\mathbf{S}(u_0,v)$ being called a v-curve, and $\mathbf{S}(u,v_0)$ a u-curve.

It is well established (see e.g. (Farin, 1993) or (Hoschek and Lasser, 1993)) that if we fix $u = u_0$ or $v = v_0$ in (11.86), the resulting curves (the isoparametric curves) are defined by (11.83). Now we try to answer the following question:

> *Which are all the surfaces $\mathbf{S}(u,v)$ such that their isoparametric curves are defined by (11.83)?*

In other words, we ask ourselves if all the surfaces whose isoparametric curves satisfy Equation (11.83) are necessarily tensor product surfaces. The answer to this question is given by the following theorem:

Theorem 11.13 (Tensor product surfaces and isoparametric curves). *All the surfaces $\mathbf{S}(u,v)$ whose u and v isoparametric curves are given by*

$$\mathbf{C}(u) = \sum_{i=0}^{n} \mathbf{P}_i f_i(u) \quad \textit{and} \quad \mathbf{C}^*(v) = \sum_{j=0}^{m} \mathbf{Q}_j g_j(v), \tag{11.90}$$

respectively where $\mathbf{P}_i$ and $\mathbf{Q}_j$ are control points and $f_i(u)$ and $g_j(v)$ belong to a given family of basis functions are of the form

$$\mathbf{S}(u,v) = \sum_{i=0}^{n}\sum_{j=0}^{m} \mathbf{P}_{ij} f_i(u) g_j(v); \tag{11.91}$$

that is, they are tensor product surfaces. ∎

Proof: We look for surfaces $\mathbf{S}(u,v)$ such that the isoparametric curves follow (11.84). This implies that for any $u = u_0$ we have:

$$\mathbf{S}(u_0,v) = \sum_{j=0}^{m} \alpha_j g_j(v)$$

where $\boldsymbol{\alpha}_j = \boldsymbol{\alpha}_j(u_0)$. Hence, the general form for $\mathbf{S}(u, v)$ is

$$\mathbf{S}(u,v) = \sum_{j=0}^{m} \boldsymbol{\alpha}_j(u) g_j(v) \tag{11.92}$$

It is evident that a similar discussion is valid for any $v = v_0$. Therefore we have

$$\mathbf{S}(u,v) = \sum_{j=0}^{m} \boldsymbol{\alpha}_j(u) g_j(v) = \sum_{i=0}^{n} \boldsymbol{\beta}_i(v) f_i(u) \tag{11.93}$$

that is

$$\boldsymbol{\alpha}(u) \bullet \mathbf{g}(v) = \boldsymbol{\beta}(v) \bullet \mathbf{f}(u) \tag{11.94}$$

where $\boldsymbol{\alpha}(u) = (\boldsymbol{\alpha}_0(u), \ldots, \boldsymbol{\alpha}_m(u))$ and $\boldsymbol{\beta}(v) = (\boldsymbol{\beta}_0(v), \ldots, \boldsymbol{\beta}_n(v))$ are unknown vector functions and $\mathbf{g}(v) = (g_0(v), \ldots, g_m(v))$ and $\mathbf{f}(u) = (f_0(u), \ldots, f_m(u))$ are known functions.

According to Corollary 11.1, from (11.94) we get:

$$\boldsymbol{\alpha}(u) = \mathbf{f}(u)\ \mathbf{D} \quad ; \quad \boldsymbol{\beta}(v) = \mathbf{g}(v)\ \mathbf{D}^T \tag{11.95}$$

where $\mathbf{D}$ is an arbitrary vector matrix. Expression (11.95) can also be written as

$$\boldsymbol{\alpha}_j(u) = \sum_{i=0}^{n} \mathbf{d}_{ij} f_i(u) \quad ; \quad \boldsymbol{\beta}_i(v) = \sum_{j=0}^{m} \mathbf{d}_{ji} g_j(v)$$

Introducing these expressions in (11.92) and (11.93) respectively, we obtain

$$\mathbf{S}(u,v) = \sum_{j=0}^{m} \sum_{i=0}^{n} \mathbf{d}_{ij} f_i(u) g_j(v) = \sum_{i=0}^{n} \sum_{j=0}^{m} \mathbf{d}_{ji} g_j(v) f_i(u),$$

which coincides with (11.91). As a conclusion, only the tensor product Bézier surfaces satisfy (11.92) and (11.93). ∎

It is obvious that similar results can be obtained no matter the family of functions we consider instead of $\mathbf{f}(u)$ and $\mathbf{g}(v)$. In particular, the previous discussion might have been introduced for instance with Bézier or B-spline surfaces.

Corollary 11.3 (Bézier tensor product surfaces). *The only parametric surfaces whose isoparametric surfaces are Bézier curves are the tensor product Bézier surfaces, and the only parametric surfaces whose isoparametric surfaces are B-spline curves are the tensor product B-spline surfaces.* ∎

11.6 Application of functional networks to fitting surfaces

Artificial neural networks have been recognized as a powerful tool for learning and simulating systems in a great variety of fields (see Castillo et al. (1999b) and references therein for a survey of this field). Since they are inspired in the behavior of the brain, they reproduce some of its most typical features, such as the ability to learn from given data. This characteristic of neural networks make them specially valuable for solving problems, in which one is interested in fitting a given set of data by using some interpolation scheme.

However, not every approximation problem admits a good representation in terms of a neural network; on the contrary, some interesting problems in computer graphics require more refined techniques for a satisfactory treatment. In this section, we apply functional networks (described in Chapter 9) as a powerful tool to be used in those cases in which its alternative description, based on neural networks, becomes inappropriate.

11.6.1 The case of parametric surfaces

In this section, the functional network paradigm is applied to fit B-spline parametric surfaces by means of Bézier surfaces. Firstly, some basic definitions to be used later on are given.

Let $\mathcal{T} = \{u_0, u_1, u_2, \ldots, u_{r-1}, u_r\}$ be a nondecreasing sequence of real numbers called *knots*. $\mathcal{T}$ is called the *knot vector*. The *ith B-spline basis function* $N_{ik}(u)$ *of order* k (or degree $k-1$) is defined by the recurrence relations

$$N_{i1}(u) = \begin{cases} 1 & \text{if} \quad u_i \le u \le u_{i+1} \\ 0 & \text{otherwise} \end{cases} \tag{11.96}$$

and

$$N_{ik}(u) = \frac{u - u_i}{u_{i+k-1} - u_i} N_{i,k-1}(u) + \frac{u_{i+k} - u}{u_{i+k} - u_{i+1}} N_{i+1,k-1}(u) \tag{11.97}$$

for $k > 1$. Then, a *B-spline curve* $\mathbf{C}(u)$ *of order* k is defined by

$$\mathbf{C}(u) = \sum_{i=0}^{m} \mathbf{P}_i N_{ik}(u) \tag{11.98}$$

where the $\{\mathbf{P}_i;\ i = 0, \ldots, m\}$ are the control points in $\mathbb{R}^3$ and the $\{N_{ik}(u)\}_i$ are the normalized B-spline basis functions of order k defined as in (11.97). As is known, in any B-spline curve its order k, the number of control points $(m+1)$ and the number of knots $(r+1)$ are related to each other by $r = m + k$ (see Anand (1993a), page 225).

With the same notation, given two knot vectors $\mathcal{U} = \{u_0, u_1, \ldots, u_r\}$ and $\mathcal{V} = \{v_0, v_1, \ldots, v_s\}$ with $r = m + k$ and $s = n + l$, a *B-spline surface* $\mathbf{S}(u, v)$ *of order* (k, l) is defined by

$$\mathbf{S}(u, v) = \sum_{i=0}^{m} \sum_{j=0}^{n} \mathbf{P}_{ij} N_{ik}(u) N_{jl}(v) \tag{11.99}$$

where the $\{\mathbf{P}_{ij}; i = 0, \ldots, m; j = 0, \ldots, n\}$ are the control points in a bidirectional net and the $\{N_{ik}(u)\}_i$ and $\{N_{jl}(v)\}_j$ are the B-spline basis functions of order k and l respectively. A more detailed discussion about B-spline curves and surfaces can be found in Piegl and Tiller (1997).

Step 1 (Statement of the problem). We are looking for the most general family of parametric surfaces $\mathbf{S}(u,v)$ such that their isoparametric curves (see Farin (1993) and Hoscheck and Lasser (1993) for a description) $u = u_0$ and $v = v_0$ are linear combinations of the sets of functions:

$$\mathbf{f}(u) = \{f_0(u), f_1(u), \ldots, f_m(u)\} \text{ and } \mathbf{f}^*(v) = \{f_0^*(v), f_1^*(v) \ldots, f_n^*(v)\},$$

respectively.

To be more precise, we look for surfaces $\mathbf{S}(u,v)$ such that they satisfy the system of functional equations

$$\mathbf{S}(u,v) \equiv \sum_{j=0}^{n} \boldsymbol{\alpha}_j(u) f_j^*(v) = \sum_{i=0}^{m} \boldsymbol{\beta}_i(v) f_i(u), \tag{11.100}$$

where the sets of coefficients

$$\{\boldsymbol{\alpha}_j(u); j = 0,1,\ldots,n\} \text{ and } \{\boldsymbol{\beta}_i(v); i = 0,1,\ldots,m\}$$

can be assumed, without loss of generality, as sets of linearly independent functions. Note that if they are not, we can rewrite Equations (11.100) in the same form but with linearly independent sets.

Step 2 (Initial topology). Based on the knowledge of the problem, the topology of the initial functional network is selected. Thus, the system of functional equations (11.100) leads to the functional network in Figure 11.21. Note that the above equations can be obtained from the network by considering the equality between the two values associated with the links connected to the output unit. We also remark that each of these values can be obtained in terms of the outputs of the preceding units by writing the outputs of the neurons as functions of their inputs, and so on.

Step 3 (Simplification). In this step, the initial functional network is simplified by using functional equations. The solution for this problem is given by Theorem 11.13. In fact, Equation (11.91) shows that the functional network in Figure 11.22 is equivalent to the functional network in Figure 11.21.

Step 4 (Uniqueness of representation). In this step, conditions for the neural functions of the simplified functional network must be obtained. For the case of Expression (11.99), two cases must be considered:

1. **The $f_i(u)$ and $f_j^*(v)$ functions are given.** Assume that there are two matrices $\mathbf{P} = \{\mathbf{P}_{ij}\}$ and $\mathbf{P}^* = \{\mathbf{P}_{ij}^*\}$ such that

$$\mathbf{S}(u,v) \equiv \sum_{i=0}^{m}\sum_{j=0}^{n} \mathbf{P}_{ij} f_i(u) f_j^*(v) = \sum_{i=0}^{m}\sum_{j=0}^{n} \mathbf{P}_{ij}^* f_i(u) f_j^*(v). \tag{11.101}$$

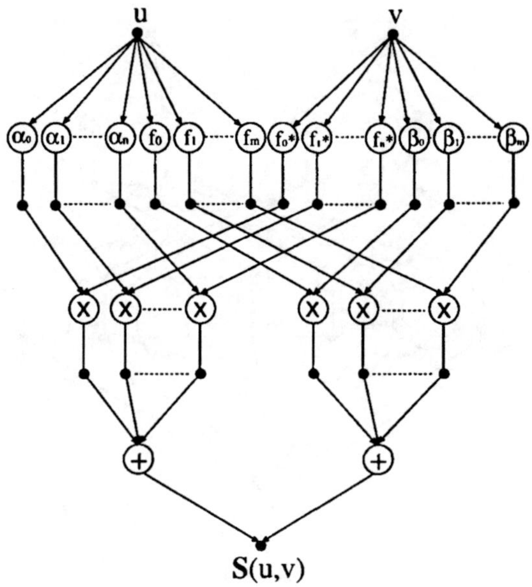

Figure 11.21: Initial functional network.

Solving the uniqueness of representation problem consists of solving equation (11.101). To this aim, we write (11.101) in the form

$$\sum_{i=0}^{m}\sum_{j=0}^{n}\left(\mathbf{P}_{ij}-\mathbf{P}^*_{ij}\right) f_i(u)f^*_j(v)=0. \tag{11.102}$$

Since the functions in the set

$$\{f_i(u)\ f^*_j(v)\mid i=0,1,\ldots,m\ ;\ j=0,1,\ldots,n\}$$

are linearly independent because the sets

$$\{f_i(u)\ |i=0,1,\ldots,m\} \text{ and } \{f^*_j(v)\mid j=0,1,\ldots,n\}$$

are linearly independent, from (11.102) we have

$$\mathbf{P}_{ij}=\mathbf{P}^*_{ij} \qquad i=0,1,\ldots,m;\ \ j=0,1,\ldots,n;$$

that is, the coefficients $\mathbf{P}_{ij}$ in (11.99) are unique.

2. **The $f_i(u)$ and $f^*_j(v)$ functions are to be learned.** In this case, assume that there are two sets of functions

$$\{f_i(u),f^*_j(v)\} \text{ and } \{\tilde{f}_i(u),\tilde{f}^*_j(v)\},$$

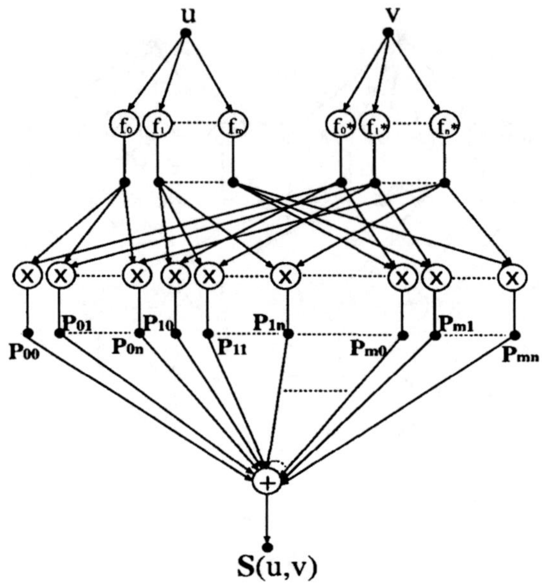

Figure 11.22: Equivalent functional network.

and two matrices $\mathbf{P}$ and $\tilde{\mathbf{P}}$ such that

$$\mathbf{S}(u,v) \equiv \sum_{i=0}^{m}\sum_{j=0}^{n}\mathbf{P}_{ij}f_i(u)f_j^*(v) = \sum_{i=0}^{m}\sum_{j=0}^{n}\tilde{\mathbf{P}}_{ij}\tilde{f}_i(u)\tilde{f}_j^*(v). \qquad (11.103)$$

Then we have

$$\sum_{i=0}^{m}\sum_{j=0}^{n}\mathbf{P}_{ij}f_i(u)f_j^*(v) - \sum_{i=0}^{m}\sum_{j=0}^{n}\tilde{\mathbf{P}}_{ij}\tilde{f}_i(u)\tilde{f}_j^*(v) = 0. \qquad (11.104)$$

According to Theorem 1 in Castillo and Iglesias (1997), the solution satisfies

$$\begin{pmatrix} \sum_{i=0}^{m} \mathbf{P}_{i0} f_i(u) \\ \sum_{i=0}^{m} \mathbf{P}_{i1} f_i(u) \\ \vdots \\ \sum_{i=0}^{m} \mathbf{P}_{in} f_i(u) \\ ------ \\ \sum_{i=0}^{m} \tilde{\mathbf{P}}_{i0} \tilde{f}_i(u) \\ \sum_{i=0}^{m} \tilde{\mathbf{P}}_{i1} \tilde{f}_i(u) \\ \vdots \\ \sum_{i=0}^{m} \tilde{\mathbf{P}}_{in} \tilde{f}_i(u) \end{pmatrix} = \begin{pmatrix} \mathbf{P}^T \\ -- \\ \mathbf{B} \end{pmatrix} \mathbf{f}^T(u) \tag{11.105}$$

$$\begin{pmatrix} f_0^*(v) \\ f_1^*(v) \\ \vdots \\ f_n^*(v) \\ -- \\ -\tilde{f}_0^*(v) \\ -\tilde{f}_1^*(v) \\ \vdots \\ -\tilde{f}_n^*(v) \end{pmatrix} = \begin{pmatrix} \mathbf{I} \\ -- \\ \mathbf{C} \end{pmatrix} (\mathbf{f}^*(v))^T \tag{11.106}$$

with

$$(\mathbf{P} \mid \mathbf{B}^T) \begin{pmatrix} \mathbf{I} \\ -- \\ \mathbf{C} \end{pmatrix} = \mathbf{0} \Leftrightarrow \mathbf{P} = -\mathbf{B}^T\mathbf{C}. \tag{11.107}$$

From (11.105) and (11.106) we get

$$\begin{aligned} \tilde{\mathbf{P}}^T \, \tilde{\mathbf{f}}^T(u) &= \mathbf{B}\mathbf{f}^T(u) \\ (\tilde{\mathbf{f}}^*(v))^T &= -\mathbf{C} \, (\mathbf{f}^*(v))^T, \end{aligned} \tag{11.108}$$

Expressions (11.105) and (11.106) give the relations between both equivalent solutions and the degrees of freedom we have.

However, if we have to learn $\mathbf{f}(u)$ and $\mathbf{f}^*(v)$ we can approximate them as:

$$\begin{aligned} \mathbf{f}(u) &= \phi(u) \, \mathbf{B} \\ \mathbf{f}^*(v) &= \psi(v) \, \mathbf{C}, \end{aligned} \tag{11.109}$$

and we get

$$\begin{aligned} &\mathbf{S}(u,v) = \mathbf{f}(u)\mathbf{P}(\mathbf{f}^*(v))^T = \\ &\phi(u)\mathbf{B}\mathbf{P}\mathbf{C}^T\psi(v)^T = \phi(u)\tilde{\mathbf{P}}\psi(v)^T, \end{aligned} \tag{11.110}$$

which is equivalent to (11.101) but with functions $\{\phi(u), \psi(v)\}$ instead of $\{\mathbf{f}(u), \mathbf{f}^*(v)\}$. Thus, this case is reduced to the first one.

Step 5 (Data collection). For the learning to be possible we need some data. In this example we have selected a set of 256 data points $\{T_{pq};\ p,q = 1,\ldots,16\}$ (in the following, the *training points*) in a regular 16×16 grid. This grid comes from the domain of different B-spline surfaces, corresponding to different choices of the control points and knot vectors.

Surface I. It is given by the control points listed in Table 11.1, $m = n = 5$, $k = l = 3$ and nonperiodic knot vectors (according to the classification used in Anand (1993a)) for both directions u and v. The corresponding surface is shown in Figure 11.23(left-up).

Surface II. It is given by the control points listed in Table 11.2, $m = n = 5$, $k = l = 3$. With the aim of checking how the knot vectors influence our results, we consider two different cases: IIa, nonperiodic knot vectors for both directions u and v (see Figure 11.23(right-up)) and IIb, periodic and nonperiodic knot vectors for u and v respectively (Figure 11.23(left-down)).

Table 11.1: Control points used to define Surface I.

(x,y,z)	(x,y,z)	(x,y,z)	(x,y,z)	(x,y,z)	(x,y,z)
(0,0,1)	(0,1,2)	(0,2,3)	(0,3,3)	(0,4,2)	(0,5,1)
(1,0,2)	(1,1,3)	(1,2,4)	(1,3,4)	(1,4,3)	(1,5,2)
(2,0,3)	(2,1,4)	(2,2,5)	(2,3,5)	(2,4,4)	(2,5,3)
(3,0,3)	(3,1,4)	(3,2,5)	(3,3,5)	(3,4,4)	(3,5,3)
(4,0,2)	(4,1,3)	(4,2,4)	(4,3,4)	(4,4,3)	(4,5,2)
(5,0,1)	(5,1,2)	(5,2,3)	(5,3,3)	(5,4,2)	(5,5,1)

Table 11.2: Control points used to define Surface II.

(x,y,z)	(x,y,z)	(x,y,z)	(x,y,z)	(x,y,z)	(x,y,z)
(0,0,1)	(0,1,2)	(0,2,3)	(0,3,3)	(0,4,2)	(0,5,1)
(1,0,2)	(1,1,3)	(1,2,4)	(1,3,2)	(1,4,3)	(1,5,2)
(2,0,1)	(2,1,4)	(2,2,5)	(2,3,5)	(2,4,4)	(2,5,3)
(3,0,3)	(3,1,4)	(3,2,5)	(3,3,1)	(3,4,2)	(3,5,3)
(4,0,2)	(4,1,2)	(4,2,4)	(4,3,4)	(4,4,3)	(4,5,2)
(5,0,1)	(5,1,2)	(5,2,3)	(5,3,3)	(5,4,2)	(5,5,1)

Surface III: It is given by the control points listed in Table 11.3, $m = 4, n = 3$, $k = 3, l = 4$ and nonperiodic knot vectors for both directions u and v (Figure 11.23(right-down)).

Table 11.3: Control points used to define Surface III.

(x,y,z)	(x,y,z)	(x,y,z)	(x,y,z)
$(-3,-3,0)$	$(-3,-1,3)$	$(-3,1,0)$	$(-3,3,3)$
$(-1,-3,3)$	$(-1,-1,7)$	$(-1,1,0)$	$(-1,3,0)$
$(1,-3,0)$	$(1,-1,0)$	$(1,1,0)$	$(1,3,3)$
$(3,-3,1)$	$(3,-1,0)$	$(3,1,3)$	$(3,3,1)$

In order to check the robustness of the proposed method, the third coordinate of the 256 three- dimensional points (x_k, y_k, z_k) was slightly modified by adding a real uniform random variable ϵ_k of mean 0 and variance 0.05. Therefore, in the following, we consider points given by (x_k, y_k, z_k^*), where

$$z_k^* = z_k + \epsilon_k \quad , \quad \epsilon_k \in (-0.05, 0, 05). \tag{11.111}$$

Such a random variable plays the role of an error measure to be used in the estimation step to learn the functional form of $\mathbf{S}(u,v)$.

Step 6 (Learning). At this point, the neural functions are estimated (learned), by using some minimization method. In the case of our example, the problem of learning the above functional network is reduced to estimating the neuron functions $x(u,v)$, $y(u,v)$ and $z(u,v)$ from a given sequence of triplets

$$\{(x_k, y_k, z_k),\ k = 1, \ldots, 256\},$$

which depend on u and v so that $x(u_k, v_k) = x_k$ and so on. To this aim we build the sum of squared errors function:

$$Q_\alpha = \sum_{k=1}^{256} \left(\alpha_k - \sum_{i=1}^{m} \sum_{j=1}^{n} a_{ij} \phi_i(u_k) \psi_j(v_k) \right)^2 , \tag{11.112}$$

where, in the present example, we must consider an error function for each variable x, y and z, and then α in the previous expression must be interpreted as three different equations, for $\alpha = x$, y and z.

The optimum value is obtained for

$$\frac{\partial Q_\alpha}{2\partial a_{rs}} = \sum_{k=1}^{256} \left(\alpha_k - \sum_{i=1}^{m} \sum_{j=1}^{n} a_{ij} \phi_i(u_k) \psi_j(v_k) \right) \times \phi_r(u_k) \psi_s(v_k) = 0 \tag{11.113}$$

$r = 1, \ldots, m; \quad s = 1, \ldots, n.$

To fit the 256 data points of our example, we have used Bernstein polynomials (see Section 11.5.5 for details) in u and v variables for the functions

$$\{\phi_i(u) = B_i^m(u) | i = 0, 1, \ldots, m\} \text{ and } \{\psi_j(v) = B_j^n(v) | j = 0, 1, \ldots, n\}.$$

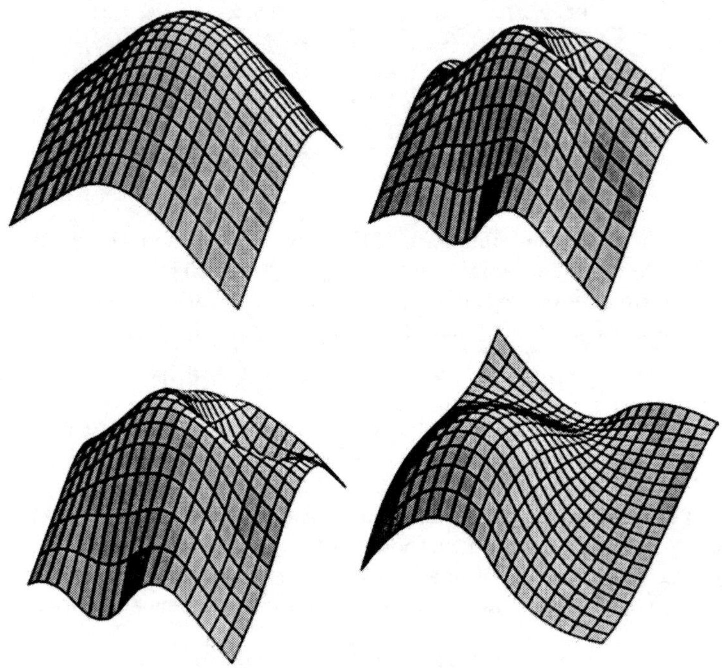

Figure 11.23: (upper-left) B-spline Surface I for $m = n = 5$, $k = l = 3$ and (nonperiodic, nonperiodic) knot vectors; (upper-right) B-spline Surface IIa for $m = n = 5$, $k = l = 3$ and (nonperiodic, nonperiodic) knot vectors; (lower-left) B-spline Surface IIb for $m = n = 5$, $k = l = 3$ and (periodic,nonperiodic) knot vectors; (lower-right) B-spline Surface III for $m = 4, n = 3$, $k = 3, l = 4$ and (nonperiodic,nonperiodic) knot vectors.

Of course, every different choice for m and n yields to the corresponding system (11.113), which must be solved. In particular, we have taken values for m and n from 2 to 6. A simple visual inspection of the data reveals that unit values for m and/or n are not adequate. On the other hand, degrees larger than 6 may lead to numerical round-off errors and are not widely used in industry.

By solving the system (11.113) for all of these cases, we obtain the control points associated with the Bézier surfaces fitting the data.

Step 7 (Model validation). At this step, a test for quality and/or the cross validation of the model is performed. Checking the obtained error is important to see whether or not the selected family of approximating functions are adequate. A cross validation of the model is also convenient.

To test the quality of the model. We have calculated the mean, the maximum and the root mean squared (RMS) errors for m and n from 2 to 6 and for the 256 training data points from the four B-splines surfaces. The obtained results for the different values of m and n are reported in Tables 11.4, 11.5 and 11.6.

Table 11.4 refers to Surface I. As the reader can appreciate, in general, errors are small indicating that the approach (which, of course, depends on the values of m and n) is reasonable. The reader will also appreciate that the best choice corresponds to $m = n = 2$, as expected, because data points come from a very smooth B-spline surface (see Figure 11.23(left-up)) of order (3,3). For $m = n = 2$ the mean and the RMS errors are 0.0023 and 0.00053 respectively. Since errors are small, the selected approximating (2,2)-degree Bézier surface, displayed in Figure 11.24(left-up), was considered adequate.

It could be argued that the good results for Surface I are due to the smoothness of the surface rather than the goodness of the method. In fact, the 36 control points used to define the surface are not apparently necessary and some of them could be removed without affecting the shape of the surface. To check this Surface II and Surface III were also considered. In addition, variants IIa and IIb were introduced to check the effect of changing a knot vector on our results. As they were very similar, only results for Surface IIa are reported here (see Table 11.5). In this case, complexity of the shape is reflected in the fact that the best approximation was obtained for the highest values of m and n. For this value, the mean and the RMS errors are respectively 0.0056 and 0.0099 for Surface IIa and 0.0057 and 0.0098 for Surface IIb.

Finally, Surface III represents a compromise between the smoothness and the complexity of the two previous surfaces. In addition, since the number of control points in the v direction is 4 and we have order 3, this surface is actually a Bézier surface in that direction. As a consequence, the best approximation is expected for $n = 3$. Table 11.6 confirms this: the best choice corresponds to $m = 6$ and $n = 3$, with values 0.0109 and 0.00219 for the mean and the RMS errors respectively. This value for m confirms that high values are more adequate to represent complex shapes and coincides with results for Surface II.

To cross validate the model. We have also used the fitted model to predict a new set of 1024 testing data points, and calculated the mean, the maximum and the root mean squared (RMS) errors, obtaining the results shown in Tables 11.4, 11.5 and 11.6. The new results confirm our previous choices for m and n in all cases. A comparison between mean and RMS error values for the training and testing data shows that, for our choice, they are comparable. Thus, we can conclude that no overfitting occurs. Note that a variance for the training data significantly smaller than the variance for the testing data is a clear indication of overfitting. This does not occur here.

Table 11.4: Mean, maximum and root mean squared errors of the 256 training points and the 1024 testing points from Surface I for different values of m and n.

TRAINING POINTS					
	$n=2$	$n=3$	$n=4$	$n=5$	$n=6$
$m=2$	**0.020** **0.0023** **0.00053**	0.016 0.004 0.00066	0.020 0.0048 0.0008	0.028 0.0053 0.0010	0.029 0.0054 0.0010
$m=3$	0.033 0.0062 0.0011	0.032 0.0066 0.0011	0.040 0.0079 0.0014	0.038 0.0083 0.0014	0.024 0.0060 0.0011
$m=4$	0.034 0.0066 0.0012	0.037 0.0071 0.0013	0.024 0.0071 0.0011	0.039 0.007 0.0013	0.040 0.0081 0.0015
$m=5$	0.030 0.0071 0.0013	0.034 0.0076 0.0014	0.031 0.0077 0.0013	0.035 0.0065 0.0011	0.054 0.0088 0.0016
$m=6$	0.041 0.0052 0.0010	0.051 0.0075 0.0014	0.041 0.0084 0.0015	0.038 0.0094 0.0016	0.050 0.0085 0.0015

TESTING POINTS					
	$n=2$	$n=3$	$n=4$	$n=5$	$n=6$
$m=2$	**0.023** **0.0024** **0.00012**	0.016 0.0038 0.00015	0.020 0.0049 0.00019	0.035 0.0057 0.00026	0.041 0.0059 0.00027
$m=3$	0.033 0.0061 0.00025	0.033 0.0065 0.00027	0.069 0.0080 0.00034	0.063 0.0086 0.00037	0.039 0.0066 0.00030
$m=4$	0.039 0.0066 0.00028	0.037 0.0070 0.00031	0.057 0.0073 0.00029	0.090 0.0082 0.00030	0.069 0.0087 0.00041
$m=5$	0.037 0.0070 0.00030	0.068 0.0077 0.00033	0.070 0.0081 0.00035	0.047 0.0068 0.00029	0.117 0.0097 0.00048
$m=6$	0.048 0.0053 0.00025	0.121 0.0079 0.00038	0.067 0.0088 0.00040	0.066 0.0100 0.00045	0.100 0.0094 0.00046

Table 11.5: Mean, maximum and root mean squared errors of the 256 training points and the 1024 testing points from Surface IIa for different values of m and n.

TRAINING POINTS					
	$n=2$	$n=3$	$n=4$	$n=5$	$n=6$
$m=2$	1.510 0.352 0.0609	1.078 0.253 0.0449	1.081 0.245 0.0440	0.949 0.234 0.0427	0.932 0.232 0.0426
$m=3$	1.207 0.316 0.0536	0.735 0.202 0.0337	0.741 0.189 0.0325	0.598 0.171 0.0302	0.609 0.169 0.0301
$m=4$	1.217 0.316 0.0526	0.746 0.188 0.0321	0.755 0.179 0.0308	0.615 0.161 0.0280	0.597 0.159 0.0281
$m=5$	0.948 0.277 0.0460	0.400 0.120 0.0190	0.405 0.105 0.0177	0.272 0.070 0.0122	0.299 0.066 0.0118
$m=6$	0.938 0.275 0.0465	0.389 0.111 0.0188	0.394 0.098 0.0165	0.248 0.061 0.0104	**0.248** **0.0564** **0.0099**

TESTING POINTS					
	$n=2$	$n=3$	$n=4$	$n=5$	$n=6$
$m=2$	1.537 0.356 0.0144	1.078 0.256 0.0106	1.081 0.249 0.0105	0.959 0.233 0.0100	0.932 0.233 0.0100
$m=3$	1.240 0.318 0.0127	0.746 0.203 0.0081	0.741 0.193 0.0079	0.678 0.172 0.0072	0.671 0.173 0.0072
$m=4$	1.261 0.323 0.0128	1.026 0.197 0.0082	1.020 0.189 0.0079	1.076 0.167 0.0073	1.084 0.167 0.0073
$m=5$	1.132 0.281 0.0112	0.441 0.126 0.0049	0.550 0.114 0.0045	0.307 0.075 0.0030	0.356 0.075 0.0031
$m=6$	1.154 0.286 0.0113	0.681 0.124 0.005	0.744 0.112 0.0048	0.779 0.073 0.0035	**0.834** **0.072** **0.0036**

Table 11.6: Mean, maximum and root mean squared errors of the 256 training points and the 1024 testing points from Surface III for different values of m and n.

TESTING POINTS					
	$n=2$	$n=3$	$n=4$	$n=5$	$n=6$
$m=2$	0.850 0.2083 0.03864	0.721 0.1677 0.03222	0.723 0.1678 0.03222	0.723 0.1679 0.03222	0.733 0.1679 0.03222
$m=3$	0.450 0.1156 0.02228	0.139 0.0329 0.00637	0.136 0.0329 0.00640	0.136 0.0330 0.00642	0.138 0.0334 0.00643
$m=4$	0.403 0.1184 0.02196	0.102 0.0265 0.00511	0.105 0.0273 0.00517	0.104 0.0268 0.00512	0.104 0.0271 0.00515
$m=5$	0.388 0.1144 0.02152	0.059 0.0138 0.00271	0.066 0.0141 0.00278	0.059 0.0138 0.00268	0.060 0.0143 0.00273
$m=6$	0.406 0.1133 0.02146	**0.056** **0.0109** **0.00219**	0.052 0.0121 0.00228	0.058 0.0129 0.00243	0.050 0.0123 0.00234

TESTING POINTS					
	$n=2$	$n=3$	$n=4$	$n=5$	$n=6$
$m=2$	0.850 0.3546 0.01153	0.725 0.2880 0.00977	0.727 0.2880 0.00977	0.725 0.2883 0.00977	0.733 0.2882 0.00977
$m=3$	0.450 0.1862 0.00640	0.139 0.0581 0.00199	0.136 0.0580 0.00200	0.136 0.0581 0.00200	0.138 0.0586 0.00200
$m=4$	0.403 0.1900 0.00629	0.107 0.0458 0.00159	0.138 0.0586 0.00200	0.112 0.0468 0.00158	0.112 0.0472 0.00159
$m=5$	0.106 0.0458 0.00159	0.062 0.0244 0.00084	0.070 0.0249 0.00086	0.060 0.0243 0.00084	0.063 0.0251 0.00085
$m=6$	0.406 0.1798 0.00607	**0.055** **0.0191** **0.00067**	0.053 0.0204 0.00068	0.058 0.0216 0.00072	0.053 0.0212 0.00072

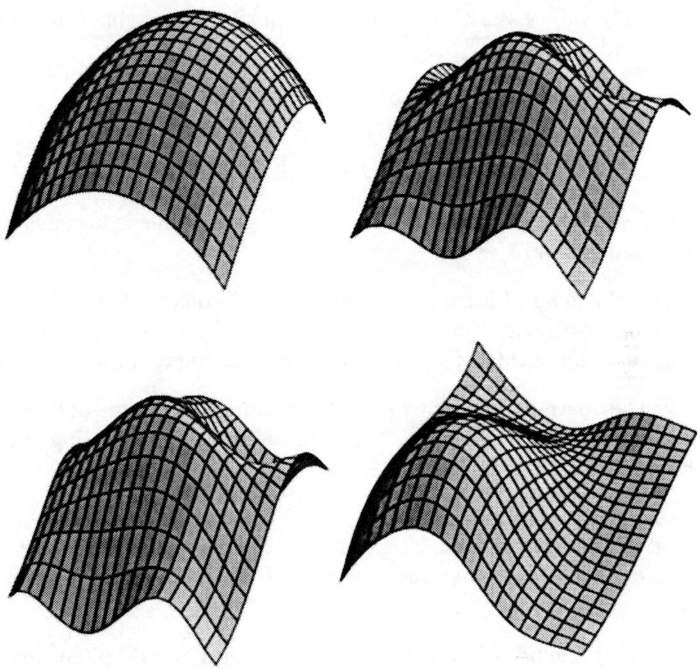

Figure 11.24: Approximations of the B-spline surfaces in Figure 11.23: (upper-left) (2,2)-degree Bézier surface approximating Surface I; (upper-right) (6,6)-degree Bézier surface approximating Surface IIa; (lower-left) (6,6)-degree Bézier surface approximating Surface IIb; (lower-right) (6,3)-degree Bézier surface approximating Surface III.

Exercises

11.1 Use the methods developed in this chapter to design and learn a functional network to reproduce the surface in the implicit form:

$$x^2y^3 + x^3y^2 + xyz = 1.$$

11.2 Generate a set of data points from the surface in the explicit form

$$z(x,y) = x^2 + y^2; \quad 0 \le x \le 1; \quad 0 \le y \le 1,$$

and then:

(a) Reproduce the above surface using polynomial functions for $u(x)$ and $v(y)$.

(b) Reproduce the above surface using sine and cosine functions for $u(x)$ and $v(y)$.

11.3 Solve the functional equation

$$f(x+y) = f(x) + f(y) + g(f(x) - f(y)) - f(x-y)$$

for f and g continuous in $\mathbb{R}_+$. This expression is a generalization of Equation (11.27).

11.4 Write the general form of a surface $\mathbf{S}(u,v)$ such that its cross sections by $u = u_0$ are Bézier curves of degree 3 and by $v = v_0$ are polynomials of degree 4. Determine the number of free parameters for such a surface.

11.5 Use the methods developed in this chapter to design and learn a functional network to reproduce the data of Table 11.1 by using the surface of Exercise 11.4.

11.6 Using the control points of Table 11.8 and the Equation (11.87), generate a set of 256 data points. Use a functional network to fit the data points by using polynomial functions. Determine the polynomial degree that best fits the data points.

11.7 Following the notation in Section 11.6.1 a rational B-spline surface $\mathbf{S}(u,v)$ of order (k,l) is defined as:

$$\mathbf{S}(u,v) = \frac{\sum_{i=0}^{m}\sum_{j=0}^{n} \mathbf{P}_{ij} w_{ij} N_{ik}(u) N_{jl}(v)}{\sum_{i=0}^{m}\sum_{j=0}^{n} w_{ij} N_{ik}(u) N_{jl}(v)}$$

where the $w_{ij} > 0$ are called the weights. Discuss whether this kind of surface can be written as a tensor-product surface.

Table 11.7: Data points used to fit the explicit surface.

x	y	z	x	y	z	x	y	z
0.000	0.000	4.990	0.000	0.125	1.820	0.000	0.250	0.005
0.000	0.375	−0.827	0.000	0.500	−1.050	0.000	0.625	−0.718
0.000	0.750	−0.027	0.000	0.875	1.390	0.000	1.000	4.040
0.125	0.000	−2.330	0.125	0.125	−1.910	0.125	0.250	0.023
0.125	0.375	1.650	0.125	0.500	2.160	0.125	0.625	1.480
0.125	0.750	−0.015	0.125	0.875	−0.738	0.125	1.000	1.600
0.250	0.000	−1.950	0.250	0.125	−1.050	0.250	0.250	−0.012
0.250	0.375	0.787	0.250	0.500	0.981	0.250	0.625	0.649
0.250	0.750	0.043	0.250	0.875	−0.694	0.250	1.000	−0.975
0.375	0.000	0.604	0.375	0.125	0.908	0.375	0.250	−0.022
0.375	0.375	−0.858	0.375	0.500	−1.100	0.375	0.625	−0.694
0.375	0.750	0.048	0.375	0.875	−0.266	0.375	1.000	−2.990
0.500	0.000	2.030	0.500	0.125	1.830	0.500	0.250	−0.027
0.500	0.375	−1.580	0.500	0.500	−1.980	0.500	0.625	−1.260
0.500	0.750	−0.025	0.500	0.875	−0.037	0.500	1.000	−4.020
0.625	0.000	0.673	0.625	0.125	0.937	0.625	0.250	0.022
0.625	0.375	−0.852	0.625	0.500	−1.060	0.625	0.625	−0.609
0.625	0.750	−0.047	0.625	0.875	−0.454	0.625	1.000	−3.740
0.750	0.000	−2.980	0.750	0.125	−1.300	0.750	0.250	−0.002
0.750	0.375	0.738	0.750	0.500	0.998	0.750	0.625	0.736
0.750	0.750	0.023	0.750	0.875	−0.998	0.750	1.000	−2.020
0.875	0.000	−6.920	0.875	0.125	−3.030	0.875	0.250	−0.038
0.875	0.375	1.590	0.875	0.500	1.920	0.875	0.625	1.150
0.875	0.750	0.036	0.875	0.875	−0.536	0.875	1.000	1.030
1.000	0.000	−7.020	1.000	0.125	−1.030	1.000	0.250	0.037
1.000	0.375	−0.973	1.000	0.500	−1.970	1.000	0.625	−1.780
1.000	0.750	-0.030	1.000	0.875	2.730	1.000	1.000	4.980

Table 11.8: Control points of a Bézier surface.

(x, y, z)	(x, y, z)	(x, y, z)	(x, y, z)
$(1, 1, 1)$	$(1, 3, 3)$	$(1, 5, 2)$	$(1, 7, 5)$
$(3, 1, 3)$	$(3, 3, 6)$	$(3, 5, 1)$	$(3, 7, 6)$
$(5, 1, 2)$	$(5, 3, 1)$	$(5, 5, 6)$	$(5, 7, 1)$
$(7, 1, 6)$	$(7, 3, 5)$	$(7, 5, 3)$	$(7, 7, 4)$

CHAPTER 12

Applications to Economics

12.1 Introduction

Functional equations have shown an incredible power to solve problems in Economics (see for example, Aczél (1966, 1975, 1984, 1987b,a, 1988), Aczél and Alsina (1984); Aczél and Eichhorn (1974), Eichhorn (1978a,b,c), Eichhorn and Kolm (1974), Eichhorn and Gehrig (1982) or Castillo et al. (1992)). They give many examples of applications, including interest formulas, price and quantity levels and indices, utility theory, production functions, aggregation problems, Fisher equation of exchange, theory of multi-sectoral growth, tax functions, etc. This chapter is devoted to developing several examples of applications of functional equations to many different problems arising from Economics. Thus, in Section 12.2 we introduce the concepts of price and quantity levels in an axiomatic form and discuss some of the problems associated with them when they are used with the Fisher equation of exchange. In Section 12.3 the axiomatic of the price indices are introduced when only prices and when both prices and quantities are used. Section 12.4 gives some interest formulas and finally, Section 12.5 is devoted to showing how functional equations can be used for modelling the demand function, that is, the sales $S(p, v)$ of a single-product as a function of the unit price p and on the advertising expenditure v. In Section 12.6 we deal with some duopoly models. Finally, in Section 12.7 we derive a taxation function from a set of conditions which include functional equations and inequalities. In particular, progressivity and independence of the type of declaration, joint or separate, for married couples is analyzed.

12.2 Price and quantity levels

Two important concepts in Economics are the concepts of price and quantity levels which are defined as follows (see Eichhorn (1978a), pp. 140-147).

Definition 12.1 (Price level). *Given* $\mathbf{p} = (p_1, p_2, \ldots, p_n)$ *the vector of prices of* n *products, we say that a non-negative real function* $P(\mathbf{p})$ *is a price level if it satisfies the following axiomatic properties:*

- *(a) Monotonicity: It is strictly increasing in any component, i.e.,*

$$\mathbf{p} > \mathbf{q} \Rightarrow P(\mathbf{p}) > P(\mathbf{q}),$$

  where $\mathbf{p} > \mathbf{q}$ *means that all the components of* $\mathbf{p}$ *and* $\mathbf{q}$ *satisfy the inequality*

- *(b) Linear homogeneity axiom: It is linearly homogeneous, that is*

$$P(\lambda\mathbf{p}) = \lambda P(\mathbf{p}); \quad \lambda \geq 0, \quad \mathbf{p} \geq \mathbf{0}.$$

■

Similarly, if $\mathbf{q} = (q_1, \ldots, q_n)$ is the vector of quantities of the above n products, we say that a function $Q(\mathbf{q})$ is a quantity level if it satisfies the same two properties above. Some price or quantity levels satisfy some of the following important properties, which allow them to be characterized:

- (c) Additivity: Any additive change of the prices yields an additive change in the price level

$$P(\mathbf{p} + \mathbf{q}) = P(\mathbf{p}) + P(\mathbf{q}).$$

- (d) Homogeneity in all $(n-1)$-tuples of prices:

$$P(\lambda p_1, \ldots, \lambda p_{j-1}, p_j, \lambda p_{j+1}, \ldots, \lambda p_n) = \lambda^{r_j} P(\mathbf{p}); \quad r_j > 0, \quad \forall j.$$

- (e) Multiplicativity:

$$P(\lambda_1 p_1, \ldots, \lambda_n p_n) = \phi(\lambda_1, \ldots, \lambda_n) P(\mathbf{p}).$$

- (f) Quasilinearity: There exist constants $a_1, \ldots, a_n$ and b with $a_1, a_2, \cdots, a_n \neq 0$ and a continuous and strictly monotonic function f such that

$$P(\mathbf{p}) = f^{-1}[a_1 f(p_1) + \ldots + a_n f(p_n) + b],$$

  where f^{-1} is the inverse of f.

In the following theorem some examples of price levels are given and characterized.

Theorem 12.1 (A class of price levels). *The class of price levels which satisfies*

- *(i) the additivity property (c) is*

$$P(\mathbf{p}) = c_1 p_1 + \ldots + c_n p_n,$$

where $c_1, \ldots, c_n$ are real arbitrary positive constants.

- *(ii) the homogeneity property (d) or the multiplicativity property (e) is*

$$P(\mathbf{p}) = C p_1^{\alpha_1} \ldots p_n^{\alpha_n}, \ \sum_{i=1}^{n} \alpha_i = 1, \ C > 0; \ \alpha_i > 0, \ i = 1, \ldots, n.$$

- *(iii) the quasi-linearity property (f) is the one above and*

$$P(\mathbf{p}) = (c_1 p_1^{\alpha} + \ldots + c_n p_n^{\alpha})^{\frac{1}{\alpha}}; \ \alpha \neq 0, \ c_i > 0, \ i = 1, \ldots, n.$$

∎

To show the importance of price and quantity levels and some of the problems associated with them, we discuss now the Fisher equation of exchange

$$MV = p_1 q_1 + \ldots + p_n q_n, \tag{12.1}$$

where M denotes the average amount of money in circulation during a certain period of time, V is the average rate of turnover of money and p_i and q_i; $i = 1, \ldots, n$, are, as before, the prices and the quantities of the n products exchanged.

The above equation can also be written as

$$p_1 q_1 + \ldots + p_n q_n = PT, \tag{12.2}$$

where P is some weighted average price and T is the sum of all q's.

The question now is whether or not this equation can be written in this form if P and T are price levels and quantity levels, respectively.

The answer to this question is negative; that is, all the pairs of functions $P(\mathbf{p})$ and $T(\mathbf{q})$ that satisfy (12.2) are not price-quantity levels (see Eichhorn (1978a), page 144).

This disappointing result led economists to redefine the concept of price and quantity levels, including prices and quantities, in the following form.

Definition 12.2 (New version of price-quantity level). *Given $\mathbf{p} = (p_1, \ldots, p_n)$ the vector of prices of n products, and $\mathbf{q} = (q_1, \ldots, q_n)$, the vector of associated quantities, we say that a function $P(\mathbf{p}, \mathbf{q})$ is a price-quantity level if it satisfies the following axiomatic properties:*

- *(a) Monotonicity: It is strictly increasing in any component of $\mathbf{p}$, i.e.,*

$$\mathbf{p} > \mathbf{r} \Rightarrow P(\mathbf{p}, \mathbf{q}) > P(\mathbf{r}, \mathbf{q}).$$

- *(b) Linear homogeneity axiom: It is linearly homogeneous, that is*

$$P(\lambda \mathbf{p}, \mathbf{q}) = \lambda P(\mathbf{p}, \mathbf{q}); \ \lambda \geq 0; \ \ \mathbf{p}, \mathbf{q} \geq 0.$$

- *(c) Commensurability axiom*

$$P\left(\lambda_1 p_1, \ldots, \lambda_n p_n; \frac{q_1}{\lambda_1}, \ldots, \frac{q_n}{\lambda_n}\right) = P(\mathbf{p}, \mathbf{q}).$$

This axiom says that any change in the units of measurement of the products keeps the price level unchanged. ∎

In a similar way we can define quantity levels.

Unfortunately, this new definition of price and quantity levels does not solve the above problem, that is, all the pairs of functions $P(\mathbf{p}, \mathbf{q})$ and $T(\mathbf{q}, \mathbf{p})$ that satisfy (12.2) are not price-quantity levels in the new sense of definition 12.2 (see Eichhorn (1978a), page 146).

12.3 Price indices

Economists, when trying to analyze the increment of the cost of living, use what they call price indices, which are defined as follows.

Definition 12.3 (Price index). *By price index we understand a positive real function $I(\mathbf{p_o}, \mathbf{p})$ such that it satisfies the following properties:*

- *(a) Monotonicity: The function $I(\mathbf{p_o}, \mathbf{p})$ is strictly increasing in $\mathbf{p}$ and strictly decreasing in $\mathbf{p_o}$, that is,*

$$\begin{aligned} \mathbf{p} > \mathbf{r} \quad &\Rightarrow I(\mathbf{p_o}, \mathbf{p}) > I(\mathbf{p_o}, \mathbf{r}), \\ \mathbf{p_o} > \mathbf{r_o} \quad &\Rightarrow I(\mathbf{p_o}, \mathbf{p}) < I(\mathbf{r_o}, \mathbf{p}). \end{aligned}$$

- *(b) Linear homogeneity: The function $I(\mathbf{p_o}, \mathbf{p})$ is linearly homogeneous of degree one with respect to $\mathbf{p}$*

$$I(\mathbf{p_o}, \lambda\mathbf{p}) = \lambda I(\mathbf{p_o}, \mathbf{p}), \quad \lambda > 0.$$

- *(c) Identity:*

$$I(\mathbf{p_o}, \mathbf{p_o}) = 1.$$

- *(d) Dimensionality: The change in the money unit leaves the price index unchanged*

$$I(\lambda\mathbf{p_o}, \lambda\mathbf{p}) = I(\mathbf{p_o}, \mathbf{p}).$$

∎

Examples of price indices are the following:

$$I(\mathbf{p_o}, \mathbf{p}) = \frac{P(\mathbf{p})}{P(\mathbf{p_o})}, \tag{12.3}$$

where $P(\mathbf{p})$ is any price level. Important particular cases are:

$$I(\mathbf{p_o}, \mathbf{p}) = \frac{c_1 p_1 + \ldots + c_n p_n}{c_1 p_{10} + \ldots + c_n p_{n0}}, \tag{12.4}$$

$$I(\mathbf{p_o},\mathbf{p}) = \left(\frac{p_1}{p_{10}}\right)^{\alpha_1} \cdots \left(\frac{p_n}{p_{n0}}\right)^{\alpha_n}; \quad \sum_{i=1}^{n} \alpha_i = 1, \;\; \alpha_i > 0, \;\; \forall i. \tag{12.5}$$

Some of the properties of the price indices, which can be derived from the axiomatic properties, are:

- (a) Proportionality:
$$I(\mathbf{p_o}, \lambda\mathbf{p_o}) = \lambda.$$
- (b) Homogeneity of degree minus one:
$$I(\lambda\mathbf{p_o}, \mathbf{p}) = \frac{1}{\lambda} I(\mathbf{p_o}, \mathbf{p}).$$
- (c) Mean value:
$$\min\left\{\frac{p_1}{p_{10}}, \ldots, \frac{p_n}{p_{n0}}\right\} \le I(\mathbf{p_o}, \mathbf{p}) \le \max\left\{\frac{p_1}{p_{10}}, \ldots, \frac{p_n}{p_{n0}}\right\}.$$

In the following paragraphs we give some characterizations of the price indices (12.3) to (12.5).

Theorem 12.2 (Characterization of a price index I). *The class of price indices that satisfy the circular property*

$$I(\mathbf{p_o}, \mathbf{p_1}) I(\mathbf{p_1}, \mathbf{p}) = I(\mathbf{p_o}, \mathbf{p})$$

is the class (12.3). ∎

Theorem 12.3 (Characterization of a price index II). *The class of price indices that satisfy the identity property and the properties*

$$\begin{aligned} (\mathbf{p_o}, \mathbf{p} + \mathbf{q}) &= I(\mathbf{p_o}, \mathbf{p}) + I(\mathbf{p_o}, \mathbf{q}), \\ \frac{1}{I(\mathbf{p_o} + \mathbf{q_o}, \mathbf{p})} &= \frac{1}{I(\mathbf{p_o}, \mathbf{p})} + \frac{1}{I(\mathbf{q_o}, \mathbf{p})}, \end{aligned}$$

is the class (12.4). ∎

Theorem 12.4 (Characterization of a price index III). *The class of price indices that satisfy the monotonicity, the linear homogeneity, the identity property and the additional property*

$$I(\alpha_1 p_{10}, \ldots, \alpha_n p_{n0}; \lambda_1 p_1, \ldots, \lambda_n p_n) = \delta(\alpha_1, \ldots, \alpha_n; \lambda_1, \ldots, \lambda_n) I(\mathbf{p_o}, \mathbf{p}),$$

with $\delta(1, \ldots, 1; 1, \ldots, 1) = 1$, *is the class (12.5).* ∎

The above Definition 12.3 is criticized because it does not take into account the amounts of the different products included in it. Thus, in some cases, economists use the following alternative definition for a price index.

Definition 12.4 (Alternative definition of price-quantity index). *By price-quantity index we understand a positive real function $I(\mathbf{p_o}, \mathbf{q_o}, \mathbf{p}, \mathbf{q})$ such that it satisfies the following properties:*

- *(a) Monotonicity: The function I is strictly increasing in $\mathbf{p}$ and strictly decreasing in $\mathbf{p_o}$, that is,*

$$\begin{array}{lll} \mathbf{p} > \mathbf{r} & \Rightarrow & wI(\mathbf{p_o}, \mathbf{q_o}, \mathbf{p}, \mathbf{q}) > I(\mathbf{p_o}, \mathbf{q_o}, \mathbf{r}, \mathbf{q}), \\ \mathbf{p_o} > \mathbf{r_o} & \Rightarrow & I(\mathbf{p_o}, \mathbf{q_o}, \mathbf{p}, \mathbf{q}) < I(\mathbf{r_o}, \mathbf{q_o}, \mathbf{p}, \mathbf{q}). \end{array}$$

- *(b) Linear homogeneity: The function I is linearly homogeneous of degree one with respect to $\mathbf{p}$*

$$I(\mathbf{p_o}, \mathbf{q_o}, \lambda\mathbf{p}, \mathbf{q}) = \lambda I(\mathbf{p_o}, \mathbf{q_o}, \mathbf{p}, \mathbf{q}), \quad \lambda > 0,$$

- *(c) Identity:*

$$I(\mathbf{p_o}, \mathbf{q_o}, \mathbf{p_o}, \mathbf{q_o}) = 1.$$

- *(d) Dimensionality: The change in the money unit leaves the price index unchanged*

$$I(\lambda\mathbf{p_o}, \mathbf{q_o}, \lambda\mathbf{p}, \mathbf{q}) = I(\mathbf{p_o}, \mathbf{q_o}, \mathbf{p}, \mathbf{q}), \quad \lambda > 0.$$

- *(e) Commensurability axiom:*

$$I\left(\lambda_1 p_{10}, \ldots, \lambda_n p_{n0}; \frac{q_{10}}{\lambda_1}, \ldots, \frac{q_{n0}}{\lambda_n}, \lambda_1 p_1 \ldots, \lambda_n p_n; \frac{q_1}{\lambda_1}, \ldots, \frac{q_n}{\lambda_n}\right)$$
$$= I(\mathbf{p_o}, \mathbf{q_o}, \mathbf{p}, \mathbf{q}).$$

■

Examples of price-quantity indices are

$$I(\mathbf{p_o}, \mathbf{q_o}, \mathbf{p}, \mathbf{q})n = \frac{\mathbf{p}\mathbf{q_o}}{\mathbf{p_o}\mathbf{q_o}} \quad (\textit{Laspeyres' index}), \tag{12.6}$$

$$I(\mathbf{p_o}, \mathbf{q_o}, \mathbf{p}, \mathbf{q}) = \frac{\mathbf{p}\mathbf{q}}{\mathbf{p_o}\mathbf{q}} \quad (\textit{Paasche's index}), \tag{12.7}$$

$$I(\mathbf{p_o}, \mathbf{q_o}, \mathbf{p}, \mathbf{q}) = \left[\frac{(\mathbf{p}\mathbf{q_o})(\mathbf{p}\mathbf{q})}{(\mathbf{p_o}\mathbf{q_o})(\mathbf{p_o}\mathbf{q})}\right]^{\frac{1}{2}} \quad (\textit{Fisher's index}), \tag{12.8}$$

$$I(\mathbf{p_o}, \mathbf{q_o}, \mathbf{p}, \mathbf{q}) = \left(\frac{p_1}{p_{10}}\right)^{\alpha_1} \cdots \left(\frac{p_n}{p_{n0}}\right)^{\alpha_n} \tag{12.9}$$
$$\sum_{i=1}^{n} \alpha_i = 1, \quad \alpha_i > 0, \forall i,$$

$$I(\mathbf{p_o}, \mathbf{q_o}, \mathbf{p}, \mathbf{q}) = \frac{[(q_{10}p_1)^\alpha + \ldots + (q_{n0}p_n)^\alpha]^{\frac{1}{\alpha}}}{[(q_{10}p_{10})^\alpha + \ldots + (q_{n0}p_{n0})^\alpha]^{\frac{1}{\alpha}}}, \quad \alpha \neq 0. \tag{12.10}$$

Some of the properties of the price indices, which can be derived from the alternative axiomatic properties, are:

- (a) Proportionality:

$$I(\mathbf{p_o}, \mathbf{q_o}, \lambda\mathbf{p_o}, \mathbf{q_o}) = \lambda.$$

- (b) Homogeneity of degree minus one:

$$I(\lambda\mathbf{p_o}, \mathbf{q_o}, \mathbf{p}, \mathbf{q}) = \frac{1}{\lambda} I(\mathbf{p_o}, \mathbf{q_o}, \mathbf{p}, \mathbf{q}).$$

- (c) Mean value:

$$\min\left\{\frac{p_1}{p_{10}}, \ldots, \frac{p_n}{p_{n0}}\right\} \le I(\mathbf{p_o}, \mathbf{q_o}, \mathbf{p}, \mathbf{q}) \le \max\left\{\frac{p_1}{p_{10}}, \ldots, \frac{p_n}{p_{n0}}\right\}.$$

For a more complete treatment of price indices see Eichhorn (1978a), pp. 152-172.

12.4 Interest rates

In Section 1.2.2 and Example 3.11 we had characterized the simple and compound interest formulas from some properties, which were written as functional equations. However, some actual bank policies differ from those previously stated. As a matter of fact, the interest rate is usually dependent on the period and/or the amount being deposited. This motivates the search for new interest formulas to be applied in the new situation of the actual economy.

As in Example 1.2.2, let $f(x,t)$ be the future value of the capital x having been invested for a period of time of duration t, and let $h(x,t) = \dfrac{f(x,t)}{x}$ be the accumulated investment rate. We consider three cases:

Case 1: These two properties suggest the following functional equation

$$h(x, t_1 + t_2) = h(x, t_1)h(x, t_2); \quad x, t \in \mathbb{R}_+, \tag{12.11}$$

where we assume $h(x, 0) = 1$ and h is continuous and increasing in both arguments. Equation (12.11) suggests that the investment rate corresponding to the sum of two periods of time is given by the product of the respective investment rates for every period of time.

Expression (12.11) is, for constant x, a Cauchy Equation (2.20), so, its general continuous solution is given by (2.21) as

$$h(x,t) = \exp[C(x)t] = p(x)^t \Rightarrow f(x,t) = xp(x)^t, \tag{12.12}$$

where $p(x)$ is a non-negative and increasing arbitrary function. We remark that:

- the cases of compound and continuous interests are particular cases with $p(x) = 1 + x$ and $p(x) = \exp(x)$, respectively.
- the case $f(x,t) = x[1 + r(x)]^t$, that is, with the interest rate being dependent on the initial investment, is included.

- the case of simple interest is not included here.

Case 2: Another possibility is to assume that the accumulated investment rate for an investment x is the accumulated investment rate for a period of duration y raised to a power which is a function of y and x, that is,

$$h(x,t) = h(y,t)^{k(y,x)}; \quad t, x, y \in \mathbb{R}_{++}, \tag{12.13}$$

which is Equation (10.43) with general (with the indicated conditions) solution (see (10.44))

$$h(x,t) = p(t)^{q(x)} = m(x)^{n(t)}; \quad k(y,x) = \frac{q(x)}{q(y)} = \frac{\log m(x)}{\log m(y)}. \tag{12.14}$$

Hence, we get

$$f(x,t) = xm(x)^{n(t)}, \tag{12.15}$$

which generalizes (12.12).

Case 3: Finally, we can assume that the accumulated investment rate for a period t_1+t_2 can be obtained from the accumulated investment rates for periods t_1 and t_2, that is,

$$h(x, t_1 + t_2) = U[h(x,t_1), h(x,t_2)], \tag{12.16}$$

which is functional Equation (10.38). Thus, its general solution is (see Model 1 in Chapter 7):

$$h(x,t) = w[tp(x)]; \quad U(x,y) = w[w^{-1}(x) + w^{-1}(y)], \tag{12.17}$$

where $w(x)$ and $p(x)$ are arbitrary functions and $w(x)$ is invertible. This leads to

$$f(x,t) = xw[tp(x)], \tag{12.18}$$

which also generalizes the interest formula (12.12). In fact, (12.11) is Equation (12.16) with $U(x,y) = xy$. Thus, the second equation of (12.17) leads to $w(x) = \exp(cx)$, which reduces (12.17) and (12.18) to (12.12).

One important particular case is given by $p(x) = x$, i.e., $f(x,t) = xw(tx)$. Note that the product tx measures the power of investment.

12.5 Demand function. Price and advertising policies

In this section we concentrate on the problem of modelling the sales $S(p,v)$ of a single-product firm such that they depend on the price p of its product and on the advertising expenditure v. In other words, we model the demand function for the product using advertising as an exogenous variable.

The novel contribution of functional equations to modelling is that we can establish some common sense properties of demand functions in terms of functional equations and, solving the resulting system, we can derive the analytical structure of these functions. This process avoids the need for guessing this structure, choosing it arbitrarily, based on convenience, on empirical considerations or easiness of mathematical manipulation, and, more important, avoiding the risk of absurd situations and inconsistencies.

The starting point for our research, and its motivation, is the model proposed by Eichhorn (1978a). He assumes that a multiplicative change in advertising expenditure yields an additive change in sales, the increment $T(p, \lambda)$ being dependent on the price p and the advertising factor increment λ. He also assumes that an additive change in the unit price yields a multiplicative change in sales, the factor $R(\pi, v)$ being dependent on the price increment π and the advertising expenditure v. However, once the system of two functional equations is solved, it turns out that R is independent of v. This suggests the use of another type of functions for R and T, such as the one described in Section 12.5.1. In this paper we assume that an additive or multiplicative change in the unit price yields an additive or multiplicative change in sales and that an additive or multiplicative change in the advertising expenditure yields an additive or multiplicative change in sales. In addition we assume that the increments depend on the price or advertising expenditure increments and either the initial prices or the initial advertising expenditure. These lead to 32 models which are completely derived in Section 12.5.1 using functional equations. Some of the resulting models are not completely specified because they depend on arbitrary functions. This means that new requirements could be established.

Some people use a linear demand function (see, for example, Anderson and Neven (1991), Allen (1992) or Boulding et al. (1994)). We shall see that in three of our models we get the linear demand, confirming the adequacy of linear demand functions; however, other valid alternative and interesting models appear too.

12.5.1 The monopoly model

Let us assume a firm such that the sales S of a single product depend on the unitary price p and on the advertising expenditure v, that is, $S = S(p, v)$. The function S cannot be arbitrary. In fact there are some common sense conditions that must be satisfied. For example, the function $S(p, v)$ must satisfy the following properties:

- **Assumption 1:** The $S(p, v)$ function is continuous in both arguments.

- **Assumption 2:** For any given v, the $S(p, v)$ function, considered as a function of p only, must be convex from below and decreasing. This implies that an increment in the unit price of the product leads to a reduction in sales, for the same advertising expenses and that its derivative decreases with p.

- **Assumption 3:** For any given p, the $S(p, v)$ function, considered as a function of v only, must be concave from below and increasing. This implies that an increment in the advertising expenses of the product leads to an increment in sales, for the same unit price.

Eichhorn's model

It is reasonable to make the following additional assumptions (see Eichhorn (1978a)) for the $S(p, v)$ function:

- **Assumption 4:** A multiplicative change in the advertising expenditure leads to an additive change in sales; that is,

$$S(p, \lambda v) = T(\lambda, p) + S(p, v), \tag{12.19}$$

where

$$\lambda \geq 0, p \geq 0, v \geq 0\,, T(1, p) = 0 \text{ and } T(\lambda, p) \text{ is increasing with } \lambda.$$

The general continuous solution of Equation (12.19) can be obtained as follows. Noting that for any p it is a Pexider functional Equation (4.3), we get (see Theorem 4.3):

$$\begin{aligned} S(p, v) &= A(p) \log v + B(p), \\ T(\lambda, p) &= A(p) \log \lambda. \end{aligned} \tag{12.20}$$

- **Assumption 5:** The sales due to an increment π in price are equal to the previous sales times a real number, which depends on π and v; that is,

$$S(p + \pi, v) = S(p, v) R(\pi, v), \tag{12.21}$$

where

$$p \geq 0, \;\; p + \pi \geq 0, \;\; v \geq 0, \;\; R(0, v) = 1 \text{ and } R(\pi, v) \text{ decreasing in } \pi.$$

The general solution of the functional Equation (12.21) can be obtained as follows: Making $p = 0$ in (12.21) we get

$$S(\pi, v) = R(\pi, v) S(0, v) = R(\pi, v) E(v), \tag{12.22}$$

where the function $E(v) = S(0, v)$ is arbitrary. Substitution of (12.22) into (12.21) leads to

$$R(p + \pi, v) = R(\pi, v) R(p, v), \tag{12.23}$$

which, for each fixed v, is a Cauchy functional equation with general solution

$$R(p, v) = \exp\left(pF(v)\right), \tag{12.24}$$

where $F(v)$ is an arbitrary function. Thus, the general solution of (12.21) is

$$\begin{aligned} S(p, v) &= E(v) \exp\left(pF(v)\right), \\ R(p, v) &= \exp\left(pF(v)\right). \end{aligned} \tag{12.25}$$

Model A4-A5

Once we have solved functional equations (12.21) and (12.19) separately, the general solution of the system (12.21)–(12.19) coincides with the general solution of the equation

$$A(p)\log v + B(p) = E(v)\exp\left(pF(v)\right), \tag{12.26}$$

which makes both solutions (12.20) and (12.25) compatible. The general continuous solution of the above equations is (see Eichhorn (1978a), page 15):

$$S(p,v) = (a + b\log v)\exp(-cp), \tag{12.27}$$

where a, b and c are arbitrary constants.

Note that in model (12.27) we have no arbitrary functions any more, but arbitrary constants. This means that the parametric model is completely specified and that we can estimate its parameters a, b and c using empirical data.

Model (12.27) shows a logarithmic increment of sales with advertising expenditures and an exponential decrease with price, in agreement with assumptions 4 and 5. One justification of this model of sales is the so-called Weber–Fechner law, that states that the stimuli of the intensity of perception is a linear function of the logarithm of the intensity of the stimulus.

Unfortunately, the resulting $R(\pi, v) = \exp(-c\pi)$ is independent of v, contrary to what was apparently suggested by Equation (12.21). The interpretation of this fact is that Equation (12.19) together with (12.21) implies this independence. In other words, no function R depending on the two arguments π and v satisfies both (12.19) and (12.21).

12.5.2 A modified Eichhorn model

In the light of the previous result, it can be argued that the function R should depend on the price p, instead of v. Thus, we can replace Assumption 5 by

- **Assumption 5a:** The sales due to an increment π in price are equal to the previous sales times a real number, which depends on π and p; that is,

  $$S(p+\pi, v) = R(\pi,p)S(p,v), \tag{12.28}$$

  where

  $$p \geq 0,\;\; p+\pi \geq 0,\;\; v \geq 0\;\; R(0,p) = 1 \text{ and } R(\pi,p) \text{ is decreasing in } \pi.$$

  The general solution of (12.28) can be obtained as follows: Making $p = 0$ in (12.28) we get

  $$S(\pi, v) = R(\pi, 0)S(0,v) = C(\pi)D(v).$$

  Back-substitution into (12.28) leads to

  $$\begin{aligned} S(p+\pi, v) &= C(p+\pi)D(v) = R(\pi,p)C(p)D(v) \\ &\Rightarrow R(\pi,p) = \frac{C(p+\pi)}{C(p)}. \end{aligned}$$

Thus, the general solution of (12.28) is

$$S(p,v) = C(p)D(v); \;\; R(\pi,p) = \frac{C(p+\pi)}{C(p)}, \tag{12.29}$$

where $C(p)$ and $D(v)$ are arbitrary functions.

Alternatively, we can assume a multiplicative, instead of an additive, change in the price p and we can question whether or not choosing between one of these assumptions influences the resulting model. We shall see that they are equivalent.

- **Assumption 5b:** The sales due to a multiplicative change (λ times) in the price are equal to the previous sales times a real number, which depends on λ and p, that is,

$$S(\lambda p,v) = R(\lambda,p)S(p,v), \tag{12.30}$$

where

$$p \geq 0, \;\; \lambda \geq 0, \;\; v \geq 0 \;\; R(1,p) = 1 \text{ and } R(\lambda,p) \text{ is decreasing in } \lambda.$$

Similarly, making $p = 1$ in (12.30) we get

$$S(\lambda,v) = R(\lambda,1)S(1,v) = C(\lambda)D(v).$$

Back-substitution into (12.30) leads to

$$\begin{aligned} S(\lambda p,v) &= C(\lambda p)D(v) \\ &= R(\lambda,p)C(p)D(v), \end{aligned}$$

which implies

$$R(\lambda,p) = \frac{C(\lambda p)}{C(p)}.$$

Thus, the general solution of (12.30) is

$$\begin{aligned} S(p,v) &= C(p)D(v), \\ R(\lambda,p) &= \frac{C(\lambda p)}{C(p)}, \end{aligned} \tag{12.31}$$

where $C(p)$ and $D(v)$ are arbitrary functions.

Note that the S functions in (12.29) and (12.31) are identical. Thus, equations (12.28) and (12.30) are equivalent. Consequently, the above mentioned two assumptions 5*a* and 5*b* lead to the same model.

Motivated by this result, we can ask whether or not using an additive instead of a multiplicative change in the advertising expenditures in Equation (12.19) influences the resulting model. Surprisingly, condition (12.19) is not equivalent to

$$S(p,v+w) = S(p,v) + T(p,w), \tag{12.32}$$

with general solution

$$S(p,v) = A(p)v + B(p). \tag{12.33}$$

Note that this is different from (12.20).

Model A4-A5a

The solution of the system (12.19)-(12.28) can be obtained by solving the functional equation

$$S(p,v) = C(p)D(v) = A(p)\log v + B(p), \tag{12.34}$$

which comes from forcing the coincidence of solutions of the forms (12.20) and (12.29).

Equation (12.34) is an equation of the form (4.13). Thus, according to (4.14) and (4.15) we can write

$$\begin{pmatrix} A(p) \\ B(p) \\ C(p) \end{pmatrix} = \begin{pmatrix} 1 \\ a \\ b \end{pmatrix} A(p); \begin{pmatrix} \log v \\ 1 \\ -D(v) \end{pmatrix} = \begin{pmatrix} 1 & 0 \\ 0 & 1 \\ c & d \end{pmatrix} \begin{pmatrix} \log v \\ 1 \end{pmatrix}$$

and

$$\begin{pmatrix} 1 & a & b \end{pmatrix} \begin{pmatrix} 1 & 0 \\ 0 & 1 \\ c & d \end{pmatrix} = \mathbf{0} \Rightarrow \begin{cases} b & = & -1/c, \\ a & = & d/c. \end{cases}$$

Then, the general solution of (12.34) is

$$B(p) = \frac{d}{c}A(p); \;\; C(p) = -\frac{A(p)}{c}; \;\; D(p) = -c\log p - d,$$

which leads to the model

$$S(p,v) = A(p)(\log v + B), \tag{12.35}$$

where the function $A(p)$ and the constant B are arbitrary. For (12.35) to satisfy Assumptions 2 and 3 above, $A(p)$ must be convex from below and decreasing. Note that $(\log v + B)$ is increasing.

Note that model (12.35) is more general than model (12.27) and that now, contrary to Eichhorn's model, $R(\pi,p) = \dfrac{A(\pi+p)}{A(p)}$ depends on both arguments, π and p, as initially stated.

We can replace (12.28) or (12.30) and (12.19) by other assumptions (see Castillo et al. (1999c)).

12.6 Duopoly Models

Duopoly models have been treated extensively in the literature (see, for example, Anderson and Neven (1991), Caplin and Nalebuff (1991), Ireland (1991), Pal (1991), Allen (1992), Dixon (1992) or Boulding et al. (1994)). In this section we investigate only two duopoly models (see Castillo et al. (2000e)).

12.6.1 Duopoly Model I

Assume now that we have two different firms that compete in the market. Assume also that the sales S of the product by firm 1 depend on the unit prices p and q and on the advertising expenditures v and w, of the two firms; that is, $S = S(p, q, v, w)$.

The function $S(p, q, v, w)$ must satisfy the following properties:

- **Assumption 1:** The $S(p, q, v, w)$ function is continuous in all arguments.
- **Assumption 2:** $S(p, q, v, w)$ is increasing in q and v.
- **Assumption 3:** $S(p, q, v, w)$ is decreasing in p and w.
- **Assumption 4:** A multiplicative change in the advertising expenditure of firm 1 leads to an additive change in sales, that is,

$$S(p, q, \lambda v, w) = S(p, q, v, w) + T(p, q, \lambda, w), \tag{12.36}$$

- **Assumption 5a:** The sales due to an increment π in price of firm 1 are equal to the previous sales times a real number, which depends on π and p, that is,

$$S(p + \pi, q, v, w) = R(\pi, p, q, w)S(p, q, v, w); \;\; p \geq 0, \;\; p + \pi \geq 0, \;\; v \geq 0, \tag{12.37}$$

where $R(0, p, q, w) = 1$.

The general solution of the system (12.36)-(12.37) is

$$S(p, q, v, w) = A(p, q, w)(\log v + B(q, w)), \tag{12.38}$$

where $A(p, , q, w)$ and $B(q, w)$ are arbitrary functions.

In addition we can make

- **Assumption 6:** The total sales of both firms is a constant K, that is,

$$S(p, q, v, w) + S(q, p, w, v) = K, \tag{12.39}$$

which using (12.38) leads to

$$A(p, q, w)(\log v + B(q, w)) + A(q, p, v)(\log w + B(p, v)) = K. \tag{12.40}$$

For fixed p and q, Equation (12.40) is of the form $\sum_{i=1}^{n} f_i(v)g_i(w)$, and then we have (see Theorem 4.5)

$$\begin{pmatrix} A(p, q, w) \\ A(p, q, w)B(q, w) \\ \log w \\ 1 \end{pmatrix} = \begin{pmatrix} a & b \\ c & d \\ 1 & 0 \\ 0 & 1 \end{pmatrix} \begin{pmatrix} \log w \\ 1 \end{pmatrix}, \tag{12.41}$$

$$\begin{pmatrix} \log v \\ 1 \\ A(q,p,v) \\ A(q,p,v)B(p,v)-K \end{pmatrix} = \begin{pmatrix} 1 & 0 \\ 0 & 1 \\ e & f \\ g & h \end{pmatrix} \begin{pmatrix} \log v \\ 1 \end{pmatrix}, \qquad (12.42)$$

where

$$\begin{pmatrix} a & c & 1 & 0 \\ b & d & 0 & 1 \end{pmatrix} \begin{pmatrix} 1 & 0 \\ 0 & 1 \\ e & f \\ g & h \end{pmatrix} = \begin{pmatrix} a+e & c+f \\ b+g & d+h \end{pmatrix}, \qquad (12.43)$$

from which we get

$$e=-a, \quad g=-b, \quad f=-c, \quad h=-d, \qquad (12.44)$$

which leads to

$$\begin{aligned} A(p,q,w) &= a(p,q)\log w + b(p,q), & (12.45)\\ A(q,p,v) &= -a(p,q)\log v - c(p,q), & (12.46)\\ B(q,w) &= \frac{c(p,q)\log w + d(p,q)}{a(p,q)\log w + b(p,q)}, & (12.47)\\ B(p,v) &= \frac{b(p,q)\log v + d(p,q) - K}{a(p,q)\log v + c(p,q)}. & (12.48) \end{aligned}$$

The compatibility of (12.45) and (12.46) leads to

$$a(p,q)\log w + b(p,q) = -a(q,p)\log w - c(q,p), \qquad (12.49)$$

which is equivalent to

$$[a(p,q)+a(q,p)]\log w + [b(p,q)+c(q,p)] = 0; \qquad (12.50)$$

that is,

$$b(p,q) = -c(q,p) \qquad (12.51)$$

and

$$a(p,q) = -a(q,p) \Rightarrow a(p,p) = 0. \qquad (12.52)$$

Similarly, the compatibility of (12.47) and (12.48) leads to

$$B(q,w) = \frac{c(p,q)\log w + d(p,q)}{a(p,q)\log w + b(p,q)} = \frac{b(q,q)\log w + d(q,q) - K}{a(q,q)\log w + c(q,q)} \qquad (12.53)$$

and

$$B(p,v) = \frac{b(p,q)\log v + d(p,q) - K}{a(p,q)\log v + c(p,q)} = \frac{c(p,p)\log v + d(p,p)}{a(p,p)\log v + b(p,p)}, \qquad (12.54)$$

and taking into account (12.52), we get

$$B(q,w) = \frac{d(q,q) - K}{c(q,q)} - \log w \tag{12.55}$$

and

$$B(p,v) = \frac{d(p,p)}{b(p,p)} - \log v, \tag{12.56}$$

and for them to be compatible we must have

$$\begin{aligned} \frac{d(q,q) - K}{c(q,q)} - \log w &= \frac{d(q,q)}{b(q,q)} - \log w \\ \Rightarrow 2\frac{d(q,q)}{c(q,q)} &= K \Rightarrow d(q,q) = K/2, \end{aligned} \tag{12.57}$$

and substitution into (12.56) leads to

$$B(q,w) = \frac{K}{2b(q,q)} - \log w. \tag{12.58}$$

Replacing now (12.45) and (12.58) in (12.38) we get

$$S(p,q,v,w) = [a(p,q)\log w + b(p,q)]\left(\log\frac{v}{w} + \frac{K}{2b(q,q)}\right), \tag{12.59}$$

which substituted into (12.39) and taking into account (12.52), gives

$$\begin{aligned} &[a(p,q)\log w + b(p,q)]\left(\log\frac{v}{w}\frac{K}{2b(q,q)}\right) \\ &\quad + (b(q,p) - a(p,q)\log v)\left(-\log\frac{v}{w} + \frac{K}{2b(p,p)}\right) = K, \end{aligned} \tag{12.60}$$

which can be written as

$$\begin{aligned} &a(p,q)\left[\log^2 v - \log^2 w\right] + \log v\left[b(p,q) - b(q,p) - \frac{Ka(p,q)}{2b(p,p)}\right] + \\ &+ \log w\left[\frac{Ka(p,q)}{2b(q,q)} - b(p,q) + b(q,p)\right] + \frac{K}{2}\left[\frac{b(p,q)}{b(q,q)} + \frac{b(q,p)}{b(p,p)} - 2\right] \equiv 0, \end{aligned} \tag{12.61}$$

and implies

$$\begin{aligned} a(p,q) &= 0, && (12.62) \\ b(p,q) &= b(q,p), && (12.63) \\ b(p,q))\,[b(p,p) + b(q,q)] &= 2b(p,p)b(q,q), \end{aligned}$$

from which we get

$$b(p,q) = \frac{1}{\alpha(p) + \alpha(q)}. \tag{12.64}$$

Thus, finally we get the model

$$S(p,q,v,w) = \frac{1}{\alpha(p)+\alpha(q)}\left[\log\frac{v}{w} + K\alpha(q)\right], \tag{12.65}$$

where $\alpha(p)$ is an arbitrary but increasing function of p.

The physical interpretation of Model (12.65) is as follows:

- If the advertisement expenditures of both firms are coincident, the sales are proportional to the ratios $\dfrac{\alpha(q)}{\alpha(p)+\alpha(q)}$ and $\dfrac{\alpha(p)}{\alpha(p)+\alpha(q)}$ for firms 1 and 2, respectively.
- The advertisement expenditures influence sales directly proportional to the logarithm of the ratio $\dfrac{v}{w}$ and inversely proportional to $\alpha(p)+\alpha(q)$.

12.6.2 Duopoly Model II

- **Assumption 7:** The sales $S(p+\alpha, q+\beta, v, w)$ of firm 1 due to increments α and β in the prices of firms 1 and 2, respectively, are the initial sales $S(p,q,v,w)$ of firm 1 times two factors which consider the associated reduction and increments due to these two changes; that is,

$$S(p+\alpha, q+\beta, v, w) = U(\alpha,p,q)V(\beta,p,q)S(p,q,v,w). \tag{12.66}$$

Making $p=q=0$ we get

$$S(\alpha,\beta,v,w) = U(\alpha,0,0)V(\beta,0,0)S(0,0,v,w), \tag{12.67}$$

which implies

$$S(p,q,v,w) = a(p)b(q)c(v,w). \tag{12.68}$$

Replacing (12.68) in (12.66) we get

$$a(p+\alpha)b(q+\beta) = U(\alpha,p,q)V(\beta,p,q)a(p)b(q), \tag{12.69}$$

and making $\beta = 0$ we obtain

$$a(p+\alpha) = U(\alpha,p,q)V(0,p,q)a(p), \tag{12.70}$$

which leads to

$$U(\alpha,p,q) = \frac{a(p+\alpha)}{a(p)e(p,q)}, \tag{12.71}$$

where $e(p,q)$ is an arbitrary function. Similarly, making $\alpha = 0$ in (12.69) we obtain

$$V(\beta,p,q) = \frac{b(q+\beta)}{b(q)d(p,q)}, \tag{12.72}$$

where

$$e(p,q) = \frac{1}{d(p,q)}. \tag{12.73}$$

- **Assumption 8:** The sales $S(p,q,v+\alpha,w+\beta)$ of firm 1 due to increments α and β in the advertisement expenditures of firms 1 and 2, respectively, are the initial sales $S(p,q,v,w)$ of firm 1 times two factors which consider the associated increments and decrements due to these two changes; that is,

$$S(p,q,v+\alpha,w+\beta)=U(\alpha,v,w)V(\beta,v,w)S(p,q,v,w). \tag{12.74}$$

The solution of this functional equation is

$$\begin{aligned} S(p,q,v,w) &= f(v)g(v)h(p,q), \\ U(\alpha,v,w) &= \frac{f(v+\alpha)}{f(v)k(v,w)}, \\ V(\beta,v,w) &= \frac{g(w+\beta)}{g(w)k(v,w)}. \end{aligned} \tag{12.75}$$

Combining now Assumptions 7 and 8 we get the system of equations

$$\begin{aligned} S(p,q,v,w) &= a(p)b(q)c(v,w), \\ S(p,q,v,w) &= f(v)g(v)h(p,q), \end{aligned} \tag{12.76}$$

which leads to the model

$$S(p,q,v,w)=a(p)b(q)f(v)g(w), \tag{12.77}$$

where the functions $a(p)$ and $g(w)$ are decreasing and the functions $b(q)$ and $f(v)$ are increasing, but otherwise arbitrary. The physical interpretation of this model is as follows:

- All the factors (prices and advertisement expenditures) act independently and contribute to the total sales of firm 1 as a factor which is less than 1 and decreasing for p and w and greater than 1 and increasing for q and v.

12.7 Taxation functions

One of the present problems in some member countries and in the European Community as a whole is the rationalization and unification of the taxation functions. It is well known that these functions are completely different from country to country and some have serious defects, in that absurd situations can occur. In fact, for some actual tax functions the net income of person A can be smaller than that of person B even though the gross income of person A is greater than that of person B and there is a coincidence in the sources of the common income. Another important problem is that associated with married couples or families, in that the tax amount can be strongly dependent on the decision of paying taxes separately or together. Thus, the normal policy is to analyze both cases and decide in favor of the smallest tax amount.

Another serious defect is related to the concept of progressive taxation. In order to have a progressive taxation, in the strong sense, it is not enough to

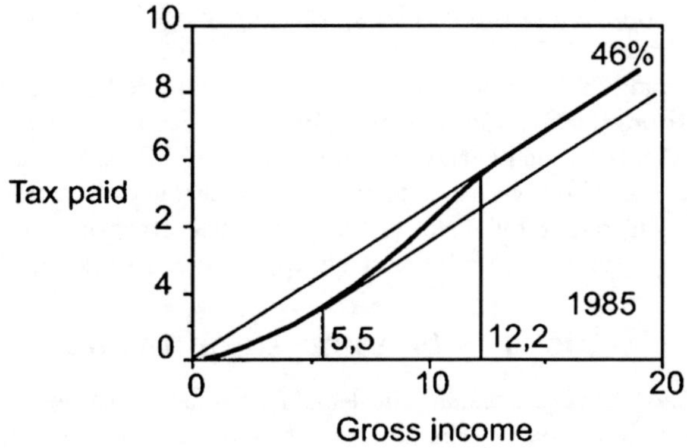

Figure 12.1: Gross income-tax paid curve associated with a faulty real taxation policy.

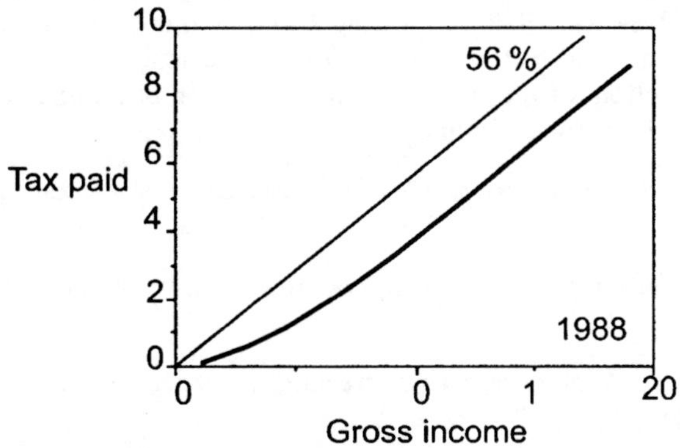

Figure 12.2: Gross income-tax paid curve associated with a progressive (in the strong sense) real taxation policy.

legislate that the larger the gross income the larger the tax paid (i.e., those having more income pay more taxes than those having less income), but also the burden of every additional unit of income should be equal to or larger than the burden of previous units. In other words, the tax function must be convex. It is worthwhile mentioning too that an increasing tax rate does not imply progressive taxation in the strong sense of the term. Figure 12.1 shows the gross income-tax paid curve associated with a real taxation policy having this defect and Figure 12.2 shows a similar curve after correction.

All the above implies that the process of deriving taxation functions must be carefully carried out if unpleasant surprises are to be avoided. In this example we show how functional equations can be satisfactorily used to solve this interesting problem.

The technique of functional equations has been applied to taxation problems in the past (Young (1987, 1988), Aczél (1987b,a), etc.). However, some of the problems were stated much earlier (see for example Mill (1973) or Stuart (1958) pp. 48-71), though they were not formulated in terms of functional equations. A very interesting paper by Aczél (1987a) calls the attention of those working in functional equations to this interesting application to taxation.

12.7.1 Restrictions to be imposed on taxation functions

To avoid all the above problems, the legislator must state very clearly all the conditions to be satisfied by the taxation function. These conditions are the result of a mixture of common sense and political decisions. Common sense is required to avoid absurd or ridiculous situations, like some of those mentioned above. Political decisions state the policies with respect to different concepts or items related to income tax. These include the number of members of the family, the source of the income (salaries or capital), the progressive character of the taxation, etc. However, it is interesting to point out that an excessive number of conditions can lead to the existence of no solution.

In the following section we present a methodology for obtaining taxation functions, which consists of the following steps:

1. Statement of the required properties to be satisfied by the taxation function.

2. Mathematical solution of the resulting system of functional equations and inequalities.

3. Selection of the remaining degrees of freedom in order to satisfy the required tax needs.

Let $s(x, y, m, p, q)$ be the tax function associated with a family unit with a salary income, x, obtained after a period of p years and a capital income, y, obtained after a period of q years, assuming it has m tax members ($m = 1$ or 2, depending on whether only husband or wife, or both receive some income, respectively. However, we assume here that m is a real number in order to allow for children and dependents to be included as non-integer values). The parameters p and q are used in order to represent irregular incomes, such as those received after irregular periods of time (more than one year). In the following examples, we assume the continuity of function s with respect to all its parameters. We also assume that the function s is defined on the following set

$$D = \{(x, y, m, p, q) \in \mathbb{R}^5 \,|\, x, y \geq 0, \quad m, p, q \geq 1\}.$$

As one example, the following conditions could be imposed on the tax function $s(x, y, m, p, q)$:

Progressivity.

$$s(x,y,m,p,q) \text{ is increasing with respect to } x, \tag{12.78}$$

$$s(x,y,m,p,q) \text{ is increasing with respect to } y, \tag{12.79}$$

$$s(x,y,m,p,q) \text{ is convex with respect to } x, \tag{12.80}$$

$$s(x,y,m,p,q) \text{ is convex with respect to } y, \tag{12.81}$$

$$s(x,y,m,p,q) \text{ is non-increasing with respect to } m. \tag{12.82}$$

Independence of the type of tax statement (joint or separate).

$$s\left(\sum_{i=1}^{m} x_i, \sum_{i=1}^{m} y_i, m, p, q\right) = \sum_{i=1}^{m} s(x_i, y_i, 1, p, q), \tag{12.83}$$

$$\left(\sum_{i=1}^{m} x_i, \sum_{i=1}^{m} y_i, m, p, q\right) \leq \sum_{i=1}^{m} s(x_i, y_i, 1, p, q), \tag{12.84}$$

$$s(x,y,m,p,q) = ms\left(\frac{x}{m}, \frac{y}{m}, 1, p, q\right). \tag{12.85}$$

Monotonicity of the net income.

$$\text{if} y_1 > y_2,\ x_1 + y_1 - s(x_1, y_1, m, p, q) > x_1 + y_2 - s(x_1, y_2, m, p, q), \tag{12.86}$$

$$\text{if} x_1 > x_2,\ x_1 + y_1 - s(x_1, y_1, m, p, q) > x_2 + y_1 - s(x_2, y_1, m, p, q). \tag{12.87}$$

Independence of the number of tax statements

$$s(x,0,m,p,q) + s(0,y,m,p,q) = s(x,y,m,p,q), \tag{12.88}$$

$$s(x,0,m,p,q_1) = ps\left(\frac{x}{p}, 0, m, 1, q_2\right) \forall q_1 q_2, \tag{12.89}$$

$$s(0,y,m,p_1,q) = qs\left(0, \frac{y}{q}, m, p_2, 1\right) \forall p_1 p_2. \tag{12.90}$$

Equal treatment for salary and capital incomes

$$s(x,y,m,p,q) = s(y,x,m,q,p), \tag{12.91}$$

$$s(x,y,m,p,p) = s(z, x+y-z, m, p, p),\ x+y \geq z. \tag{12.92}$$

The first five conditions refer to the progressivity of the tax function. Conditions (12.78) and (12.79) state that the larger the salary or capital incomes the larger the tax to be paid. Conditions (12.80) and (12.81) imply progressivity; that is, they state that the burden of every additional unit of income must be equal to or larger than the burden of previous units. Condition (12.82) states that the larger the number of tax members in the family unit, the smaller the tax to be paid.

Conditions (12.83),(12.84) and (12.85) refer to joint and separate tax statements of married couples. Equation (12.83) expresses the coincidence of the associated tax amounts for identical total income of the family unit. Inequality (12.84) enforces the joint declaration (as a couple) to lead to smaller, or at the most, equal, tax amounts than the separate tax statements. It must be understood that the equal sign must be attainable, i.e., a particular case must exist in which both tax statements lead to the same tax amount. Finally, Equation (12.85) enforces the coincidence only in the case of equal incomes of all members. This condition, together with conditions (12.80), (12.81) and (12.88) will be shown to be equivalent to condition (12.84).

Inequalities (12.86) and (12.87) imply the monotonicity of the net income. They state that the larger the gross income the larger the net income, regardless of whether it comes from salary or capital sources.

Equation (12.88) refers to the independence of the number of tax statements and it states that the tax amount must depend only on the total income and not on whether they are declared in one or several years.

Equations (12.89) and (12.90) establish the independence of the time when the incomes are generated, i.e., the tax amount must be dependent on the total income, but not on the time of its generation.

Finally, Equation (12.91) makes no distinction between salary and capital incomes and Equation (12.92) establishes the same tax amounts when the total income is the same, regardless of its source.

The relation of conditions (12.80), (12.81), (12.84) and (12.88) to condition (12.85) is shown in theorem 12.5. We first give the following lemma.

Lemma 12.1 *The general solution of the functional Equation (12.88) is*

$$\begin{aligned} s(x,y,m,p,q) &= a(x,m,p,q)+b(y,m,p,q), \\ a(0,m,p,q)+b(0,m,p,q) &= 0, \end{aligned} \tag{12.93}$$

where a and b are arbitrary continuous functions.

Proof: With $a(x,m,p,q) = s(x,0,m,p,q)$ and $b(y,m,p,q) = s(0,y,m,p,q)$ arbitrary continuous functions, the first part of (12.93) holds. If we now make $x = y = 0$ in (12.88) we get $s(0,0,m,p,q) = 0$ and then we obtain the second part of (12.93). ∎

Theorem 12.5 (Taxation functions). *If the equal sign of (12.84) is attainable and if conditions (12.80), (12.81) and (12.88) are satisfied, then (12.85) holds.* ∎

Proof: Taking into account (12.93), condition (12.84) can be written as

$$\begin{aligned} &s(x,y,m,p,q) \le \sum_{i=1}^{m} s(x_i,y_i,1,p,q) = \sum_{i=1}^{m} [a(x_i,1,p,q)+b(y_i,1,p,q)], \\ &x_i,y_i \ge 0, \;\; i=1,2,\ldots,m, \;\; x=\sum_{i=1}^{m} x_i, \;\; y=\sum_{i=1}^{m} y_i, \end{aligned} \tag{12.94}$$

where the functions a and b, according to (12.80) and (12.81), must be convex with respect to x and y, respectively, and then, they must satisfy

$$\left.\begin{array}{l}\sum\limits_{i=1}^{m}\alpha_i a(x_i,1,p,q)\geq a\left(\sum\limits_{i=1}^{m}\alpha_i x_i,1,p,q\right)\\ \sum\limits_{i=1}^{m}\alpha_i b(y_i,1,p,q)\geq b\left(\sum\limits_{i=1}^{m}\alpha_i y_i,1,p,q\right)\end{array}\right\};\ \sum_{i=1}^{m}\alpha_i=1,\ \ \alpha_i>0,\ \ \forall i, \tag{12.95}$$

which for $\alpha_i = \dfrac{1}{m}$ becomes

$$\begin{array}{l}\sum\limits_{i=1}^{m}a(x_i,1,p,q)\geq ma\left(\dfrac{1}{m}\sum\limits_{i=1}^{m}x_i,1,p,q\right)=ma\left(\dfrac{x}{m},1,p,q\right),\\ \sum\limits_{i=1}^{m}b(y_i,1,p,q)\geq mb\left(\dfrac{1}{m}\sum\limits_{i=1}^{m}y_i,1,p,q\right)=mb\left(\dfrac{y}{m},1,p,q\right).\end{array} \tag{12.96}$$

If the equality in (12.94) is attainable, then $s(x, y, m, p, q)$ must be the minimum of the right hand side of (12.94). However, from (12.96) we can write

$$\begin{array}{l}\sum\limits_{i=1}^{m}\left[a(x_i,1,p,q)+\ b(y_i,1,p,q)\right] \ -\\ \qquad\geq ma\left(\dfrac{x}{m},1,p,q\right)+mb\left(\dfrac{y}{m},1,p,q\right)=ms\left(\dfrac{x}{m},\dfrac{y}{m},1,p,q\right),\end{array} \tag{12.97}$$

and then the minimum is $ms\left(\dfrac{x}{m},\dfrac{y}{m},1,p,q\right)$. This proves that (12.85) holds. ∎

12.7.2 Compatible tax functions

In this section we study two different cases:

- Case 1: Same taxes for joint and separate tax statements.
- Case 2: Joint statement tax amount smaller than that for separate statements.

Case 1: Same taxes for joint and separate statements

In this case, we find the most general tax function that satisfies the set of restrictions (12.78) to (12.83) and (12.86) to (12.90) and then we add either condition (12.91) or condition (12.92). We start with several lemmas.

Lemma 12.2 *The general continuous solution of functional Equation (12.83) is*

$$s(x, y, m, p, q) = w_1(p, q)x + w_2(p, q)y + w_3(p, q)m, \tag{12.98}$$

where w_1, w_2 and w_3 are arbitrary continuous functions.

Proof: In effect, making $x_1 = x$, $y_1 = y$ and $x_i = y_i = 0$; $(i = 2, 3, \ldots, m)$ in (12.83) we obtain

$$s(x, y, m, p, q) = s(x, y, 1, p, q) + (m-1)s(0, 0, 1, p, q). \tag{12.99}$$

Substituting Expression (12.99) into (12.83) we get

$$s\left(\sum_{i=1}^{m} x_i, \sum_{i=1}^{m} y_i, 1, p, q\right) + (m-1)s(0, 0, 1, p, q) = \sum_{i=1}^{m} s(x_i, y_i, 1, p, q), \tag{12.100}$$

and calling

$$w(x, y, p, q) = s(x, y, 1, p, q) - s(0, 0, 1, p, q), \tag{12.101}$$

we arrive at

$$w\left(\sum_{i=1}^{m} x_i, \sum_{i=1}^{m} y_i, p, q\right) = \sum_{i=1}^{m} w(x_i, y_i, p, q), \tag{12.102}$$

which is a generalized Cauchy equation. Thus, because of the continuity of $s(x, 0, m, p, q)$ and $s(0, y, m, p, q)$, its general continuous solution is

$$w(x, y, p, q) = w_1(p, q)x + w_2(p, q)y. \tag{12.103}$$

From (12.99), (12.101) and (12.103) we get

$$s(x, y, m, p, q) = w_1(p, q)x + w_2(p, q)y + ms(0, 0, 1, p, q), \tag{12.104}$$

and calling

$$w_3(p, q) = s(0, 0, 1, p, q), \tag{12.105}$$

we finally obtain (12.98). ∎

Lemma 12.3 *If s(x,y,m,p,q) satisfies Equation (12.89) it can be written as*

$$s(x, 0, m, p, q) = xd\left(\frac{x}{p}, m\right), \tag{12.106}$$

where d is an arbitrary continuous function.

Proof: Making $q_1 = q_2 = q$ in (12.89) we get

$$s(x, 0, m, p, q) = ps\left(\frac{x}{p}, 0, m, 1, q\right), \tag{12.107}$$

hence

$$s(x, 0, m, p, q) = x\frac{p}{x}\, s\left(\frac{x}{p}, 0, m, 1, q\right) = xe\left(\frac{x}{p}, m, q\right), \tag{12.108}$$

where $e(y, m, q) = \dfrac{s(y, 0, m, 1, q)}{y}$ is an arbitrary continuous function. A new substitution of (12.108) into (12.89) leads to

$$xe\left(\frac{x}{p}, m, q_1\right) = xe\left(\frac{x}{p}, m, q_2\right) \;;\; \forall q_1 \,,\, q_2 \;\Rightarrow\; e(x, m, q) = d(x, m), \tag{12.109}$$

from which we get (12.106). ∎

Similarly, Equation (12.90) is equivalent to

$$s(0,y,m,p,q) = yg\left(\frac{y}{q},m\right), \tag{12.110}$$

where g is an arbitrary continuous function.

Lemma 12.4 *The general solution of the system (12.88)–(12.89)–(12.90) is*

$$s(x,y,m,p,q) = xd\left(\frac{x}{p},m\right) + yg\left(\frac{y}{q},m\right). \tag{12.111}$$

Proof: From (12.93), (12.106) and (12.110) we get

$$s(x,0,m,p,q) = a(x,m,p,q) + b(0,m,p,q) = xd\left(\frac{x}{p},m\right), \tag{12.112}$$

$$s(0,y,m,p,q) = a(0,m,p,q) + b(y,m,p,q) = yg\left(\frac{y}{q},m\right). \tag{12.113}$$

Finally, from (12.112), (12.113) and (12.93) we obtain (12.111). ∎

Lemma 12.5 *The general continuous solution of the system (12.83)–(12.88)–(12.89)–(12.90) is*

$$s(x,y,m,p,q) = Ax + By. \tag{12.114}$$

Proof: Combining now (12.98) with (12.111) we obtain

$$w_1(p,q)x + w_2(p,q)y + w_3(p,q)m = d\left(\frac{x}{p},m\right)x + g\left(\frac{y}{q},m\right)y, \tag{12.115}$$

and making $x = y = 0$ we obtain $w_3(p,q) = 0$ and with $y = 0$ we get

$$w_1(p,q) = d\left(\frac{x}{p},m\right) \Rightarrow w_1(p,q) = d(x,m) = A, \tag{12.116}$$

where A is an arbitrary constant.
Similarly,

$$w_2(p,q) = g\left(\frac{y}{q},m\right) \Rightarrow w_2(p,q) = g(y,m) = B, \tag{12.117}$$

where B is another arbitrary constant. With this the solution of the system (12.83)–(12.88)–(12.89)–(12.90) becomes (12.114). ∎

Conditions (12.78), (12.79), (12.86) and (12.87) imply

$$0 < A < 1, \;\; 0 < B < 1, \tag{12.118}$$

and conditions (12.80), (12.81) and (12.82) are obviously satisfied.

Thus, we have proved the following theorem:

Theorem 12.6 (Equal join and separate statements). *The most general tax function which is continuous in x and y and satisfies conditions (12.78) to (12.83) and (12.86) to (12.90) is*

$$s(x,y,m,p,q) = Ax + By, \quad 0 < A < 1, \quad 0 < B < 1. \tag{12.119}$$

■

This shows that the same taxes for joint and separate statements imply constant, though not necessarily equal, tax rates for salary and capital incomes.

Corollary 12.1 (Particular case). *If in addition we enforce condition (12.91) or condition (12.92) we get*

$$s(x,y,m,p,q) = A(x+y), \tag{12.120}$$

which implies identical tax rates for salary and capital incomes. ■

Because this tax function can be politically unsatisfactory (it is not progressive), we solve case 2 in the following section.

Case 2: Joint statement tax amount smaller than that for separate statements

In this section we obtain the most general tax function that satisfies conditions (12.78) to (12.82) and (12.86) to (12.90) with (12.84) or (12.85) and later we add either condition (12.91) or (12.92).

Lemma 12.1 (Tax function). *The general solution of (12.85) is*

$$s(x,y,m,p,q) = mr\left(\frac{x}{m}, \frac{y}{m}, p, q\right), \tag{12.121}$$

where $r(x,y,p,q)$ is an arbitrary continuous function. ■

Lemma 12.6 *The general solution of the system (12.85)-(12.88)-(12.89)-(12.90) is*

$$s(x,y,m,p,q) = m\left[pu\left(\frac{x}{pm}\right) + qv\left(\frac{y}{qm}\right)\right]; \quad u(0) = v(0) = 0. \tag{12.122}$$

Proof: Combining (12.111) with (12.121) we obtain

$$s(x,y,m,p,q) = xd\left(\frac{x}{p}, m\right) + yg\left(\frac{y}{q}, m\right) = mr\left(\frac{x}{m}, \frac{y}{m}, p, q\right), \tag{12.123}$$

and making first $y=0$, $q=1$ and $p=1$, and later $x=0$, $p=1$ and $q=1$, we get

$$d(x,m) = \frac{m}{x}r\left(\frac{x}{m}, 0, 1, 1\right) = \frac{m}{x}u\left(\frac{x}{m}\right), \tag{12.124}$$

$$g(y, m) = \frac{m}{y} r\left(0, \frac{y}{m}, 1, 1\right) = \frac{m}{y} v\left(\frac{y}{m}\right), \tag{12.125}$$

which when substituted into (12.123) leads to the first part of (12.122). But (12.88) also implies $0 = s(0, 0, m, p, q) = pu(0) + qv(0)$ for all p, q, that is, $u(0) = v(0) = 0$. ∎

On the other hand, conditions (12.78) and (12.87) imply

$$u(x) \text{ and } x - u(x) \text{ are increasing,} \tag{12.126}$$

and (12.79) and (12.86)

$$v(y) \text{ and } y - v(y) \text{ are increasing.} \tag{12.127}$$

Condition (12.82) leads to

$$\text{the functions } mu\left(\frac{x}{m}\right) \text{ and } mv\left(\frac{y}{m}\right) \text{ clearly being non-increasing in } m. \tag{12.128}$$

Finally, condition (12.80) is equivalent to

$$u(x) \text{ being convex,} \tag{12.129}$$

and condition (12.81) to

$$v(x) \text{ being convex.} \tag{12.130}$$

Thus, we have proved the following theorem.

Theorem 12.7 (Joint less than separate tax functions). *The most general tax function that satisfies conditions (12.78) to (12.82) and (12.86) to (12.90) with (12.84) or (12.85) is*

$$s(x, y, m, p, q) = m\left[pu\left(\frac{x}{pm}\right) + qv\left(\frac{y}{qm}\right)\right], \tag{12.131}$$

where u and v are functions such that

$$\begin{array}{l} u(0) = v(0) = 0, \\ u(x) \text{ and } x - u(x) \text{ are increasing,} \\ v(y) \text{ and } y - v(y) \text{ are increasing,} \\ u(x) \text{ is continuous and convex,} \\ v(x) \text{ is continuous and convex,} \\ \mu\left(\frac{x}{m}\right) \text{ and } mv\left(\frac{y}{m}\right) \text{ are non-increasing in } m, \end{array} \tag{12.132}$$

but otherwise arbitrary. ∎

The physical interpretation of the tax function (12.131) is as follows. Incomes due to salaries and incomes due to capital must be separated when calculating the tax amount. In addition, both lead to progressive tax functions u and v, respectively. The progressive character of functions $u(x)$ and $v(y)$ is only limited by conditions (12.132). On the other hand, the tax amount associated with every member of a family unit is that corresponding to the mean value (gross income per individual and year).

This tax function can also be used to encourage a rise in the birthrate and take into account the number of children because the parameter m can be increased. In other words, m can be the sum of as many units as tax persons in the family plus some fractions to consider sons and daughters or dependents. In the light of all the above and the tax function (12.131), we can say that it is not a mere coincidence that tax functions used in some countries of the European Community have a separate consideration for salary and capital incomes or that they use mean values for the incomes of family units. These decisions are not arbitrary but the consequence of a compatibility imposed by some system of common sense and political conditions.

Corollary 12.2 (Particular case). *If in addition, we impose condition (12.91) to tax function (12.131) we get*

$$s(x,y,m,p,q) = m\left[pu\left(\frac{x}{pm}\right) + qu\left(\frac{y}{qm}\right)\right]; \quad u(0) = v(0) = 0. \tag{12.133}$$

■

Proof: Compatibility of (12.91) and (12.131) leads to

$$p\left[u\left(\frac{x}{pm}\right) - v\left(\frac{x}{pm}\right)\right] = q\left[u\left(\frac{y}{qm}\right) - v\left(\frac{y}{qm}\right)\right] = 0 \Rightarrow u(x) = v(x), \tag{12.134}$$

which implies

$$u(x) = v(x), \tag{12.135}$$

and then (12.133) holds. ■

Lemma 12.2 (). *The general solution of (12.92) is*

$$s(x,y,m,p,p) = t(x+y,m,p), \tag{12.136}$$

where t is an arbitrary continuous function. ■

Proof: In effect, making $z = 0$ in (12.92) we get

$$s(x,y,m,p,p) = s(0,x+y,m,p,p) = t(x+y,m,p), \tag{12.137}$$

but (12.137) satisfies (12.92) for any arbitrary continuous t. ■

Corollary 12.3 (Particular case). *If we impose condition (12.92) to (12.131) we get the tax function*

$$s(x,y,m,p,q) = A(x+y). \tag{12.138}$$

■

Proof: Combining (12.136) with (12.131) leads to

$$m\left[pu\left(\frac{x}{pm}\right)+pv\left(\frac{y}{pm}\right)\right]=t(x+y,m,p), \tag{12.139}$$

and making first $x=0$ and then $y=0$ we obtain

$$mpv\left(\frac{y}{pm}\right)=t(y,m,p);mpu\left(\frac{x}{pm}\right)=t(x,m,p), \tag{12.140}$$

from which we get

$$v\left(\frac{x}{pm}\right)=u\left(\frac{x}{pm}\right)\Rightarrow v(x)=u(x). \tag{12.141}$$

Substituting (12.141) into (12.139) and making $m=p=1$ leads to Pexider's equation

$$u(x)+u(y)=t(x+y,1,1)=s(x+y)\Rightarrow u(x)=Ax+B, \tag{12.142}$$

but $u(0)=0$ implies $B=0$ and then (12.138) holds. ■

Once the structure of tax functions is known it only remains to select the free parameters and arbitrary functions in order to get the required tax amounts.

The most important conclusions of the above results are:

1. A general methodology has been presented for obtaining tax functions and consists of the following steps:
   - Statement of all the conditions to be satisfied by the tax functions,
   - solution of the resulting system of functional equations and inequalities and
   - selection of the arbitrary constants and functions in order to satisfy all the tax needs.
2. The only tax functions leading to the same tax amounts for separate and joint tax statements of married couples are those with constant, though not necessarily equal, tax rates for salary and capital incomes.
3. A tax function (12.131) has been given such that a joint statement by married couples leads to a tax amount smaller than the separate tax statement amount. However, there is a coincidence if both members of the couple receive the same incomes.
4. All obtained tax functions showed the possibility of separating salary and capital incomes.
5. The technique of functional equations has proved to be a very powerful tool in dealing with the problem of obtaining compatible tax functions.

Exercises

12.1 Propose other alternatives for an interest formula satisfying equation in Example 5.3 closer to the actual bank policies. Discuss whether or not they are reasonable.

12.2 Propose alternative assumptions to the model in Section 12.5.1.

12.3 Solve the following system of functional equations and give it an economic interpretation in terms of a cost-advertising demand model.

$$S(p+\pi, v) = S(p, v)R(\pi, v)$$

and

$$S(P, \lambda v) = R(\lambda, p)S(p, v)$$

12.4 Modify the assumptions in the two models in Section 12.6 and solve the corresponding system of equations.

12.5 One interesting practical problem is to find the optimal advertising policy for a firm. This means determining the price and the advertising expenditure which maximize the earnings. The benefit of the firm is given by the function

$$G(p, v) = pS(p, v) - K[S(p, v)] - v, \tag{12.143}$$

where $K(x)$ is the cost function, which gives the cost of producing x units of product.

Make some assumptions for the K and S functions in terms of functional equations, and then obtain the optimal solution for the firm.

12.6 Is the linear convex combination of a set of price indices a price index?

CHAPTER 13

Applications to Probability and Statistics

13.1 Introduction

One of the areas in which functional equations have been most utilized is in Probability and Statistics. In this chapter we give several applications to characterize families of distributions. In Section 13.2 we characterize all bivariate distributions with normal conditionals and we discover that, surprisingly, not only the bivariate normal has this property. In Section 13.3 we characterize bivariate distributions with gamma conditionals. In Section 13.4 we characterize other important distributions. Section 13.5 deals with the problem of linear regressions with conditionals in location-scale families. In Section 13.6 we estimate a multinomial model using functional equations. In Section 13.7 all random variables resulting from the sum of a random number of random variables is dealt with. In Section 13.8 the problem of conjugate distributions is analyzed. In Section 13.9 the maximum stability problem is analyzed. Finally, in Section 13.10 all reproductive parametric families of one and two parameters are obtained.

The reader interested in a complete coverage of these problems can consult Arnold et al. (1992).

13.2 Bivariate distributions with normal conditionals

In this section we deal with a problem stated by Castillo and Galambos (1987a). Let us assume an absolutely continuous bivariate random variable (X, Y) whose joint, marginal and conditional densities are denoted by

$$f_{(X,Y)}(x, y), g(x), h(y), f_{X|Y}(x|y)$$

and $f_{Y|X}(y|x)$, respectively. Hence, it is evident that

$$f_{(X,Y)}(x,y) = f_{X|Y}(x|y)h(y) = f_{Y|X}(y|x)g(x). \tag{13.1}$$

In this example we assume bivariate distributions with normal conditionals, i.e.,

$$\begin{aligned} f_{Y|X}(y|x) &\div \frac{\exp\left\{-\frac{1}{2}\left[\frac{y-a(x)}{b(x)}\right]^2\right\}}{b(x)}, \\ f_{X/Y}(x|y) &\div \frac{\exp\left\{-\frac{1}{2}\left[\frac{x-d(y)}{c(y)}\right]^2\right\}}{c(y)}, \end{aligned} \tag{13.2}$$

where $b(x) > 0, c(y) > 0$. Note that $a(x)$ and $d(y)$ define the regression lines and $b(x)$ and $c(y)$ the conditional standard deviations.

Upon substitution into (13.1), taking logarithms, setting

$$u(x) = \log[g(x)/b(x)]; \quad v(y) = \log[h(y)/c(y)], \tag{13.3}$$

and rearranging terms, it follows that

$$\begin{aligned} [2u(x)b^2(x) - a^2(x)]c^2(y) + b^2(x)[d^2(y) - 2v(y)c^2(y)] & \\ -y^2c^2(y) + x^2b^2(x) + 2a(x)yc^2(y) - 2xb^2(x)d(y) &= 0, \end{aligned} \tag{13.4}$$

which is an equation of the form (4.13).

According to (4.14), the solution satisfies

$$\begin{bmatrix} 2u(x)b^2(x) - a^2(x) \\ b^2(x) \\ 1 \\ x^2b^2(x) \\ 2a(x) \\ xb^2(x) \end{bmatrix} = \begin{bmatrix} a_{11} & a_{12} & a_{13} \\ 1 & 0 & 0 \\ a_{31} & a_{32} & a_{33} \\ 0 & 0 & 1 \\ a_{51} & a_{52} & a_{53} \\ 0 & 1 & 0 \end{bmatrix} \begin{bmatrix} b^2(x) \\ xb^2(x) \\ x^2b^2(x) \end{bmatrix} \tag{13.5}$$

and

$$\begin{bmatrix} c^2(y) \\ d^2(y) - 2v(y)c^2(y) \\ -y^2c^2(y) \\ 1 \\ yc^2(y) \\ -2d(y) \end{bmatrix} = \begin{bmatrix} 1 & 0 & 0 \\ b_{24} & b_{25} & b_{26} \\ 0 & 0 & -1 \\ b_{44} & b_{45} & b_{46} \\ 0 & 1 & 0 \\ b_{64} & b_{65} & b_{66} \end{bmatrix} \begin{bmatrix} c^2(y) \\ yc^2(y) \\ y^2c^2(y) \end{bmatrix}, \tag{13.6}$$

where

$$\begin{bmatrix} a_{11} & 1 & a_{31} & 0 & a_{51} & 0 \\ a_{12} & 0 & a_{32} & 0 & a_{52} & 1 \\ a_{13} & 0 & a_{33} & 1 & a_{53} & 0 \end{bmatrix} \begin{bmatrix} 1 & 0 & 0 \\ b_{24} & b_{25} & b_{26} \\ 0 & 0 & -1 \\ b_{44} & b_{45} & b_{46} \\ 0 & 1 & 0 \\ b_{64} & b_{65} & b_{66} \end{bmatrix} = \mathbf{0}.$$

Thus, the system of solutions of Equation (13.4) becomes

$$a(x) = \frac{-(A + Bx + Cx^2)}{(D + 2Ex + Fx^2)}; \quad d(y) = \frac{-(H + By + Ey^2)}{(J + 2Cy + Fy^2)};$$

$$b^2(x) = \frac{1}{(D + 2Ex + Fx^2)}; \quad c^2(y) = \frac{1}{(J + 2Cy + Fy^2)};$$

$$u(x) = -\tfrac{1}{2}\left[G + 2Hx + Jx^2 - \frac{(A + Bx + Cx^2)^2}{(D + 2Ex + Fx^2)}\right];$$

$$v(y) = -\tfrac{1}{2}\left[G + 2Ay + Dy^2 - \frac{(H + By + Ey^2)^2}{(J + 2Cy + Fy^2)}\right];$$

$$g(x) = (D + 2Ex + Fx^2)^{-1/2}\exp\left\{-\frac{G + 2Hx + Jx^2 - \frac{(A+Bx+Cx^2)^2}{(D+2Ex+Fx^2)}}{2}\right\};$$

$$h(y) = (J + 2Cy + Fy^2)^{-1/2}\exp\left\{-\frac{G + 2Ay + Dy^2 - \frac{(H+By+Ey^2)^2}{(J+2Cy+Fy^2)}}{2}\right\};$$

$$f(x,y) = \frac{1}{\sqrt{2\pi}}\exp\left\{-\frac{G}{2}\right\}\bullet \tag{13.7}$$

$$\bullet\exp\left\{-\frac{1}{2}[2Hx+2Ay+Jx^2+Dy^2+2Bxy+2Cx^2y+2Exy^2+Fx^2y^2]\right\}.$$

For the function $f(x,y)$ above to be a probability density function the sets of arbitrary constants $\{A, B, C, D, E, F, G, H, J\}$ must satisfy one of the following sets of conditions:

(*i*) $F = E = C = 0, \; D > 0, \; J > 0, \; B^2 < DJ,$
(*ii*) $F > 0, \; FD > E^2, \; JF > C^2.$

Model (i) is the bivariate normal model and model (ii) has the following properties:

- Regression lines need not be straight lines
- Marginal distributions are not normal
- The mode is at the intersection of regression lines

Figure 13.1 shows a classical bivariate normal model with its corresponding marginals and regression lines in the upper part and a non-normal but normal conditionals distribution with its corresponding marginals and regression lines (lower part).

Figure 13.2 shows a two-mode (non-normal) normal conditionals distribution with its corresponding marginals and regression lines.

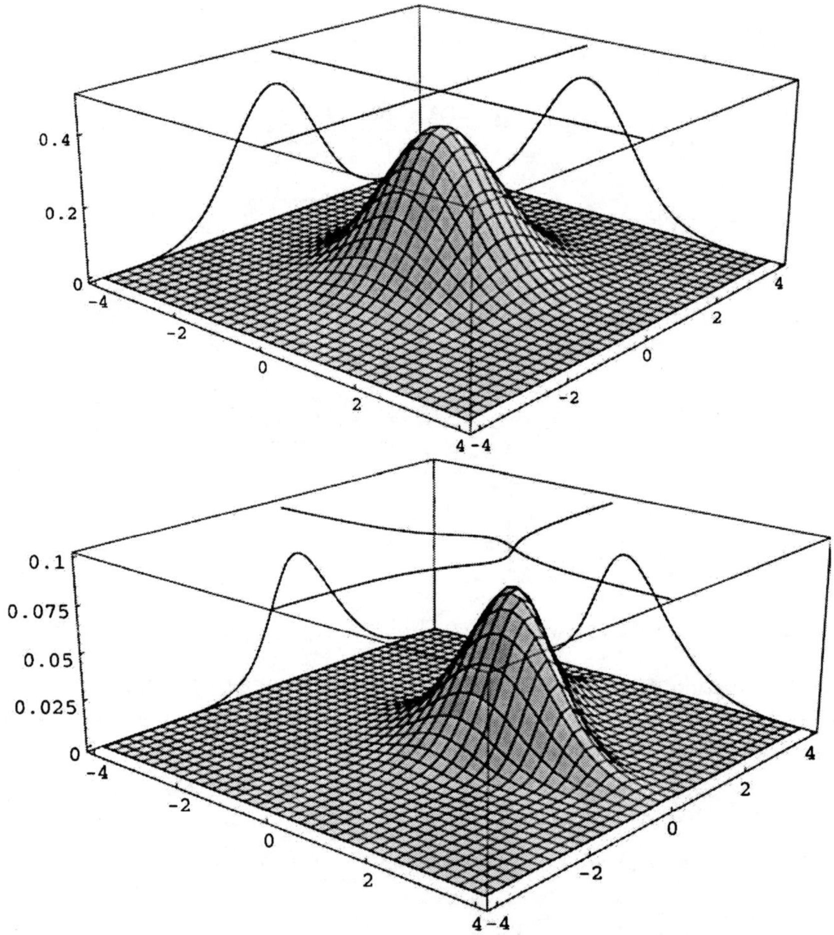

Figure 13.1: A classical bivariate normal model with its corresponding marginals and regression lines (upper figure). A non-normal but normal conditionals distribution with its corresponding marginals and regression lines (lower figure).

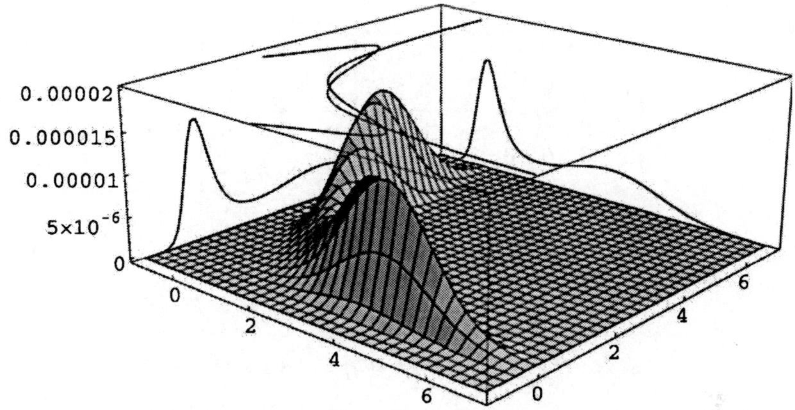

Figure 13.2: A two-mode (non-normal) normal conditionals distribution with its corresponding marginals and regression lines.

13.2.1 Case of independence

For the $f(x, y)$ function in (13.7) to lead to the case of independence we must have

$$2Hx + 2Ay + Jx^2 + Dy^2 + 2Bxy + 2Cx^2y + 2Exy^2 + Fx^2y^2 = r(x) + s(y);$$

that is,

$$[2Hx + Jx^2 - r(x)] + [2Ay + Dy^2 - s(y)] + Bxy + Bxy$$
$$+ \left[2Cy + \frac{F}{2}y^2\right]x^2 + \left[2Ex + \frac{F}{2}x^2\right]y^2 = 0,$$

which is of the form (4.13). Then, (4.14) and (4.15) become

$$\begin{bmatrix} 2Hx + Jx^2 - r(x) \\ 1 \\ x \\ x \\ x^2 \\ 2Ex + \frac{F}{2}x^2 \end{bmatrix} = \begin{bmatrix} a & b & c \\ 1 & 0 & 0 \\ 0 & 1 & 0 \\ 0 & 1 & 0 \\ 0 & 0 & 1 \\ 0 & 2E & \frac{F}{2} \end{bmatrix} \begin{bmatrix} 1 \\ x \\ x^2 \end{bmatrix},$$

$$\begin{bmatrix} 1 \\ 2Ay + Dy^2 - s(y) \\ By \\ By \\ 2Cy + \frac{F}{2}y^2 \\ y^2 \end{bmatrix} = \begin{bmatrix} 1 & 0 & 0 \\ m & n & p \\ 0 & B & 0 \\ 0 & B & 0 \\ 0 & 2C & \frac{F}{2} \\ 0 & 0 & 1 \end{bmatrix} \begin{bmatrix} 1 \\ y \\ y^2 \end{bmatrix},$$

where

$$\begin{bmatrix} a & 1 & 0 & 0 & 0 & 0 \\ b & 0 & 1 & 1 & 0 & 2E \\ c & 0 & 0 & 0 & 1 & \frac{F}{2} \end{bmatrix} \begin{bmatrix} 1 & 0 & 0 \\ m & n & p \\ 0 & B & 0 \\ 0 & B & 0 \\ 0 & 2C & \frac{F}{2} \\ 0 & 0 & 1 \end{bmatrix} = 0.$$

But this implies

$$a = -m \; ; \; n = p = b = c = B = C = E = F = 0;$$

$$r(x) = 2Hx + Jx^2 + m; \quad s(y) = 2Ay + Dy^2 - m,$$

which shows that independence is possible only for the bivariate normal model.

13.3 Bivariate distributions with gamma conditionals

We derive the most general bivariate density with gamma conditionals as shown by Castillo et al. (1990b). Let $a > -1$ and $p > 0$. A random variable X is called *gamma*, and is denoted by $X \sim G(a+1, p)$, if its probability density function is

$$f_X(x) = \frac{p^{a+1}}{\Gamma(a+1)} x^a \exp(-px) \text{ if } x \geq 0. \tag{13.8}$$

If in Expression (13.1) we assume that all conditional distributions are gamma, i.e.,$X|Y = y \sim G[e(y)+1, f(y)]$; $Y|X = x \sim G[b(x)+1, c(x)]$, where $e(y), f(y), b(x)$ and $c(x)$ are real functions, we have

$$a(x)g(x)y^{b(x)} \exp[-c(x)y] = d(y)h(y)x^{e(x)} \exp[-f(y)x], \tag{13.9}$$

where

$$a(x) = \frac{c(x)^{b(x)+1}}{\Gamma[b(x)+1]}; \quad d(y) = \frac{f(y)^{e(y)+1}}{\Gamma[e(y)+1]}, \tag{13.10}$$

and $b(x), c(x), e(y), f(y), g(x)$ and $h(y)$ are the unknown functions, which are restricted by

$$b(x) > -1, \;\; c(x) > 0 \text{ if } x \geq 0; \;\; e(y) > -1; \;\; f(y) > 0 \text{ if } y \geq 0. \tag{13.11}$$

Taking logarithms in (13.9) we get the functional equation

$$\log[a(x)g(x)] + b(x)\log y - c(x)y - \log[d(y)h(y)] - e(y)\log x + xf(y) = 0, \tag{13.12}$$

which is of the form (4.13).
If we now use Theorem 4.5 and take into consideration that the systems $\{1, x, \log x\}$

and $\{1, y, \log y\}$ are systems of linearly independent functions, the general solution of (13.12) can be written as

$$\begin{pmatrix} \log[a(x)g(x)] \\ b(x) \\ -c(x) \\ -1 \\ -\log x \\ x \end{pmatrix} = \mathbf{A} \begin{pmatrix} 1 \\ x \\ \log x \end{pmatrix}; \quad \begin{pmatrix} 1 \\ \log y \\ y \\ \log[d(y)h(y)] \\ e(y) \\ f(y) \end{pmatrix} = \mathbf{B} \begin{pmatrix} 1 \\ y \\ \log y \end{pmatrix}, \tag{13.13}$$

where $\mathbf{A}$ and $\mathbf{B}$ are matrices, the transposes of which are

$$\mathbf{A}' = \begin{pmatrix} a_{11} & a_{21} & a_{31} & -1 & 0 & 0 \\ a_{12} & a_{22} & a_{32} & 0 & 0 & 1 \\ a_{13} & a_{23} & a_{33} & 0 & -1 & 0 \end{pmatrix}; \quad \mathbf{B}' = \begin{pmatrix} 1 & 0 & 0 & b_{44} & b_{54} & b_{64} \\ 0 & 0 & 1 & b_{45} & b_{55} & b_{65} \\ 0 & 1 & 0 & b_{46} & b_{56} & b_{66} \end{pmatrix}, \tag{13.14}$$

and we must have

$$\mathbf{A}'\mathbf{B} = \mathbf{0}, \tag{13.15}$$

which leads to

$$\begin{array}{lll} a_{11} = b_{44} = G, & a_{31} = b_{45} = D, & a_{21} = b_{46} = A, \\ a_{12} = -b_{64} = H, & a_{32} = -b_{65} = E, & a_{22} = -b_{66} = B, \\ a_{13} = b_{54} = J, & a_{33} = b_{55} = F, & a_{23} = b_{56} = C, \end{array}$$

and then

$$\begin{array}{rcl rcl} b(x) & = & A + Bx + C\log x; & c(x) & = & -(D + Ex + F\log x); \\ e(y) & = & J + Fy + C\log y; & f(y) & = & -(H + Ey + B\log y); \\ a(x)g(x) & = & e^{(G+Hx+J\log x)}; & d(y)h(y) & = & e^{(G+Dy+A\log y)}, \end{array}$$

so finally we get

$$\begin{array}{rcl} f(x,y) & = & a(x)g(x)y^{b(x)}\exp[-c(x)y] \\ & = & \exp[G + Hx + Dy + Exy + J\log x + A\log y \\ & & +Bx\log y + Fy\log x + C\log x\log y], \end{array} \tag{13.16}$$

where A, B, C, D, E, F, G, H and J are arbitrary constants.

For (13.16) to be a probability density function some extra conditions must be satisfied by the arbitrary constants. Table 13.1 gives the only five possible models.

The marginal probability density functions are

$$g(x) = \frac{\Gamma[A+1+Bx+C\log(x)]\exp[G+Hx+J\log(x)]}{[-D-Ex-F\log(x)]^{A+1+Bx+C\log(x)}}, \quad x > 0$$

$$h(y) = \frac{\Gamma[J+1+Fy+C\log(y)]\exp[G+Dy+A\log(y)]}{[-H-Ey-B\log(y)]^{J+1+Fy+C\log(y)}}, \quad y > 0$$

Table 13.1: Feasible sets of parameters for the Gamma–Gamma model.

<table>
<tr><th></th><th>$C=0$
$E<0$</th><th>$C<0$
$E<0$</th><th>$C=0$
$E=0$</th></tr>
<tr><td>$B>0$
$F>0$</td><td>MODEL 4
$A>-1$
$D<F-F\log\left(\frac{-F}{E}\right)$
$H<B-B\log\left(\frac{-B}{E}\right)$
$J>-1$</td><td>MODEL 5
$A>C-1-C\log\left(\frac{-C}{B}\right)$
$D<F-F\log\left(\frac{-F}{E}\right)$
$H<B-B\log\left(\frac{-B}{E}\right)$
$J>C-1-C\log\left(\frac{-C}{F}\right)$</td><td>imp.</td></tr>
<tr><td>$B>0$
$F=0$</td><td>MODEL 3b
$A>-1$
$D<0$
$H<B-B\log\left(\frac{-B}{E}\right)$
$J>-1$</td><td>impossible</td><td>imp.</td></tr>
<tr><td>$B=0$
$F>0$</td><td>MODEL 3a
$A>-1$
$D<F-F\log\left(\frac{-F}{E}\right)$
$H<0$
$J>-1$</td><td>impossible</td><td>imp.</td></tr>
<tr><td>$B=0$
$F=0$</td><td>MODEL 2
$A>-1$
$D<0$
$H<0$
$J>-1$</td><td>impossible</td><td>M. 1
$A>-1$
$D<0$
$H<0$
$J>-1$</td></tr>
</table>

and the conditional expected values and variances become

$$
\begin{aligned}
E(X|Y=y) &= -\frac{J+1+Fy+C\log y}{H+Ey+B\log y}, \\
E(Y|X=x) &= -\frac{A+1+Bx+C\log x}{D+Ex+F\log x}, \\
Var(X|Y=y) &= \frac{J+1+Fy+C\log y}{(H+Ey+B\log y)^2}, \\
Var(Y|X=x) &= \frac{A+1+Bx+C\log x}{(D+Ex+F\log x)^2}.
\end{aligned}
$$

Other applications of Equation (4.13) to the characterization of bivariate distributions by its marginals can be seen in Arnold et al. (1992).

13.4 Other equations

In this section we give the general measurable solution of a functional equation which has applications to the characterization of the normal distribution.

13.4.1 One characterization of the normal distribution

Let X and Y be independent random variables with density functions f and g, respectively. If the random variables $U = X + Y$ and $V = X - Y$ are independent, then X and Y are normal distributions with the same variance (see Aczél (1966), page 109).

Proof: If $h(x, y)$ is the joint probability density function of (X, Y), then the density function, $p(u, v)$, of (U, V) is

$$p(u,v) = \frac{1}{2}h\left(\frac{u+v}{2}, \frac{u-v}{2}\right); \quad u, v \in \mathbb{R},$$

but, because of the independence of X and Y and the required independence of U and V, we must have

$$p_1(u)p_2(v) = \frac{1}{2}f\left(\frac{u+v}{2}\right)g\left(\frac{u-v}{2}\right); \quad u, v \in \mathbb{R},$$

and using the transformation

$$x = \frac{u+v}{2}; \quad y = \frac{u-v}{2}; \quad q_1(x) = \sqrt{2}p_1(x); \quad q_2(x) = \sqrt{2}p_2(x),$$

we obtain the functional equation

$$f(x)g(y) = q_1(x+y)q_2(x-y); \quad x, y \in \mathbb{R},$$

which is of the form (4.37) and then

$$f(x) = \alpha_1 \exp(a_1x + b_1x^2); \quad g(x) = \alpha_2 \exp(a_2x + b_1x^2).$$

For $f(x)$ to be a probability density function we must have

$$\int_{-\infty}^{\infty} \alpha_1 \exp(a_1x + b_1x^2)dx = 1 \Rightarrow \frac{\alpha_1}{\sqrt{-b_1}}\sqrt{\pi}\exp\left(-\frac{a_1^2}{4b_1}\right) = 1,$$

and the similar condition for $g(x)$. With this, and calling

$$\mu_1 = -\frac{a_1}{2b_1}; \quad \mu_2 = -\frac{a_2}{2b_1}; \quad \sigma = \sqrt{-\frac{1}{2b_1}},$$

it can be written as

$$f(x) = \frac{1}{\sqrt{2\pi\sigma^2}} \exp\left[-\frac{(x-\mu_1)^2}{2\sigma^2}\right]; \quad g(x) = \frac{1}{\sqrt{2\pi\sigma^2}} \exp\left[-\frac{(x-\mu_2)^2}{2\sigma^2}\right],$$

which proves the result. ■

Proof: (Alternative proof): Let φ_X and φ_Y be the characteristic functions of X and Y, respectively, then, because X and Y on one hand and $X+Y$ and $X-Y$ on the other hand are independent, we can write

$$\begin{aligned}\varphi_{X+Y}(t) = \varphi_X(t)\varphi_Y(t); \quad \varphi_{X-Y}(t) = \varphi_X(t)\varphi_Y(-t);\\ \varphi_{2X}(t) = \varphi_X(2t) = \varphi_{X+Y}(t)\varphi_{X-Y}(t) = \varphi_X(t)^2\varphi_Y(t)\varphi_Y(-t);\\ \varphi_{2Y}(t) = \varphi_Y(2t) = \varphi_{X+Y}(t)\varphi_{X-Y}(-t) = \varphi_X(t)\varphi_Y(t)^2\varphi_X(-t),\end{aligned}$$

and then

$$\frac{\varphi_X(2t)}{\varphi_X(-2t)} = \frac{\varphi_X(t)^2}{\varphi_X(-t)^2}; \quad \frac{\varphi_Y(2t)}{\varphi_Y(-2t)} = \frac{\varphi_Y(t)^2}{\varphi_Y(-t)^2},$$

and calling

$$\rho_1(t) = \frac{\varphi_X(t)}{\varphi_X(-t)}; \quad \rho_2(t) = \frac{\varphi_Y(t)}{\varphi_Y(-t)},$$

we obtain

$$\left.\begin{array}{l}\rho_1(2t) = \rho_1(t)^2\\ \rho_2(2t) = \rho_2(t)^2\end{array}\right\} \Rightarrow \left\{\begin{array}{l}\rho_1(t) = \exp(C_1t)\\ \rho_2(t) = \exp(C_2t)\end{array}\right\} \quad C_1 \text{ and } C_2 \text{ complex.}$$

But this implies

$$\left.\begin{array}{l}\varphi_X(2t) = \varphi_X(t)^2\varphi_Y(t)^2\exp(-C_2t)\\ \varphi_Y(2t) = \varphi_X(t)^2\varphi_Y(t)^2\exp(-C_1t)\end{array}\right\} \Rightarrow \frac{\varphi_X(t)}{\varphi_Y(t)} = \exp\left[\frac{t}{2}(C_1 - C_2)\right],$$

and then we have

$$\varphi_X(2t) = \varphi_X(t)^4\exp(-C_1t) \Rightarrow \log\varphi_X(2t) = 4\log\varphi_X(t) - C_1t + 2K\pi i,$$

which can be written as

$$\psi_X(2t) = 4\psi_X(t) - C_1t + 2K\pi i \Rightarrow \quad \psi_X(t) = A_1t^2 + \frac{C_1}{2}t - \frac{2K\pi i}{3},$$

and then

$$\varphi_X(t) = \exp\left(A_1t^2 + \frac{C_1}{2}t - \frac{2K\pi i}{3}\right),$$

but, taking into account that for any characteristic function $\varphi(0) = 1$, we get $K = 0$. Thus,

$$\varphi_X(t) = \exp\left(A_1t^2 + \frac{C_1}{2}t\right) \Rightarrow X \approx N\left(\frac{C_1}{2i}, -2A_1\right).$$

Similarly, we get

$$\varphi_Y(t) = \exp\left(A_1t^2 + \frac{C_2}{2}t\right) \Rightarrow Y \approx N\left(\frac{C_2}{2i}, -2A_1\right),$$

which concludes the proof.

■

13.5 Linear regressions with conditionals in location-scale families

In this example we characterize all bivariate distributions with conditionals in location-scale families (Arnold et al. (1992)). In other words, we solve the functional equation

$$f\left(\frac{x-a-by}{\sigma_1}\right)g\left(\frac{y-\mu_2}{\sigma_2}\right)=h\left(\frac{y-c-dx}{\sigma_3}\right)k\left(\frac{x-\mu_4}{\sigma_4}\right), \qquad (13.17)$$

where k and g, f and h are the marginal and the conditional probability density functions, respectively; a, b, c, and d are arbitrary constants; and we have assumed that $\sigma_1, \sigma_2, \sigma_3$ and σ_4 are finite and different from zero.

Our problem consists of finding all functions f, g, h and k that satisfy (13.17). We distinguish the following cases:

1. **Case 1**: $b = d = 0$. In this case, Equation (13.17) can be written as

$$\frac{f\left(\dfrac{x-a}{\sigma_1}\right)}{k\left(\dfrac{x-\mu_4}{\sigma_4}\right)}=\frac{h\left(\dfrac{y-c}{\sigma_3}\right)}{g\left(\dfrac{y-\mu_2}{\sigma_2}\right)}=C, \qquad (13.18)$$

   where C is a constant. Thus, we have the trivial solution of (13.17); i.e., the independence model.

2. **Case 2**: $b \neq 0$ or $d \neq 0$. Then (13.17) can be written as

$$f(Ax+By+C)g(Dy+E)=h(Fx+Gy+H)k(Mx+N), \qquad (13.19)$$

   where $A, D, G, M \neq 0$. Without loss of generality we can assume $A = D = G = M = 1$. If we now make the transformation

$$\begin{aligned} u &= x+By+C,\\ v &= y+E, \end{aligned} \qquad (13.20)$$

   with inverse

$$\begin{aligned} x &= u-Bv+DE-C,\\ y &= v-E, \end{aligned} \qquad (13.21)$$

   Equation (13.19) becomes

$$f(u)g(v)=h(A^*u+B^*v+C^*)k(u+D^*v+E^*), \qquad (13.22)$$

   where

$$\begin{aligned} A^* &= F,\\ B^* &= 1-FB,\\ C^* &= FBE-FC+H-E,\\ D^* &= -B,\\ E^* &= BE-C+N. \end{aligned} \qquad (13.23)$$

If we now make the new transformation (note that $A^*D^* - B^* = -1$)

$$\begin{aligned} u^* &= u - u_0; \quad u_0 = \frac{B^*E^* - C^*D^*}{A^*D^* - B^*}, \\ v^* &= v - v_0; \quad v_0 = \frac{C^* - A^*E^*}{A^*D^* - B^*}, \end{aligned} \tag{13.24}$$

Expression (13.22) becomes

$$f(u^* + u_0)g(v^* + v_0) = h(A^*u^* + B^*v^*)k(u^* + D^*v^*), \tag{13.25}$$

and calling

$$f^*(u^*) = f(u^* + u_0), \quad g^*(v^*) = g(v^* + v_0), \tag{13.26}$$

we finally get the functional equation

$$f^*(u^*)g^*(v^*) = h(A^*u^* + B^*v^*)k(u^* + D^*v^*), \tag{13.27}$$

which solution is given by Theorem 4.10:

$$\begin{aligned} f^*(x) &= \alpha_1 \exp(a_1 x + b_1 x^2), \\ g^*(x) &= \alpha_2 \exp\left(a_2 x - \frac{B^*D^*}{A^*} b_1 x^2\right), \\ h(x) &= \beta_1 \alpha_1 \alpha_2 \exp\left[(a_2 - a_1 D^*)x + \frac{D^*}{A^*} b_1 x^2\right], \\ k(x) &= \frac{1}{\beta_1} \exp\left[(a_1 B^* - a_2 A^*)x - B^* b_1 x^2\right], \end{aligned} \tag{13.28}$$

where a_1, a_2, b_1 are real constants and $\alpha_1, \alpha_2, \beta_1$ non-null real constants, and we have taken into consideration that $A^*D^* - B^* = -1$. Using now (13.24) and (13.26) leads to

$$\begin{aligned} f(x) &= \alpha_1 \exp\left[a_1(x - B^*E^* + C^*D^*) + b_1(x - B^*E^* + C^*D^*)^2\right], \\ g(x) &= \alpha_2 \exp\left[a_2(x - C^* + A^*E^*) - \frac{B^*D^*}{A^*} b_1 (x - C^* + A^*E^*)^2\right], \\ h(x) &= \beta_1 \alpha_1 \alpha_2 \exp\left[(a_2 - a_1 D^*)x + \frac{D^*}{A^*} b_1 x^2\right], \\ k(x) &= \frac{1}{\beta_1} \exp\left[(a_1 B^* - a_2 A^*)x - B^* b_1 x^2\right], \end{aligned} \tag{13.29}$$

which shows that only uniform and normal distributions are possible.

Table 13.2 shows a discussion of different cases depending on the values of the parameters a_1, a_2 and b_1.

Cases 2, 4, 6 and 8 correspond to normal bivariate models; case 1 is the independent uniform model; and case 7 is the bivariate exponential model. Finally, models 3 and 5 do not lead to a bivariate distribution because of the incompatibility of ranges between conditionals and marginals. No more bivariate distributions satisfying (13.17) are possible.

Table 13.2: Feasible solutions

#	a_1	a_2	b_1	f(x)	g(x)	h(x)	k(x)	Comments
1	0	0	0	U	U	U	U	Independent
2	0	0	$\neq 0$	N	N	N	N	-
3	0	$\neq 0$	0	U	E	E	E	Impossible
4	0	$\neq 0$	$\neq 0$	N	N	N	N	-
5	$\neq 0$	0	0	E	U	E	E	Impossible
6	$\neq 0$	0	$\neq 0$	N	N	N	N	-
7	$\neq 0$	$\neq 0$	0	E	E	E	E	Independent
8	$\neq 0$	$\neq 0$	$\neq 0$	N	N	N	N	-
U=Uniform, E=Exponential, N=Normal								

13.6 Estimation of a multinomial model

One of the most important problems to deal with when working with probability based expert systems is the parameter estimation based on incomplete data. In medical diagnosis, a typical problem consists of estimating a model when a sample of patients is known. For each patient a set of symptoms, not necessarily equal for all patients, is known. Hence, we must deal with incomplete data (some symptoms are unknown for some patients).

In the following paragraphs we present a method of estimation, based on the maximum likelihood principle, which can be reduced to a functional Equation (see Castillo et al. (1990d)).

For the sake of simplicity we shall assume a discrete bidimensional random variable (X_1, X_2) such that X_i $(i = 1, 2)$ only takes the values 0 or 1. However, the derived conclusions are still valid for the multidimensional case and even for random variables taking more than 2 different values.

Let us assume that we observe a sample with $n_1(i)$ $(i = 0, 1)$, elements with $X_1 = i$ and X_2 unknown, $n_2(i)$ $(i = 0, 1)$, elements with $X_2 = i$ and X_1 unknown and $n_{12}(i, j)$ elements with $X_1 = i$ and $X_2 = j$. Then, the likelihood function of the sample becomes

$$\begin{aligned} V &= (a+b)^{n_1(0)}(c+d)^{n_1(1)}(a+c)^{n_2(0)}(b+d)^{n_2(1)} \\ & \quad a^{n_{12}(0,0)}b^{n_{12}(0,1)}c^{n_{12}(1,0)}d^{n_{12}(1,1)} \end{aligned} \tag{13.30}$$

where $a = P(0,0)$, $b = P(0,1)$, $c = P(1,0)$, $d = P(1,1)$ and $P(x, y)$ is the probability mass function of (X_1, X_2) and

$$a + b + c + d = 1. \tag{13.31}$$

Taking logarithms in (13.30) we get

$$\begin{aligned} L = \log(V) &= n_1(0)\log(a+b) + n_1(1)\log(c+d) + n_2(0)\log(a+c) \\ &\quad + n_2(1)\log(b+d) + n_{12}(0,0)\log a + n_{12}(0,1)\log b \\ &\quad + n_{12}(1,0)\log c + n_{12}(1,1)\log d. \end{aligned} \tag{13.32}$$

By considering the following Lagrange auxiliary function

$$Q = L + \lambda(a+b+c+d-1), \tag{13.33}$$

the maximum likelihood estimators can be obtained from the system of equations

$$\begin{aligned} \frac{\partial Q}{\partial a} &= \frac{n_1(0)}{a+b} + \frac{n_2(0)}{a+c} + \frac{n_{12}(0,0)}{a} + \lambda = 0, \\ \frac{\partial Q}{\partial b} &= \frac{n_1(0)}{a+b} + \frac{n_2(1)}{b+d} + \frac{n_{12}(0,1)}{b} + \lambda = 0, \\ \frac{\partial Q}{\partial c} &= \frac{n_1(1)}{c+d} + \frac{n_2(0)}{a+c} + \frac{n_{12}(1,0)}{c} + \lambda = 0, \\ \frac{\partial Q}{\partial d} &= \frac{n_1(1)}{c+d} + \frac{n_2(1)}{b+d} + \frac{n_{12}(1,1)}{d} + \lambda = 0. \end{aligned} \tag{13.34}$$

Making the following change of variables

$$a^* = -\lambda a, \quad b^* = -\lambda b, \quad c^* = -\lambda c, \quad d^* = -\lambda d, \tag{13.35}$$

we can take $\lambda = -1$ and if we denote

$$\begin{array}{llll} x = n_1(0); & y = n_1(1); & z = n_2(0); & p = n_2(1); \\ q = n_{12}(0,0); & r = n_{12}(0,1); & s = n_{12}(1,0); & t = n_{12}(1,1), \end{array} \tag{13.36}$$

the system (13.34) can be written as

$$\begin{aligned} \frac{x}{a^*+b^*} + \frac{z}{a^*+c^*} + \frac{q}{a^*} &= 1, \\ \frac{x}{a^*+b^*} + \frac{p}{b^*+d^*} + \frac{r}{b^*} &= 1, \\ \frac{y}{c^*+d^*} + \frac{z}{a^*+c^*} + \frac{s}{c^*} &= 1, \\ \frac{y}{c^*+d^*} + \frac{p}{b^*+d^*} + \frac{t}{d^*} &= 1, \end{aligned} \tag{13.37}$$

where now the unknowns are a^*, b^*, c^* and d^*. Let

$$a^* = h(x, y, z, p, q, r, s, t) \tag{13.38}$$

be the function giving the a^* solution of (13.37). Note that the arguments in this function are the rest of parameters appearing in (13.37). Then we have

$$\begin{aligned} b^* &= h(x, y, p, z, r, q, t, s), \\ c^* &= h(y, x, z, p, s, t, q, r), \\ d^* &= h(y, x, p, z, t, s, r, q). \end{aligned} \tag{13.39}$$

It is important to note that, due to the symmetry of equations in (13.37), the four unknowns a^*, b^*, c^* and d^* can be obtained by means of the same function h only by making the adequate permutations.

The first equation in (13.37) can be written as

$$\begin{aligned}&\frac{x}{h(x,y,z,p,q,r,s,t)+h(x,y,p,z,r,q,t,s)}\\&+\frac{z}{h(x,y,z,p,q,r,s,t)+h(y,x,z,p,s,t,q,r)}+\frac{q}{h(x,y,z,p,q,r,s,t)}=1,\end{aligned} \tag{13.40}$$

which is a functional equation equivalent to the system (13.37) because all its equations lead to the same functional equation. Hence, the solution of (13.40) is equivalent to the solution of (13.37).

Equation (13.40) is not easy to solve. Thus, with the purpose of illustrating the method of solution, we solve only the particular case: $z = p = 0$; that is, the case when the symptom X_1 is known for all patients. In this case (13.40) becomes

$$\frac{x}{f(x,y,q,r,s,t)+f(x,y,r,q,t,s)}+\frac{q}{f(x,y,q,r,s,t)}=1, \tag{13.41}$$

where

$$f(x,y,q,r,s,t)=h(x,y,0,0,q,r,s,t). \tag{13.42}$$

Due to the fact that (13.41) depends only on x and q and because the variables x and q appear only as arguments in the first, third and fourth places of function f, we conclude that this function is independent of y, t and s. Using this reasoning or by making $y = y_0$, $t = t_0$ and $s = s_0$, Equation (13.41) can be written as

$$\frac{x}{f(x,q,r)+f(x,r,q)}+\frac{q}{f(x,q,r)}=1, \tag{13.43}$$

and interchanging r and q in (13.43) we obtain

$$\frac{x}{f(x,r,q)+f(x,q,r)}+\frac{r}{f(x,r,q)}=1, \tag{13.44}$$

and from (13.43) and (13.44) we get

$$\frac{q}{f(x,q,r)}=\frac{r}{f(x,r,q)}. \tag{13.45}$$

Substitution now of $f(x,q,r)$, obtained from (13.45), into (13.44) leads to

$$f(x,r,q)=\frac{x}{1+\frac{q}{r}}+r=\frac{r(r+q+x)}{r+q}. \tag{13.46}$$

With this, if $z = p = 0$ we get

$$
\begin{aligned}
a^* &= f(x,q,r) &= \frac{q(x+q+r)}{q+r}, \\
b^* &= f(x,r,q) &= \frac{r(x+q+r)}{q+r}, \\
c^* &= f(y,s,t) &= \frac{s(y+s+t)}{s+t}, \\
d^* &= f(y,t,s) &= \frac{t(y+s+t)}{s+t},
\end{aligned}
\tag{13.47}
$$

and then, considering (13.31) and (13.35), finally leads to

$$
\begin{aligned}
a &= \frac{q(x+q+r)}{n(q+r)}, \\
b &= \frac{r(x+q+r)}{n(q+r)}, \\
c &= \frac{s(y+s+t)}{n(s+t)}, \\
d &= \frac{t(y+s+t)}{n(s+t)},
\end{aligned}
\tag{13.48}
$$

where

$$n = x + y + q + r + s + t \tag{13.49}$$

is the sample size including complete and incomplete cases.

Note that this is the solution of the system (13.31)-(13.37) for the particular case $z = p = 0$.

13.7 Sum of a random number of discrete random variables

Let R be a random variable such that

$$R = X_1 + \ldots + X_N, \tag{13.50}$$

where N is a discrete random variable with probability generating function $g_N(s)$ and the X_i $(i = 1, 2, \ldots)$ are independent and identically distributed discrete random variables with probability generating function $g_X(s)$. Then, the probability generating function (real version) of R is given by

$$g_R(s) = g_N[g_X(s)].$$

Then, the most general single parameter family of random variables that can be generated by the process (13.50) using single parameter families of X and N random variables satisfies the functional equation

$$F[G(s,y),z] = K[s, N(y,z)],$$

where $F[s,z], G(s,y)$ and $K(s,x)$ are the probability generating functions of the families associated with N, X and R, respectively.

In this case we cannot guarantee that all the regularity conditions in Corollary 6.1 hold. However, the invertibility of F with respect to the first argument for a fixed value of the second holds because the random variable N is integer. We can also force the invertibility of G with respect to its first argument if the random variables X take integer values. Thus, Corollary 6.1 can be used in order to find solutions to the above problem, but we cannot conclude whether or not we get the general continuous solution.

According to (6.28), we get

$$\begin{array}{rclrcl} F(s,z) &=& k[f(s)+g(z)]; & G(s,y) &=& f^{-1}[p(s)+q(y)]; \\ K(s,z) &=& k[p(s)+n(z)]; & N(y,z) &=& n^{-1}[q(y)+g(z)]. \end{array}$$

For F, G and K to be probability generating functions we must have

$$\begin{array}{lll} F(1,z) = k[f(1)+g(z)] = 1 & \Rightarrow & f(1) = \pm\infty, \quad k(\pm\infty) = 1, \\ G(1,y) = f^{-1}[p(1)+q(y)] = 1 & \Rightarrow & p(1) = \pm\infty, \quad f^{-1}(\pm\infty) = 1, \\ K(1,z) = k[p(1)+n(z)] = 1 & \Rightarrow & p(1) = \pm\infty, \quad k(\pm\infty) = 1. \end{array}$$

13.7.1 Particular case

If the random variables X, N and R belong to the same single parameter family we must have

$$F[F(s,y),z] = F[s, N(y,z)],$$

which is the transformation Equation (6.74). We can get invertibility of F with respect to its first argument if we work with integer X. However, we must add extra conditions in order to have invertibility with respect to its second argument. Thus, one solution is

$$F(x,y) = k[k^{-1}(x)+n(y)]; \quad N(x,y) = n^{-1}[n(x)+n(y)].$$

For $F(x,y)$ to be a probability generating function we must have

$$F(1,y) = 1 \Rightarrow k[k^{-1}(1)+n(y)] = 1 \Rightarrow k(\pm\infty) = 1.$$

One simple particular case is

$$k(s) = \frac{a+s}{s}; \quad n(z) = z; \quad za \le 0; \quad F(s,z) = \frac{(z+a)s-z}{zs+a-z},$$

with associated probability mass function

$$P(x) = \begin{cases} \dfrac{z}{z-a} & \text{if } x = 0, \\ \dfrac{a^2}{(a-z)^2}\left(\dfrac{z}{z-a}\right)^{x-1} & \text{otherwise.} \end{cases}$$

13.8 Bayesian conjugate distributions

Let us assume that the random variable X belongs to a single parameter family with likelihood function $M(x,\theta)$, where $\theta \in \Theta$ is the parameter, and that θ has an "a priori" probability density function $F(y,\theta)$; i.e., it belongs to a single parameter family with parameter y. If we force the "a posteriori" distribution to belong to the same family, we must have

$$F[G(x,y),\theta] = M(x,\theta)F(y,\theta), \tag{13.51}$$

where the function G gives the value of the new parameter as a function of the sample value x and the old parameter value y.

Functional Equation (13.51) is a particular case of (7.79) with $N = F$ and $H(x,y) = xy$. We cannot guarantee that all the regularity conditions in Theorem 7.10 hold here; however, conditions $H_1(x,y) \neq 0$ and $H_2(x,y) \neq 0$ hold if the random variables X and θ are unlimited on both sides. Thus, (7.80) does not necessarily lead to the general solution. However, taking into account (7.80), we get

$$\begin{array}{l} F(x,y) = l[f(y)g^{-1}(x) + \alpha(y) + \beta(y)] = n^{-1}[f(y)k(x) + \beta(y)], \\ G(x,y) = g[h(x) + k(y)], \; M(x,y) = m^{-1}[f(y)h(x) + \alpha(y)], \\ H(x,y) = l[m(x) + n(y)] = xy, \end{array} \tag{13.52}$$

and then we have

$$l^{-1}(xy) = m(x) + n(y); \quad x, y \in \mathbb{R}_{++},$$

which is Equation (4.3), with the general continuous or strictly monotonic solution (see Theorem 4.3)

$$l^{-1}(x) = A\log(BCx); \;\; m(x) = A\log(Bx); \;\; n(x) = A\log(Cx),$$

then

$$F(x,y) = \frac{1}{BC}\exp\left[\frac{f(y)g^{-1}(x) + \alpha(y) + \beta(y)}{A}\right] = \frac{1}{C}\exp\left[\frac{f(y)k(x) + \beta(y)}{A}\right], \tag{13.53}$$

which leads to

$$g^{-1}(x) - k(x) = \frac{A\log B - \alpha(y)}{f(y)} = D \;\; \Rightarrow \;\; \begin{cases} g^{-1}(x) = k(x) + D, \\ \alpha(y) = A\log B - Df(y), \end{cases}$$

and then (13.52) becomes

$$\begin{array}{l} F(x,y) = \dfrac{1}{C}\exp\left[\dfrac{f(y)k(x) + \beta(y)}{A}\right]; \;\; M(x,y) = \exp\left[\dfrac{f(y)(h(x) - D)}{A}\right]; \\ G(x,y) = k^{-1}\left[h(x) + k(y) - D\right]. \end{array}$$

which satisfies (13.51).

Finally, making the following change of functions

$$f^*(x) = \exp\left[\frac{f(x)}{A}\right]; \ \beta^*(x) = \frac{\exp\left[\beta(x)/A\right]}{C}; \ h^*(x) = h(x) - D,$$

we get

$$\begin{array}{ll} F(x,y) = f^*(y)^{k(x)}\beta^*(y); & M(x,y) = f^*(y)^{h^*(x)}; \\ G(x,y) = k^{-1}[h^*(x) + k(y)], & \end{array} \tag{13.54}$$

which are solutions of (13.51).

13.9 Maximum stability

Let us assume two independent random variables X and Y with cumulative distribution functions $F_X(x)$ and $G_Y(y)$, respectively. Then the cumulative distribution function of the random variable $Z = \max(X,Y)$ is $H_Z(z) = F_X(z)G_Y(z)$. If we want X, Y and Z to belong to the same single parameter family, we must have

$$F[z, G(a,b)] = F(z,a)F(z,b), \tag{13.55}$$

where the function G gives the value of the new parameter.

Equation (13.55) is a particular case of (7.79) with $M = N = F$ and $H(x,y) = xy$. In this case we can easily find cases in which the conditions in Theorem 7.10 hold. For example, if the range of the random variables is $[0,\beta]$, we can choose the domain $A^2 \times (0,\beta]$, where A is the domain of a and b. Then, $F(z,x) \neq 0, \forall x$ and $F(\beta,x) = 1, \forall x$. If, in addition, we assume that $F_1(z,a) \neq 0$, all regularity conditions hold.

Calling $F^t(x,y) = F(y,x)$, (13.55) becomes

$$F^t\left[G(a,b), z\right] = F^t(a,z)\, F^t(b,z), \tag{13.56}$$

which is a special case of (13.51) with $M = F^t$. So Example 13.8 yields

$$F^t(x,y) = f^*(y)^{k(x)}\,\beta^*(y) = f^*(y)^{h^*(x)}; \ \ G(a,b) = k^{-1}(h^*(a) + k(b)).$$

That is,

$$F(x,y) = f^*(x)^{k(y)-D}; \ \ G(x,y) = k^{-1}(k(x) + k(y) - D), \tag{13.57}$$

since $k(x_1) - k(x_2) = h^*(x_1) - h^*(x_2) \Rightarrow h^*(x) = k(x) - D$ for some constant D. Finally, substitution into (13.55) gives $D = 0$ and we obtain the solution

$$F(x,y) = f^*(x)^{h^*(y)}; \ \ G(x,y) = h^{*-1}[h^*(x) + h^*(y)],$$

where $f(x)$ is an arbitrary positive function.

We can apply this solution to the case of fatigue life of longitudinal elements. In this case, the survivor function $Q(x,y)$ of an element of length y plays the same role as $F(x,y)$ before. But now we must have $G(x,y) = x + y$, which implies $h(x) = cx$ and leads to

$$Q(x,y) = f(x)^{cy}.$$

13.10 Reproductivity

In this section we characterize reproductive families of distributions. We consider two cases: one- and two-parameter families.

13.10.1 Reproductivity in single parameter families

We say that a family of random variables is reproductive under convolution if the sum of independent random variables of the family belongs to the family; that is, if the family is closed under sums. We know that the characteristic function of the sum of two independent random variables is the product of the characteristic functions of the components. Thus, reproductivity for single parameter families can be written as the functional equation

$$F[G(x,y),t] = F(x,t)F(y,t),$$

where F (characteristic function) is a complex function of two real variables. The function $G(x,y)$ is a real function of two real variables and shows how the parameter of the sum can be obtained as a function of the parameters of the two random variables being added.

This equation, taking into account that

$$F(x,t) = F_1(x,t)\ \exp[iF_2(x,t)],$$

where $F_1(x,t)$ and $F_2(x,t)$ are the modulus and the argument of $F(x,t)$ respectively, can be written as

$$\begin{aligned} F_1[G(x,y),t] &= F_1(x,t)F_1(y,t),\\ F_2[G(x,y),t] &= F_2(x,t) + F_2(y,t). \end{aligned} \tag{13.58}$$

The first of these equations was solved in Example 13.9. Thus, we can write some solutions as

$$G(x,y) = h^{-1}[h(x) + h(y)];\ \ F_1(x,t) = \exp[r(t)h(x)]. \tag{13.59}$$

To solve the second equation, we follow a similar process as that used in Example 13.9. We consider this equation as a particular case of (7.79) with $F = M = N = F_2$ and $H(x,y) = x + y$. Not all regularity conditions in Theorem 7.10 hold here for the two equations in (13.58). Thus, we use it in order to find some solutions to our problem, but we cannot guarantee them to be the general solution. Thus, from (7.80) we get

$$\begin{aligned} F_2(x,y) &= l[f(y)g^{-1}(x) + \alpha(y) + \beta(y)] = m^{-1}[f(y)h(x) + \alpha(y)]\\ &= n^{-1}[f(y)k(x) + \beta(y)],\\ G(x,y) &= g[h(x) + k(y)],\\ H(x,y) &= l[m(x) + n(y)] = x + y, \end{aligned}$$

and then we have

$$l^{-1}(x + y) = m(x) + n(y),$$

which is Equation (4.1), with general continuous-at-a-point solution (see Theorem 4.1)

$$l^{-1}(x) = Ax + B + C; \quad m(x) = Ax + B; \quad n(x) = Ax + C,$$

then

$$\begin{aligned} F_2(x,y) &= \frac{f(y)g^{-1}(x) + \alpha(y) + \beta(y) - B - C}{A} = \frac{f(y)h(x) + \alpha(y) - B}{A} \\ &= \frac{f(y)k(x) + \beta(y) - C}{A}, \end{aligned}$$

which leads to

$$g^{-1}(x) - k(x) = \frac{B - \alpha(y)}{f(y)} = D \Rightarrow \qquad \begin{cases} g^{-1}(x) = k(x) + D, \\ \alpha(y) = B - Df(y). \end{cases}$$

$$h(x) - k(x) = \frac{\beta(y) - \alpha(y) + B - C}{f(y)} = E$$

$$\Rightarrow \begin{cases} h(x) = k(x) + E, \\ \beta(y) - \alpha(y) + B - C = Ef(y). \end{cases}$$

and then we can write

$$\begin{aligned} F_2(x,y) &= \frac{f(y)[k(x) + E - D]}{A}, \\ G(x,y) &= k^{-1}[k(x) + k(y) + E - D], \end{aligned}$$

and making the following change of functions

$$f^*(x) = \frac{f(x)}{A}; \quad k^*(x) = k(x) + E - D,$$

we finally get the following solution of the second equation in (13.58):

$$F_2(x,y) = f^*(y)k^*(x); \quad G(x,y) = {k^*}^{-1}[k^*(x) + k^*(y)]. \tag{13.60}$$

Equaling the two expressions for $G(x,y)$ in (13.59) and (13.60) we get

$$G(x,y) = h^{-1}[h(x) + h(y)] = {k^*}^{-1}[k^*(x) + k^*(y)] \Rightarrow ah(x) = k^*(x),$$

where a is an arbitrary constant.

Thus, finally we obtain

$$\begin{aligned} F(x,t) &= F_1(x,t)\ \exp[iF_2(x,t)] \\ &= \exp[r(t)h(x)]\exp[if^*(t)ah(x)] \\ &= \exp[r(t) + iaf^*(t)]^{h(x)} \\ &= S(t)^{h(x)}, \end{aligned}$$

where $S(t)$ is an arbitrary complex function of real variable and $h(x)$ is a real arbitrary strictly monotonic and continuously differentiable function of real variable.

In Table 13.3 we give some examples of one-parameter reproductive families together with their characteristic functions and their associated $S(t)$ and $h(x)$ functions.

Table 13.3: Some single parameter reproductive families.

Family	$F(x,t)$	$S(t)$	$h(x)$
Binomial	$[p\exp(it)+q]^x$	$p\exp(it)+q$	x
Negative binomial	$\left[\dfrac{pe^{it}}{1-qe^{it}}\right]^x$	$\dfrac{pe^{it}}{1-qe^{it}}$	x
Gamma	$\left[1-\dfrac{it}{\lambda}\right]^{-x}$	$\left[1-\dfrac{it}{\lambda}\right]^{-1}$	x
Normal $N(0,x)$	$\exp(-x^2t^2/2)$	$\exp(-t^2/2)$	x^2
Chi-square	$(1-2it)^{-x/2}$	$(1-2it)^{-1/2}$	x

13.10.2 Reproductivity in two-parameter families

In this example we extend the problem analyzed in Section 13.10.1 to the case of two-parameter families. Thus, we need to solve the functional equation

$$\mathbf{F}[\mathbf{G}(\mathbf{x},\mathbf{y}),t] = \mathbf{F}(\mathbf{x},t) * \mathbf{F}(\mathbf{y},t), \tag{13.61}$$

where $\mathbf{F}$ (the characteristic function) is a complex function of a real three-dimensional vector, the function $\mathbf{G}(\mathbf{x},\mathbf{y})$ is a real bidimensional vector function of two real bidimensional vectors and shows how the vector parameter of the sum can be obtained as a function of the vector parameters, $\mathbf{x}$ and $\mathbf{y}$, of the two random variables being added and "*" denotes the complex product.

This equation is a particular case of Equation (8.21) with $\mathbf{F} = \mathbf{K} = \mathbf{L}$ and $\mathbf{H}$ the complex product; that is, according to (8.22), its general (with conditions (a) to (d)) solution is given by

$$\begin{aligned} \mathbf{F}[\mathbf{z},u] &= \mathbf{w}[\mathbf{C}(u)\mathbf{r}(\mathbf{z}) + \mathbf{a}(u) + \mathbf{b}(u)] \\ &= \mathbf{s}^{-1}[\mathbf{C}(u)\mathbf{p}(\mathbf{z}) + \mathbf{a}(u)] = \mathbf{t}^{-1}[\mathbf{C}(u)\mathbf{q}(\mathbf{z}) + \mathbf{b}(u)], \\ \mathbf{G}(\mathbf{x},\mathbf{y}) &= \mathbf{r}^{-1}[\mathbf{p}(\mathbf{x}) + \mathbf{q}(\mathbf{y})], \\ \mathbf{H}(\mathbf{u},\mathbf{v}) &= \mathbf{w}[\mathbf{s}(\mathbf{u}) + \mathbf{t}(\mathbf{v})]. \end{aligned} \tag{13.62}$$

Note that the multiparametric case ($n > 2$) cannot satisfy condition (b) in Theorem 8.10.

Due to the fact that the complex product is associative, we get (see Aczél (1966)):

$$\begin{aligned} \mathbf{s} &= \mathbf{Dr} + \mathbf{P}, \\ \mathbf{t} &= \mathbf{Dr} + \mathbf{Q}, \\ \mathbf{w}^{-1} &= \mathbf{Dr} + \mathbf{P} + \mathbf{Q}, \\ \mathbf{r}(\rho_1,\theta_1) + \mathbf{r}(\rho_2,\theta_2) &= \mathbf{r}(\rho_1\rho_2, \theta_1 + \theta_2), \end{aligned}$$

where $\mathbf{P}, \mathbf{Q} \in \mathbb{R}^2$ are arbitrary constants, $\mathbf{D}$ is a 2×2 non-singular matrix and the $\rho's$ and $\theta's$ refer to the modula and the arguments of the corresponding complex numbers.

This equation can be written as

$$r_i(\rho_1, \theta_1) + r_i(\rho_2, \theta_2) = r_i(\rho_1\rho_2, \theta_1 + \theta_2), \quad i = 1, 2, \tag{13.63}$$

where the subindices refer to the real and imaginary components.

Successively making $\rho_1 = \rho_2 = 1$ and $\theta_1 = \theta_2 = 0$ we get

$$\left.\begin{array}{l} r_i(1, \theta_1) + r_i(1, \theta_2) = r_i(1, \theta_1 + \theta_2) \\ r_i(\rho_1, 0) + r_i(\rho_2, 0) = r_i(\rho_1\rho_2, 0) \end{array}\right\}; \quad i = 1, 2, \tag{13.64}$$

and calling

$$\left.\begin{array}{l} g_i(\theta) = r_i(1, \theta) \\ f_i(\rho) = r_i(\rho, 0) \end{array}\right\}; \quad i = 1, 2, \tag{13.65}$$

we obtain

$$\left.\begin{array}{l} g_i(\theta_1) + g_i(\theta_2) = g_i(\theta_1 + \theta_2) \\ f_i(\rho_1) + f_i(\rho_2) = f_i(\rho_1\rho_2) \end{array}\right\}; \quad i = 1, 2, \tag{13.66}$$

which are Cauchy's equations (3.7) and (3.9), respectively. Thus

$$\left.\begin{array}{l} g_i(\theta) = a_{2i}\theta \\ f_i(\rho) = a_{1i} \log \rho \end{array}\right\}; \quad i = 1, 2. \tag{13.67}$$

But, then we have

$$r_i(\rho, \theta) = r_i(\rho 1, 0 + \theta) = f_i(\rho) + g_i(\theta) = a_{1i} \log \rho + a_{2i}\theta; \quad i = 1, 2 \tag{13.68}$$

which in matrix form becomes

$$\mathbf{r}(x, y) = \begin{pmatrix} a_{11} \log x + a_{21}y \\ a_{12} \log x + a_{22}y \end{pmatrix}, \tag{13.69}$$

which implies

$$\mathbf{w}(x, y) = \mathbf{r}^{-1}(x, y) = \begin{pmatrix} \alpha_1 \exp(b_{11}x + b_{21}y) \\ b_{12}x + b_{22}y + \alpha_2 \end{pmatrix},$$

where we have assumed

$$b_{11}b_{22} - b_{21}b_{12} \neq 0. \tag{13.70}$$

Now, from (13.62) we get

$$\begin{aligned} \mathbf{C}(u)\mathbf{r}(\mathbf{x}) + \mathbf{a}(u) + \mathbf{b}(u) - \mathbf{P} - \mathbf{Q} &= \mathbf{C}(u)\mathbf{p}(\mathbf{x}) + \mathbf{a}(u) - \mathbf{P} \\ &= \mathbf{C}(u)\mathbf{q}(\mathbf{x}) + \mathbf{b}(u) - \mathbf{Q}, \end{aligned}$$

which implies

$$\begin{array}{c}\mathbf{C}(u)[\mathbf{p}(\mathbf{x}) - \mathbf{r}(\mathbf{x})] = \mathbf{b}(u) - \mathbf{Q},\\ \mathbf{C}(u)[\mathbf{q}(\mathbf{x}) - \mathbf{r}(\mathbf{x})] = \mathbf{a}(u) - \mathbf{P},\end{array}$$

and

$$\begin{array}{ll}\mathbf{p}(\mathbf{x}) = \mathbf{k}_1 + \mathbf{r}(\mathbf{x}); & \mathbf{b}(u) = \mathbf{C}(u)\mathbf{k}_1 + \mathbf{Q},\\ \mathbf{q}(\mathbf{x}) = \mathbf{k}_2 + \mathbf{r}(\mathbf{x}); & \mathbf{a}(u) = \mathbf{C}(u)\mathbf{k}_2 + \mathbf{P},\end{array}$$

where $\mathbf{k}_1$ and $\mathbf{k}_2$ are constant vectors.

Substituting now back into (13.62) leads to

$$\begin{array}{rcl}\mathbf{F}[\mathbf{z}, u] &=& \mathbf{w}\mathbf{C}(u)[\mathbf{r}(\mathbf{z}) + \mathbf{k}] + \mathbf{P} + \mathbf{Q},\\ \mathbf{G}(\mathbf{x}, \mathbf{y}) &=& \mathbf{r}^{-1}[\mathbf{r}(\mathbf{x}) + \mathbf{r}(\mathbf{y}) + \mathbf{k}],\end{array}$$

where $\mathbf{k} = \mathbf{k}_1 + \mathbf{k}_2$, and making $\mathbf{m}(\mathbf{x}) = \mathbf{r}(\mathbf{x}) + \mathbf{k}$ they can be written as

$$\begin{array}{rcl}\mathbf{F}[\mathbf{z}, u] &=& \mathbf{w}[\mathbf{C}(u)\mathbf{m}(\mathbf{z}) + \mathbf{P} + \mathbf{Q}],\\ \mathbf{G}(\mathbf{x}, \mathbf{y}) &=& \mathbf{m}^{-1}\,[\mathbf{m}(\mathbf{x}) + \mathbf{m}(\mathbf{y})].\end{array}$$

Then, we finally get

$$\mathbf{F}(\mathbf{z}, u) = \begin{pmatrix}\exp\left[(b_{11}\mathbf{C}_1(u) + b_{21}\mathbf{C}_2(u))\mathbf{m}(\mathbf{z})\right]\\ (b_{12}\mathbf{C}_1(u) + b_{22}\mathbf{C}_2(u))\,\mathbf{m}(\mathbf{z})\end{pmatrix}, \tag{13.71}$$

where the elements in the matrix are the modulus and the argument of the characteristic function, respectively, and $\mathbf{C}_i(u)$ is the i-th row of the matrix $\mathbf{C}(u)$.
Equation (13.71) can also be written as

$$\mathbf{F}(\mathbf{z}, u) = \exp\left[(\mathbf{A}_1(u) + i\mathbf{A}_2(u))\mathbf{m}(\mathbf{z})\right], \tag{13.72}$$

where $\mathbf{A}_i(u)$ $(i = 1, 2)$, according to (13.71), are two arbitrary, but independent real n-vector functions.

In Table 13.4 we give $\mathbf{F}(\mathbf{z}, u), \mathbf{m}(\mathbf{z}), \mathbf{A}_1(u)$ and $\mathbf{A}_2(u)$ for the normal and Chi-square families of distributions.

Thus, if the necessary conditions hold, we have proved that if $\mathbf{F}(\mathbf{z}, c) = u$ and $\dfrac{\partial \mathbf{F}(\mathbf{z}, 0)}{\partial t} = u$ have unique inverses and $\mathbf{F}$ and $\mathbf{G}$ are differentiable, then the general (with the indicated additional conditions) solution of (13.61) is (13.72).

Exercises

13.1 Assume two independent random variables X and Y with cumulative distribution functions $F_X(x)$ and $G_Y(y)$, respectively. Then the cumulative distribution function of the random variable $Z = \min(X, Y)$ is $H_Z(z) = 1 - (1 - F_X(z))(1 - F_Y(z))$.

Table 13.4: Two two-parameter reproductive families

Family	Normal	Chi-square
$\mathbf{z}$	$\begin{pmatrix}\mu \\ \sigma\end{pmatrix}$	$\begin{pmatrix}\lambda \\ n\end{pmatrix}$
$\mathbf{F}(\mathbf{z},u)$	$\exp\left[iu\mu - \frac{u^2\sigma^2}{2}\right]$	$(1-2iu)^{-\frac{n}{2}}\exp\left[\frac{iu\lambda}{1-2iu}\right]$
$\mathbf{m}(\mathbf{z})$	$\begin{pmatrix}\mu \\ \sigma^2\end{pmatrix}$	$\begin{pmatrix}\lambda \\ n\end{pmatrix}$
$\mathbf{A}_1(u)$	$\begin{pmatrix}0 & -\frac{u^2}{2}\end{pmatrix}$	$\begin{pmatrix}-\frac{2u^2}{1+4u^2} & -\frac{\log(1+4u^2)}{4}\end{pmatrix}$
$\mathbf{A}_2(u)$	$\begin{pmatrix}u & 0\end{pmatrix}$	$\begin{pmatrix}\frac{u}{1+4u^2} & \frac{arc\tan(2u)}{2}\end{pmatrix}$

(a) Write the functional equation for X, Y and Z to belong to a common single parameter family and solve it.

(b) Find some particular cases.

(c) Check whether or not the Weibull distribution is one of such families.

13.2 We say that a random variable has a Pareto distribution and denote it as $P(\sigma, \alpha)$, if its probability density function is

$$f(x,\alpha) = \frac{\alpha}{\sigma}\left(1+\frac{x}{\sigma}\right)^{-(\alpha+1)} I(x>0),$$

where $\alpha, \sigma > 0$.

Obtain the most general bivariate family with Pareto conditionals; i.e.:

$$X|Y=y \sim P(\sigma_1(y), \alpha)$$

and

$$Y|X=x \sim P(\sigma_2(x), \alpha).$$

13.3 Find all joint bivariate densities with conditionals of the form:

$$f(x|y) = \lambda(y)e^{-\lambda(y)x}, \quad x>0,$$

such that $\lambda(y) > 0$, and

$$f(y|x) = \frac{\alpha}{\sigma(x)}\left(1+\frac{y}{\sigma(x)}\right)^{-(\alpha+1)}, \quad y>0,$$

with $\sigma(x) > 0$ and $\alpha > 0$.

13.4 Write the functional equation associated with the compatibility of two given conditional distributions of a bivariate density; i.e.,write the functional conditions for two given conditional families to come from a joint density.

13.5 Given a family of conditional densities

$$f_{X|Y}(x|y) = a(x,y), \quad x \in S(X), \quad y \in S(Y), \tag{13.73}$$

and a regression function

$$E(Y|X = x) = \psi(x), \quad x \in S(X) \tag{13.74}$$

State the functional equations that give an answer to the following problems:

(a) Are $a(x,y)$ and $\psi(x)$ compatible in the sense that there will exist a joint density function $f_{X,Y}(x,y)$ with $a(x,y)$ as its corresponding family of conditional densities and with $\psi(x)$ as its regression function of Y on X?

(b) Assuming that $a(x,y)$ and $\psi(x)$ are compatible, under what conditions do they determine a unique joint density?

(c) Given $a(x,y)$, identify the class of all compatible functions ψ.

13.6 The basic idea of a probability paper plot, of a two-parameter family of distributions $F(x;a,b)$, consists of modifying the random variable X scale to the $U = h(X)$ scale and the probability scale P to the $V = g(P)$ scale in such a manner that the cdfs become a family of straight lines. In this way, when the cdf is drawn a linear trend is an indication of the sample coming from the corresponding family.

Obtain the functional equation of all transformations $h(X)$ and $g(P)$ and all two-parameter families of distributions $F(x;a,b)$ such that they can be represented on a probability plot.

13.7 A random variable is said to be a finite mixture if its pdf is of the form

$$f(x) = \pi_1 f_1(x) + \ldots + \pi_k f_k(x)$$

where $\pi_i \geq 0$ and $\pi_1 + \ldots + \pi_k = 1$ and $f_1(x), \ldots, f_k(x)$ are known linearly independent pdfs.

Obtain the most general pdf $f(x,y)$ with conditionals

$$f(x|y) = \pi_1(y) f_1(x) + \ldots + \pi_k(y) f_k(x)$$

and

$$f(y|x) = \tilde{\pi}_1(x) g_1(y) + \ldots + \tilde{\pi}_r(x) g_r(y),$$

where $\{f_i(x), i = 1, \ldots, k\}$ and $\{g_j(y); i = 1, \ldots, n\}$ are sets of linearly independent pdfs.

Bibliography

ABEL, N. H. (1823). Méthode générale pour trouver des functions d'une seule quantité variable lorsqu'une propieté de ces fonctions est exprimée par une equation entre deux variables. *Mag. Naturvidenskab*, 1:1–10.

ABEL, N. H. (1826a). Untersuchungen der functionen zweier unabhangigen veranderlichen grossen x und y, wie $f(x, y)$ welche die eigenschaft haben, dass $f[z, f(x, y)]$ eine symmetrische function von x, y und z ist. *J. Reine Angew. Math.*, 1:11–5.

ABEL, N. H. (1826b). Untersuchungen uber die eihe $1 + (m/1)x + (m(m-1))((1.2)x^2 + \ldots$. *J. Reine Angew. Math.*, 1:311–319.

ACZÉL, J. (1966). *Lectures on functional equations and their applications*. Vol. 19, Mathematics in Science and Engineering. Academic Press.

ACZÉL, J. (1975). On a system of functional equations determining price and productivity indices. *Utilitas Math.*, 7:345–362.

ACZÉL, J. (1984). On history, applications and theory of functional equations (introduction). In J. Aczél, ed., *Functional Equations: History, Applications and Theory*, chap. 1, pp. 3–12. Reidel Publishing Company, Dordrecht/Boston/Lancaster.

ACZÉL, J. (1987a). Scale-invariant equal sacrifice in taxation and conditional functional equations. *Aequationes Math.*, 32:336–349.

ACZÉL, J. (1987b). *A Short Course on Functional Equations Based upon Recent Applications to the Social and Behavioral Sciences*. Reidel Publishing Company.

ACZÉL, J. (1988). Measurement in economics: theory and applications of economic indices. *Physica, Heidelberg*, pp. 3–17.

ACZÉL, J. and ALSINA, C. (1984). Contributions to production theory, natural resources, economic indices and related topics. *Methods of Operations Research*, 487.

ACZÉL, J. and DHOMBRES, J. (1989). *Functional Equations in several variables*. Cambridge University Press.

ACZÉL, J. and EICHHORN, W. (1974). Systems of functional equations determining price and productivity indices. *Utilitas Math.*, 5:213–226.

ALEGRE, J., ARCARONS, J., BOLANCÉ, C., and DÍAZ, L. (1995). *Ejercicios y problemas de Econometría*. Editorial AC, Madrid.

ALLEN, B. (1992). Price and quantity competition in homogeneous duopoly markets. *Economics Letters*, 38:417–422.

ALSINA, C. (1981a). Some functional equations in the space of uniform distribution functions. *Aequationes Math.*, 22:153–164.

ALSINA, C. (1981b). Triangle functions and composition of probability distribution functions. *Stochastica*, 5:25–32.

ALSINA, C. and BONNET, E. (1979). On sums of dependent uniformly distributed random variables. *Stochastica*, 3:33–43.

ANAND, V. (1993a). *Computer Graphics and Geometric Modeling for Engineers*. John Wiley and Sons, New York.

ANAND, V. B. (1993b). *Computer Graphics and Geometric Modeling for Engineers*. Chapter 13, John Wiley and Sons.

ANDERSON, J. A. and ROSENBERG, E. (1988). *Neurocomputing: Foundations of Research*. The MIT Press: Cambridge.

ANDERSON, S. P. and NEVEN, D. J. (1991). Cournot competition yields spatial agglomeration. *International Economic Review*, 32:793–808.

ARNOLD, B., CASTILLO, E., and SARABIA, J. M. (1992). *Conditionally Specified Distributions*. Springer Verlag.

ARNOLD, B. C., CASTILLO, E., and SARABIA, J. M. (1993). Multivariate distributions with generalized pareto conditionals. *Statistics and Probability Letters*, 17:361–368.

BAKER, J. A. (1974). On the functional equation $f(x)\,g(y) = \prod_{i=1}^{n} h_i(a_i\,x + b_i\,y)$. *Aequationes Math.*, 11:154–162.

BARNSLEY, M. F. (1990). *Fractals Everywhere*. Academic Press, New York, Second Edition.

BARNSLEY, M. F. and HURD, L. P. (1993). *Fractal Image Compression*. A. K. Peters, Massachussetts.

BENZ, W. (1993). On a general principle in geometry that leads to functional equations. *Aequationes Math.*, 46:3–10.

BOGDANOFF, J. and KOZIN, F. (1987). Effect of length on fatigue life of cables. *J. of Engineering Mechanics*, 113:925–940.

BOISSONAT, J. D. (1985). Surface reconstruction from planar cross-sections. In *Proceedings of the IEEE Conference on Computer Vision and Pattern Recognition*, pp. 393–397. Detroit, MI.

BOULDING, W., LEE, E., , and STAELIN, R. (1994). Mastering the mix: Do advertising, promotion, and sales force activities lead to differentiation? *Journal of Marketing Research*, 31:159–172.

BU-QING, S. and DING-YUAN, L. (1989). *Computational Geometry, Curve and Surface Modeling*. AcademicPress, San Diego.

CAPLIN, A. and NALEBUFF, B. (1991). Aggregation and imperfect competition: on the existence of equilibrium. *Econometrica*, 59:25–59.

CASTILLO, E. (1988). *Extreme Value Theory in Engineering*. Academic Press, New York.

CASTILLO, E. (1996). *Algunas aplicaciones de las ecuaciones funcionales*. Universidad de Cantabria. Servicio de publicaciones, Apertura del Curso Académico 1996/97.

CASTILLO, E. (1998). Functional networks. *Neural Processing Letters*, 7:151–159.

CASTILLO, E., COBO, A., GÓMEZ-NESTERKIN, R., and HADI, A. S. (2000a). A general framework for functional networks. *Networks*, 35(1):70–82.

CASTILLO, E., COBO, A., GUTIÉRREZ, J. M., and PRUNEDA, R. E. (1999a). *Functional Networks with Applications. A Neural-Based Paradigm*. Kluwer Academic Publishers.

CASTILLO, E., COBO, A., GUTIÉRREZ, J. M., and PRUNEDA, R. E. (1999b). Working with differential, functional and difference equations using functional networks. *Applied Mathematical Modeling*, 23:89–107.

CASTILLO, E., COBO, A., GUTIÉRREZ, J. M., and PRUNEDA, R. E. (2000b). Functional networks. A new neural network based methodology. *Computer-Aided Civil and Infrastructure Engineering*, 15:90–106.

CASTILLO, E., DÁVILA, M. R., and RUIZ, M. R. (1989). Resolución numérica de ecuaciones diferenciales mediante una ecuación funcional equivalente. In *XIV Jornadas Hispano-Lusas de Matemáticas*. Tenerife, Spain.

CASTILLO, E., FERNÁNDEZ, A., RUIZ, J. R., and SARABIA, J. M. (1990a). Statistical models for analysis of fatigue life of long elements. *J. of Eng. Mech., ASCE*, 116:1036–1049.

CASTILLO, E., FERNÁNDEZ-CANTELI, A., ESSLINGER, V., and THURLIMANN, B. (1985). Statistical model for fatigue analysis of wires, strands and cables. *Iabse Periodica*, 1:1–40.

CASTILLO, E. and GALAMBOS, J. (1987a). Bivariate distributions with normal conditionals. In *Proceedings of the IASTED International Symposium on Simulation, Modeling and Development SMD -87*, pp. 59–62. El Cairo, Egypt.

CASTILLO, E. and GALAMBOS, J. (1987b). Lifetime regression models based on a functional equation of physical nature. *Journal of Applied Probabilit*, 24:160–169.

CASTILLO, E., GALAMBOS, J., and SARABIA, J. M. (1987). The selection of the domain of attraction of an extreme value distribution from a set of data. *Extreme Value Theory Proc. Oberwolfach 1987, Lectures Notes in Statistics, Springer Verlag*, 51:181–190.

CASTILLO, E., GALAMBOS, J., and SARABIA, J. M. (1990b). Caracterización de modelos bivariantes con distribuciones condicionadas tipo gamma. *Estadística Española*, 32:439–450.

CASTILLO, E. and GUTIÉRREZ, J. M. (1998). Nonlinear time series modeling and prediction using functional networks. extracting information masked by chaos. *Physics Letters A*, 244:71–84.

CASTILLO, E., GUTIÉRREZ, J. M., COBO, A., and CASTILLO, C. (2000c). A minimax method for learning functional networks. *Neural Processing Letters*, 11(1):39–49.

CASTILLO, E., GUTIÉRREZ, J. M., COBO, A., and CASTILLO, C. (2000d). Some learning methods in functional networks. *Computer Aided Civil and Infrastructure Engineering*, 15:427–439.

CASTILLO, E., GUTIÉRREZ, J. M., HADI, A. S., and LACRUZ, B. (2001). Some applications of functional networks in statistics and engineering. *Technometrics*, 43(1):10–24.

CASTILLO, E. and IGLESIAS, A. (1995). Some applications of functional equations to the characterization of families of surfaces. In D. Lasser, ed., *Proceedings Cagd 94*, pp. 153–169. Shaker-Verlag, Kaiserlautern.

CASTILLO, E. and IGLESIAS, A. (1997). Some characterizations of families of surfaces using functional equations. *ACM Transactions on Graphics*, 16:296–318.

CASTILLO, E., RUIZ, M. R., and ALVAREZ, E. (1990c). Aplicaciones de las ecuaciones funcionales a sistemas expertos basados en probabilidad. In *XV Jornadas Hispano-Lusas de Matemáticas*. Evora, Portugal.

CASTILLO, E., RUIZ, M. R., and ALVAREZ, E. (1990d). Estimación de los parámetros de un modelo multinomial mediante ecuaciones funcionales. In *XV Jornadas Hispano-Lusas de Matemáticas*. Evora, Portugal.

CASTILLO, E., RUIZ, M. R., and COBO, A. (1990e). Caracterización del producto complejo mediante ecuaciones funcionales. In *XV Jornadas Hispano-Lusas de Matemáticas*, p. . Evora, Portugal.

CASTILLO, E. and RUIZ-COBO, R. (1992). *Functional Equations in Science and Engineering*. Marcel Dekker, New York.

CASTILLO, E., RUIZ-COBO, R., and ALSINA, C. (1992). Una metodología para la obtención de tarifas impositivas congruentes. *Hacienda Pública Española*, 122:27–36.

CASTILLO, E., SARABIA, J. M., and MERCEDES GONZALEZ, A. (2000e). Some demand functions in a duopoly market with advertising. In *Functional equations and inequalities*, vol. 518 of *Math. Appl.*, pp. 31–54. Kluwer Acad. Publ., Dordrecht.

CASTILLO, E., SARABIA, J. M., SARABIA, J. M., and GONZÁLEZ, A. M. (1999c). Some models for demand functions with advertising. *CEJOR Cent. Eur. J. Oper. Res.*, 7(2):71–92.

CAUCHY, A. L. (1821). *Cours d'Analyse de l'Ecole Polytechnique, Vol. I, Analyse Algébrique*. Debure, Paris.

CHANDRU, V. and KOCHAR, B. S. (1987). Analytic techniques for geometric intersection problems. In G. E. Farin, ed., *Geometric Modeling: Algorithms and new Trends*, pp. 305–318. SIAM.

CHIONH, E. W. and GOLDMAN, R. N. (1992). Implicitizing rational surfaces with base points by applying perturbations and the factors of zero theorem. In T. Lynche and L. L. Schumaker, eds., *Mathematical Methods in Computer Aided Geometric Design*, pp. 101–110. Academic Press.

COONS, S. A. (1964). *Surfaces for Computer Aided Design*. MIT, Mechanical Engineering Department, Design Division.

COX, D. R. (1972). Regression models and life-tables. *J. R. Statist. Soc.*, 2:187–202.

CUOMO, K. and OPPENHEIM, A. (1993). Circuit implementation of synchronized chaos with applications to communications. *Phys. Rev. Lett.*, 71:65–68.

D'ALEMBERT, L. (1747). Recherches sur la courbe qui forme une corde tendue mise en vibration. i. *His. Acad. Berlin*, pp. 214–219.

D'ALEMBERT, L. (1750). Adittion au mémoire sur la courbe qui forme une corde tendue mise en vibration. *His. Acad. Berlin*, pp. 355–360.

D'ALEMBERT, L. (1769). Mémoire sur les principes de mécanique. *His. Acad. Paris*, pp. 278–286.

DHOMBRES, J. G. and GER, R. (1975). Conditional Cauchy equations. *Glasnik Mat. Ser. III*, 13:39–62.

DIXON, H. D. (1992). The competitive outcome as the equilibrium in an Edgeworthian price-quantity model. *The Economic Journal*, 102:301–309.

EICHHORN, W. (1978a). *Functional equations in Economics*. Addison-Wesley Publishing Co.

EICHHORN, W. (1978b). Inequalities and functional equations in the theory of the price index. In *Proc. First International Conf, on General Inequalities*, pp. 23–28. Oberwolfach, 1976.

EICHHORN, W. (1978c). What is an economic index? In *Proc. International Sympos.*, pp. 23–28. Karlsruhe, 1976.

EICHHORN, W. (2002). Solving certain systems of nonlinear differential equations by algebraic methods. *CEJOR Cent. Eur. J. Oper. Res.*, 10(3):229–236.

EICHHORN, W. and GEHRIG, W. (1982). Measurement of inequality in economics. In *Modern Applied Mathematics- Optimization and Operations Research*, pp. 657–693. North Holland, Amsterdam, Nueva York.

EICHHORN, W. and KOLM, S. (1974). Technical progress, neutral inventions and cobb-douglas. In *Production Theory*. Springer Verlag, Berlin.

EKOULE, A. B., PEYRIN, F. C., and ODET, C. L. (1991). A triangulation algorithm form arbitrary shaped multiple planar contours. *ACM Transactions on Graphics*, 10:182–199.

FARIN, G. (1993). *Curves and Surfaces for Computer Aided Geometric Design*. AcademicPress, San Diego.

FARIN, G. E., ed. (1987). *Geometric Modeling: Algorithms and New Trends*. Proceedings of the Conference on Geometric Modeling and Robotics SIAM.

FARIN, G. E. (1990). *Curves and Surfaces for Computer Aided Geometric Design*. Academic Press, Second Edition.

FAROUKI, R. T. (1986). The characterization of parametric surface sections. *Computer Vision, Graphics and Image Processing*, 33:209–236.

FAUX, I. D. and PRATT, M. J. (1985). *Computational Geometry for Design and Manufacture*. John Wiley, New York.

FREEMAN, J. A. and SKAPURA, D. M. (1991). *Neural Networks: Algorithms, Applications, and Programming Techniques*. Addison-Wesley, Reading, MA.

GALAMBOS, J. (1987). *The asymptotic theory of extreme order statistics.* Robert E. Krieger, Malabar, Florida York.

GALILEO, G. (1638). *Discorsi e dimostrazioni intorno a due nouve scienze.* Leyden.

GENEST, G. and ZIDEK, J. V. (1986). Combining probability distributions: a critique and an annotated bibliography. *Stat. Science*, 1:114–148.

GÓMEZ-BAYÓN, J. (1984). *Modelos estadísticos compatibles para el estudio de la resistencia a fatiga de elementos simples y tendones.* Ph.D. thesis, Universidad de Cantabria, Santander.

GÓMEZ-NESTERKÍN, R. (1997). Modelación y predicción mediante redes funcionales. *Revista Electrónica Foro Red Mat. Facultad de Ciencias, UNAM*, 2.

GORDON, W. J. (1993). Sculptured surface definition via blending-function methods fundamental developments of computer-aided geometric modeling. In *Fundamental Developments of Computer-Aided Geometric Design*, pp. 117–134. Academic Press.

GRASSBERGER, P. and PROCACCIA, I. (1983). *Physical Review Letters*, 50:346.

GUTIÉRREZ, J., IGLESIAS, A., and RODRÍGUEZ, M. A. (1996). A multifractal analysis of ifsp measures: Application to fractal image generation. *Fractals*, 4:17–27.

GUTIÉRREZ, J., IGLESIAS, A., RODRÍGUEZ, M. A., and RODRÍGUEZ, V. A. (1997). Efficient rendering in fractal images. *The Mathematica Journal*, 7:7–14.

HAMEL, S. M. (1990). A noise reduction method for chaotic systems. *Phys. Letters A.*, 148:421–428.

HERTZ, J., KROGH, A., and PALMER, R. G. (1991). *Introduction to the Theory of Neural Computation.* Addison Wesley, Redwood City, CA.

HOFFMANN, C. M. (1989). *Geometric and Solid Modeling: An Introduction.* Morgan Kaufmann.

HOITSMA, D. H. and ROCHE, M. (1983). The computation of all plane-surface intersections for cad/cam applications. *Computer Aided Geometric Modeling, NASA C.P. 2272*, pp. 15–18.

HOLMES, P. (1979). The computation of all plane-surface intersections for cad/cam applications. *Philos. T. Roy. Soc.*, 292:419–.

HOSCHECK, J. and LASSER, D. (1993). *Fundamental Developments of Computer Aided Geometric Design.* A. K. Peters, Wellesley, MA.

HOSCHEK, J. and LASSER, D. (1993). *Fundamentals of CAGD*. A. K. Peters.

HUTCHINSON, J. E. (1981). Fractals and self-similarity. *Indiana Univ. Math. J.*, 30:713–747.

IRELAND, N. J. (1991). Welfare and non-linear pricing in a cournot oligopoly. *The Economic Journal*, 101:949–957.

JENSEN, J. L. W. V. (1906). Sur les fonctions convexes et les inégalités entre les valeurs moyennes. *Act. Math.*, 30:175–193.

KAHLIG, P. (1990). A general evaporation formula derived by dimensional analysis. *Annales Geophysicae*, 8:167–170.

KEPLER, J. (1624). *Chilias logarithmorum ad totidem numeros rotundos*. Marburg.

KUCZMA, M. (1978). Functional equations on restricted domains. *Aequationes Mathematicae*, 18:1–34.

KUCZMA, M., CHOCZEWSKI, B., and GER, R. (1990). *Iterative functional equation*. Encyclopedia of Mathematics and its applications, Cambridge Univ. Press.

LAGRANGE (1788). *Mécanique analitique*.

LAGRANGE (1799). *Lecons sur le calcul des fonctions*. Paris.

LAJKÓ, K. (1973). Uber die allgemeinen losungen der funktionalgleichung $f(x) + f(y) - f(xy) = h(x + y - xy)$. *Plub. Math. Debrecen*, 19:68–76.

LEE, R. B. and FREDERICKS, D. A. (1984). Intersection of parametric surfaces and a plane. *IEEE Computer Graphics and Applications*, 4:48–51.

LEGENDRE, A. M. (1791). *Eléments de géometrie. Note II*. Didot, Paris.

LESTER, J. (1986). Martin's theorem for euclidean-space and a generalization to the perimeter case. *J. Geom.*, 27:29–35.

LORENZ, E. (1963). Deterministic nonperiodic flow. *J. Atmos. Sci.*, 20:130–.

MILL, J. S. (1973). *Principles of political economy*. J. W. parker, West Strand, London, 1848, Second Edition, A. M. Kelley, Clifton.

MONREAL, A. and SANTOS, M. (1998). On some functional equations arising in computer graphics. *Aequationes Math.*, 55:61–72.

MORTENSON, M. E. (1985). *Geometric Modeling*. John Wiley and Sons.

MURALI, K. and LAKSHMANAN, M. (1998). *Phys. Lett. A*, 30:303–.

NAPIER, J. (1614). *Cannon of Logarithms*.

NAPIER, J. (1617). *Rabdologia*. Edinburgh.

NAPIER, J. (1620). *Mirifici Logarithmorum Canonis Constructio*. Edinburgh.

ORESME, N. (1347). *Questiones super geometriam Euclidis*. Manuscript, Paris.

ORESME, N. (1352). *Tractatus de configurationibus qualitatum et motuum*. Manuscript, Paris.

PAL, D. (1991). Cournot doupoly with two production periods and cost differentials. *Journal of Economic Theory*, 55:441–448.

PECORA, L. M. (1993). *Chaos in Communications*. SPIE Proceedings Vol. 2038.

PÉREZ, G. and CERDEIRA, H. A. (1995). Extracting messages masked by chaos. *Phys. Rev. Lett.*, 74:1970–1973.

PETERSEN, C. S. (1983). *Contours of Three and Four Dimensional Surfaces*. Ph.D. thesis, Master Thesis, Dept. Mathematics, Univ. Utah, Salt Lake City, UT.

PEXIDER, J. V. (1903). Notiz uber funktionaltheoreme. *Monatsh. Math. Phys.*, 14:293–301.

PICCIOTTO, R. (1970). *Tensile fatigue characteristics of sized polyester/viscose yarn and their effect on weaving performance*. Master's Thesis, North Carolina State Univ.

PIEGL, L. and TILLER, W. (1997). *The NURBS Book*. Springer Verlag, Berlin Heidelberg.

RASSIAS, J. M. (1994). *Geometry, Analysis and Mechanics*. World Scientific Singapore.

RICHART, F. E., HALL, J. R., and WOODS, R. D. (1970). *Vibrations of Soils and Foundations*. Prentice Hall International Series in Theoretical and Applied Mechanics, Englewood Cliffs, New Jersey.

ROGERS, D. and ADAMS, J. A. (1990). *Mathematical Elements for Computer Graphics*. McGraw-Hill, New York.

SARABIA, J. M. and CASTILLO, E. (1989). Una familia triparamétrica para aproximar distribuciones de extremos. In *XIV Jornadas Hispano-Lusas de Matemáticas*, pp. 1055–1059.

SEDERBERG, T. W. (1983). *Implicit and parametric curves and surfaces for Computer Aided Geometric Design*. Ph.D. thesis, Ph. D. Thesis, Purdue University.

SEDERBERG, T. W., ANDERSON, D. C., and GOLDMAN, R. N. (1984). Implicit representation of parametric curves and surfaces. *Computer Vision, Graphics and Image Processing*, 28:72–74.

STUART, A. J. C. (1958). *On progressive taxation.* La Hague, 1889. English translation in 'Classics in the theory of public finance', Macmillan, London-New York.

YOUNG, H. P. (1987). Progressive taxation and the equal sacrifice principle. *J. Public. Econom.*, 32.

YOUNG, H. P. (1988). On certain related functional equations. *J. Econom. Theory*, 44:321–335.

Index

Mathematics in Science and Engineering
Edited by C.K. Chui, Stanford University

Recent titles:
I. Podlubny, *Fractional Differential Equations*

Lightning Source UK Ltd.
Milton Keynes UK
UKOW031335260213

206846UK00001B/49/P

9 780444 517883